Plant genetic resources of Ethiopia

Plant genetic resources of Ethiopia

Edited by

J. M. M. ENGELS

Plant Genetic Resources Centre/Ethiopia, Addis Ababa, Ethiopia

J. G. HAWKES

School of Continuing Studies, University of Birmingham, UK

MELAKU WOREDE

Plant Genetic Resources Centre/Ethiopia, Addis Ababa, Ethiopia

CAMBRIDGE UNIVERSITY PRESS

Cambridge
New York Port Chester
Melbourne Sydney

CAMBRIDGE UNIVERSITY PRESS
Cambridge, New York, Melbourne, Madrid, Cape Town, Singapore, São Paulo

Cambridge University Press
The Edinburgh Building, Cambridge CB2 8RU, UK

Published in the United States of America by Cambridge University Press, New York

www.cambridge.org
Information on this title: www.cambridge.org/9780521384568

First published 1991
This digitally printed version 2008

A catalogue record for this publication is available from the British Library

ISBN 978-0-521-38456-8 hardback
ISBN 978-0-521-06553-5 paperback

Contents

Contributors

ABEBE, Dawit. c/o Coordinating Office for Traditional Medicine, PO Box 5117, Addis Ababa, Ethiopia.

ALEMAYEHU, Fekadu. Institute of Agricultural Research, Holetta Research Centre, PO Box 2003, Addis Ababa, Ethiopia.

AMEHA, Mesfin. Institute of Agricultural Research, Jima Research Centre, PO Box 2003, Addis Ababa, Ethiopia.

BELAYNEH, Hiruy. Institute of Agricultural Research, Holetta Research Centre, PO Box 2003, Addis Ababa, Ethiopia.

DEMISSIE, Abebe. Plant Genetic Resources Centre/Ethiopia, PO Box 30726, Addis Ababa, Ethiopia.

de VLETTER, J. German Agency for Technical Cooperation (GTZ), PO Box 60054, Addis Ababa, Ethiopia.

DOGGETT, H. 38a Cottenham Road, Histon, Cambridge CB4 4ES, UK.

EDWARDS, Sue B. c/o Asmara University, PO Box 1220, Asmara, Ethiopia.

ENGELS, Jan M. M. IBPGR Regional Coordinator for South and Southeast Asia, c/o NBPGR, Pusa Campus, New Delhi 110012, India (permanent address: Wörbesgarten 6a, 6239 Eppstein-Ehlhalten, Federal Republic of Germany).

FEYISSA, Regassa. Plant Genetic Resources Centre/Ethiopia, PO Box 30726, Addis Ababa, Ethiopia.

GEBRE, Hailu. Institute of Agricultural Research, Holetta Research Centre, PO Box 2003, Addis Ababa, Ethiopia.

GEBRE EGZIABHER, Tewolde Berhan. Asmara University, PO Box 1220, Asmara, Ethiopia.

GEBRE-MARIAM, Hailu. Institute of Agricultural Research,

Holetta Research Centre, PO Box 2003, Addis Ababa, Ethiopia.

GOETTSCH, E. Scharweg 72, Kiel, Federal Republic of Germany.

HAGOS, Estifanos. c/o Coordinating Office for Traditional Medicine, PO Box 5117, Addis Ababa, Ethiopia.

HANSON, Jean. International Livestock Centre for Africa, PO Box 5689, Addis Ababa, Ethiopia.

HAWKES, John G. University of Birmingham, c/o School of Continuing Studies, PO Box 363, Birmingham B15 2TT, UK.

KEBEDE, Yilma. Institute of Agricultural Research, Holetta Research Centre, PO Box 2003, Addis Ababa, Ethiopia.

KETEMA, Seyfu. Debre Zeit Agricultural Research Centre/AUA, PO Box 32, Debre Zeit, Ethiopia.

LAZIER, John R. 187 King Street, West Cobourg, Ontario, Canada KgA 2M8.

MEHRA, K. L. c/o National Bureau of Plant Genetic Resources, New Delhi 110012, India.

MEKBIB, Hailu. Plant Genetic Resources Centre/Ethiopia. PO Box 30726, Addis Ababa, Ethiopia.

MENGISTU, Alemayehu. c/o Dr John Lazier, PO Box 5689, Addis Ababa, Ethiopia.

MENGISTU, Solomon. International Livestock Centre for Africa, PO Box 5689, Addis Ababa, Ethiopia.

SENDEK, Enyat. Plant Genetic Resources Centre/Ethiopia, PO Box 30726, Addis Ababa, Ethiopia.

SMITHSON, John B. Centro Internacional de Agricultura Tropical, Apartado Aereo 6713, Cali, Colombia.

TESEMMA, Tesfaye. Debre Zeit Agricultural Research Centre/AUA, PO Box 32, Debre Zeit, Ethiopia.

TULLU, Abebe. Debre Zeit Agricultural Research Centre/AUA, PO Box 32, Debre Zeit, Ethiopia.

WOREDE, Melaku. Plant Genetic Resources Centre/Ethiopia, PO Box 30726, Addis Ababa, Ethiopia.

List of acronyms used in this volume

AAASA	Association for the Advancement of Agricultural Sciences in Africa
ADD	Agricultural Development Department of MOA (Ethiopia)
AVRDC	Asian Vegetable Research and Development Center
CADU	Chilalo Agricultural Development Unit
CATIE	Centro Agronómico Tropical de Investigación y Enseñanza (Costa Rica)
CEPGL	Communaute Economique des Pays des Grands Lacs
CGIAR	Consultative Group on International Agricultural Research
CIAT	Centro Internacional de Agricultura Tropical (International Center of Tropical Agriculture) (Colombia)
CIBC	Commonwealth Institute for Biological Control
CIMMYT	Centro Internacional de Mejoramiento de Maíz y Trigo (International Centre for Maize and Wheat Improvement) (Mexico)
CIP	Centro Internacional de la Papa (International Potato Centre) (Peru)
CSIRO	Commonwealth Scientific and Industrial Research Organisation (Australia)
ESC	Ethiopian Seed Corporation
FAL	Forschungsanstalt für Landwirtschaft (Federal Republic of Germany)
FAO	Food and Agricultural Organization of the United Nations
FNE	Forage Network in Ethiopia

GTZ	Deutsche Gesellschaft für Technische Zusammenarbeit (German Agency for Technical Cooperation)
IAR	Institute of Agricultural Research (Ethiopia)
IARCs	International Agricultural Research Centres
IBPGR	International Board for Plant Genetic Resources
ICAR	Indian Council of Agricultural Research
ICARDA	International Center for Agricultural Research in the Dry Areas (Syria)
ICRISAT	International Crops Research Institute for the Semi-Arid Tropics (India)
IDRC	International Development Research Center (Canada)
IITA	International Institute of Tropical Agriculture (Nigeria)
ILCA	International Livestock Centre for Africa
ILRAD	International Laboratory for Research on Animal Diseases (Kenya)
IRAZ	Institut de Recherche Agricole et Zootechnique
IRGC	International Rice Germplasm Center
IRRI	International Rice Research Institute (Philippines)
IRTP	International Rice Testing Program
ISNAR	International Service for National Agricultural Research
ISTA	International Seed Testing Association
IUCN	International Union for the Conservation of Nature
IUFRO	International Union of Forestry Research Organisations
MOA	Ministry of Agriculture (Ethiopia)
NAS	National Academy of Sciences
NBPGR	National Bureau of Plant Genetic Resources (India)
NIHORT	National Horticultural Research Institute (Nigeria)
NOAA	National Oceanographic and Atmospheric Administration (USA)
NPGS	National Plant Germplasm System (USA)
OECD	Organization for Economic Cooperation and Development
ORSTOM	Office de la Recherche Scientifique et Technique d'Outre-mer (France)
OXFAM	Oxford Committee for Famine Relief
PGRC/E	Plant Genetic Resources Centre/Ethiopia
RRC	Relief and Rehabilitation Commission (Ethiopia)
SADCC	Southern African Development Coordination Conference
SAREC	Swedish Agency for Research Cooperation
SIDA	Swedish International Development Authority

UNDP	United Nations Development Programme
UNEP	United Nations Environment Programme
UNESCO	United Nations Educational, Scientific and Cultural Organization
USDA	United States Department of Agriculture
USDA/GRIN	United States Department of Agriculture/Genetic Resources Information Network
WARDA	West African Rice Development Association
WWF	World Wide Fund for Nature (formerly the World Wildlife Fund)

Preface

Plant genetic resources constitute the building blocks of all modern plant breeding. They form the raw material from which new varieties have been systematically bred to meet the growing need for more food. These traditional genepools are an invaluable asset to the welfare of mankind and should be preserved, both for current use and for posterity. Loss of genetic diversity is detrimental to crop improvement programmes. To prevent this loss countries in all parts of the world are endeavouring to conserve and utilize these precious materials. The plant genetic resources (PGR), thus, must be systematically collected, characterized, evaluated, documented and conserved, for effective utilization. This is all the more important, now, since agriculture is becoming more and more industrialized, thus leading to a narrowing of the genetic base, demanding genetic uniformity and causing vulnerability to pest and disease attacks, all of which pose high risks to sustainable agricultural production systems. The variability in landraces, and the primitive cultivars held by traditional farming societies, can provide the world with appropriate raw material to stop such unwanted processes. It is therefore essential to conserve it at all costs.

The urgency and the need to collect and conserve plant genetic wealth was advocated some three decades back by the Food and Agricultural Organization of the United Nations (FAO), and since the 1960s the network of activities in this context has spread considerably. Since the establishment of the International Board for Plant Genetic Resources (IBPGR) in 1974 much has been done to create a better awareness in plant genetic resources activities and to help developing countries with training and equipment to this end. The International Agricultural Research Centres have also been equally instrumental in conserving and utilizing the genetic resources of several major crops and offering the enhanced materials to Third World national breeding programmes. The more recent concern of the FAO Commission on Plant Genetic Resources is equally laudable in this direction.

With the above international developments, national programmes have also gradually been strengthened, for instance in Brazil, Ethiopia, India, Indonesia and the Philippines. The Plant Genetic Resources Centre/Ethiopia (PGRC/E) is an excellent example of such a functional national programme of a country which in this case is a well-known centre of diversity and domestication for several world crops.

The PGRC/E was established in 1976 and with the assistance of the government of the Federal Republic of Germany has developed into a full-fledged genebank. PGRC/E, in collaboration with the Deutsche Gesellschaft für Technische Zusammenarbeit (GTZ), organized the First International Symposium on the Conservation and Utilization of Ethiopian Germplasm in Addis Ababa, Ethiopia, from 13 to 16 October 1986. Sixty-one participants from Ethiopia and 29 from overseas took part in this symposium.

This book presents a synthesis of the activities on plant genetic resources in Ethiopia and the richness of its plant wealth, laying emphasis on economic plants of traditional use. The work is dealt with in four sections, each of which highlights Ethiopia as a centre of diversity of crop plants and their wild relatives, laying stress on their collection and conservation and giving detailed accounts of activities in the evaluation and utilization of these national assets. The international linkages of PGRC/E with International Agricultural Research Centres and other regional and national programmes have been summarized in the last chapter.

It is felt that this book will add to the existing literature on plant genetic resources and will be useful not only to scientists, but also to teachers, policy makers and conservationists. It might in particular be useful to other national programmes which are being developed at present and which might require an easy reference source.

The editors are grateful to the contributors, to the secretarial staff of PGRC/E, to Mrs Caryl Sheffield who was responsible for the copy editing of the majority of the chapters and to the various organizations which have supported the Organizing Committee of the symposium in various ways. Special mention should be made here of the GTZ who financed the production of the proceedings of the symposium and of IBPGR for their contribution to the printing of the cover. Undoubtedly, the book is an outcome of the fruitful cooperation and the sincere efforts of all those involved.

J. M. M. Engels
J. G. Hawkes
Melaku Worede

Part I

General introduction

1

An Ethiopian perspective on conservation and utilization of plant genetic resources

MELAKU WOREDE

Introduction

The Ethiopian region is characterized by a wide range of agro-climatic conditions, which account for the enormous diversity of biological resources that exist in the country. Probably the most important of these resources is the immense genetic diversity of the various crop plants grown in the country.

The indigenous landraces of the crop plant species, their wild relatives and the wild and weedy species which form the basis of Ethiopia's plant genetic resources, are highly prized for their potential value as sources of important variations for crop improvement programmes.

Populations of these forms of plant species also represent sources having the greatest potential for genetic diversity and can therefore serve as invaluable means to fill the gaps that still exist in the available base of genetic diversity in the world collection of many major crop species. Among the most important traits which are believed to exist in these materials are earliness, disease and pest resistance, nutritional quality, resistance to drought and other stress conditions, and characteristics especially useful in low-input agriculture.

Scientists from many parts of the world have identified highly desirable genetic characteristics in relatively few germplasm collections of various crop species and they are currently being utilized intensively in a number of breeding programmes. Preservation of the indigenous stock has a particular significance in the country's breeding programmes as characters of resistance and adaptation needed by breeders to solve acute national problems exist in these materials.

Much of the existing diversity is in constant danger of being irretrievably lost as the normally lower-yielding indigenous landraces are being rapidly replaced by introduced or improved cultivars. Seeds imported as food grain by relief agencies pose an even more serious threat, as shown during the recent drought when farmers were forced as a result of the food shortages to eat their own seeds in order to survive or to sell them for consumption purposes. Entire ecosystems are being demolished with advances in agriculture and changes in land use.

Scientists have long realized the dangers of genetic erosion and have stressed the need for timely action to salvage the country's still abundant genetic resources. The accumulated efforts of these scientists and those of various concerned national and international organizations led to the establishment, in 1976, of the Plant Genetic Resources Centre (PGRC/E) in Ethiopia. The Centre has the following major objectives:

- to promote the collection, evaluation, documentation and scientific study of crop germplasm in Ethiopia, East Africa and adjacent regions;
- to preserve germplasm by long-term storage and maintenance in order to make valuable germplasm available to breeding programmes;
- to provide germplasm for breeding programmes aimed at the development of such characters as higher yield, better quality, and disease and pest resistance;
- to introduce new crop germplasm into Ethiopia by means of exchange with other institutions.

The story pertaining to the establishment of PGRC/E and the role played by the International Board for Plant Genetic Resources (IBPGR) and the German Agency for Technical Cooperation (GTZ) in attaining this goal are documented in a previous report (Worede, 1983a).

In this review an overall perspective of the various crop genetic resources activities of the Centre is presented. Other aspects covered in some detail include problems related to the Centre's germplasm collections with particular reference to past and present situations as well as utilization. PGRC/E efforts to coordinate ongoing and projected activities at national and international levels are also discussed.

Exploration and collection

In previous years, several scientific expeditions were made in Ethiopia by a number of international (and national) explorers and many species identified as being worthy of collection and preservation for their genetic diversity. Much of the work, however, was confined to roadsides and less remote areas and in many instances the collections were biased to meet specific needs. Organized missions to collect germplasm throughout Ethiopia started with the creation of the genebank in 1976. More than 115 successful collecting missions have since been undertaken, involving nearly all administrative regions of the country and covering a relatively broad range of crop types and agro-ecological zones. In the field collecting operation, priority has been given to those species of greatest economic and social importance which are most threatened by genetic erosion. Areas chosen for collecting are those where the danger of germplasm loss due to expansion of new varieties, natural disasters and changes in land use is greatest.

Apart from the conventional germplasm explorations conducted routinely by the Centre, systematic collecting of large samples of indigenous landraces in drought-prone areas has been launched in collaboration with the Ethiopian Seed Corporation (ESC). These samples are being stored at strategic seed reserve centres and subsequently redistributed to farmers as required. This is being done primarily to avoid drastic losses of genetic variability in situations where the farmer either is forced to eat his own seed or is replacing traditional varieties with imported seeds distributed through relief agencies.

Germplasm collecting in Ethiopia is, therefore, developing into a major national and international effort responding to emergency situations and the need to build up comprehensive germplasm collections for sustained provision of such materials to breeding programmes. As large numbers of samples accumulate, however, questions of efficiency in the collecting and maintenance of the material will inevitably arise.

To avoid such problems, measures for rational planning of future collecting activities to explore areas of high genetic diversity are already being undertaken. A case in point is the series of expeditions (Seegeler, 1986) carried out recently to study genetic diversity of oil plants in Ethiopia, which resulted in the identification of areas for comprehensive collecting of such crops in the future. Such an undertaking is a long-term proposition, however, and will require a more

Table 1. *Germplasm collecting at PGRC/E (June, 1986)*

Crop type	Total number of collected and donated accessions	%	Number of accessions collected by PGRC/E	%	PGRC/E collections as percentage of total number of accessions
Cereals	28 849	73.1	8219	56.4	28
Oilseeds	4490	11.4	2355	16.2	52
Legumes	4170	10.6	2890	19.8	69
Spices	749	1.9	520	3.6	69
Coffee	702	1.7	140	0.9	20
Medicinal	62	0.1	61	0.4	98
Others	452	1.2	397	2.7	88
Total	39 474	100.0	14 582	100.0	—

extensive survey of large areas representing a broad range of agro-ecological conditions and the collection of pertinent data over many years.

Seed conservation

Indigenous landraces, mostly populations, form the bulk of the existing germplasm collections currently maintained by PGRC/E (Table 1). Such materials are maintained as active or base collections at +4 °C and −10 °C, respectively, following established seed-processing procedures (Krauss, 1983). Seed viability is maintained through regeneration of material in the field at ecologically appropriate locations.

Field genebanks

The main focus of the living collection is on coffee. It is now universally agreed that Ethiopia is the primary centre of diversification for *Coffea arabica* and perhaps the only region, covering the area bordering southern Sudan and part of Uganda, where the species occurs spontaneously. The genetic diversity that exists is tremendous and this has great significance for the economy of the country and the rest of the coffee growing world.

In realization of the urgent need for effective measures to preserve and utilize the existing variability, which at present is being disastrously eroded, a special effort is being made to conserve coffee in its natural growing environment. This includes conservation of the semi-cultivated coffee in areas where the forest coffee occurs spontaneously, and where large variation exists, and maintenance of the forest coffee in its natural ecosystem in certain protected areas, the so-called genetic reserves (Worede, 1982). A field genebank, comprising some 700 accessions, is being established within the Kefa administrative region. In the future this genebank will be extended into other appropriate sites as the size of the collection continues to increase.

Other living collections include *Phytolacca* spp., commonly known as 'endod', *Ensete ventricosum* and several spices and root crops, maintained at different sites in the country in collaboration with existing agricultural research and other relevant scientific institutions.

Situation with existing collections

In previous years, germplasm collections were maintained mostly in small holdings by plant breeders who specialized in a few major food and other crops. In such situations, many of the landraces and wild types were probably excluded unless they exhibited characters of immediate breeding interest.

The genebank's present holdings (see Table 1) are an assemblage of old collections that were acquired through transfer from several breeding centres in Ethiopia, donations by various national and international organizations, and material collected in the field. The collections that were acquired through transfer from breeding centres and other sources are, for the most part, deficient in documented records. Many of these old collections are probably not representative of the populations from which they were sampled and will have been subjected to subsequent natural and artificial selection. Poor storage conditions and improper maintenance of such material may also have resulted in losses of seed viability.

Germplasm collections in Ethiopia are relatively secure at present under the improved storage conditions at the Centre. Ethiopian materials are represented in world collections, with numerous samples being exchanged among genebanks and breeders (Mengesha, 1975; Worede, 1983a). Duplication is inevitable but this serves as a safeguard against losses and allows populations to be maintained and studied under more diverse agro-ecological conditions. This may, however, be of little value unless all passport and evaluation data are made available by the various centres holding such collections.

Evaluation and documentation

Scientific studies of germplasm at PGRC/E focus primarily on recording information which is essential in breeding activities. Such studies include characterization and preliminary and further evaluation of germplasm and these are carried out in collaboration with breeders and other scientists. Appropriate descriptors, developed jointly with the plant breeders, are utilized together with those provided by IBPGR whenever possible.

Further evaluation of material deals mainly with in-depth screening in the laboratory or greenhouse and in the field for disease/pest resistance and adaptation to stress conditions. Screening under low-input conditions is also included.

With landraces which are represented by highly heterogeneous populations, maintenance or handling of accessions is often difficult and is associated with a number of problems (e.g. genetic drift, contamination of material by foreign pollen) which occur during multiplication and evaluation. With the self-pollinating material, which is composed of a large number of distinct genotypes, some compromise is often needed to overcome the practical problems which such diversity poses to the evaluator. At PGRC/E, this is done by subdividing samples into distinct agro-morphotypes to form the various components, which are then multiplied/characterized separately with their own accession numbers. With the cross-pollinating types, subdividing is often not necessary as evaluation is dealt with at the population level. Certain isolation techniques such as multiple bagging are, however, applied in developing the various components.

Data generated in the field and in the laboratory during evaluation are collected and processed through the computer-assisted documentation system at the Centre for dissemination to breeders or other users (Engels, 1983).

Genetic resources information available from PGRC/E may, in the future, be incorporated into a regional or global network of information through the integration of pertinent data into an international data bank (lead institutes, world directory, etc.).

Utilization of germplasm

The genebank in Ethiopia is designed primarily to provide a sustained supply of germplasm material required by plant scientists for the development of improved or new, superior crop varieties. Its major activities are therefore geared to meet this requirement and are aligned to follow each other in a logical sequence.

In many genebanks there are certain gaps in those activities linking breeding with other related aspects which are the responsibility of the germplasm user. This problem has been largely overcome in Ethiopia as a result of integration of the Centre's utilization oriented activities

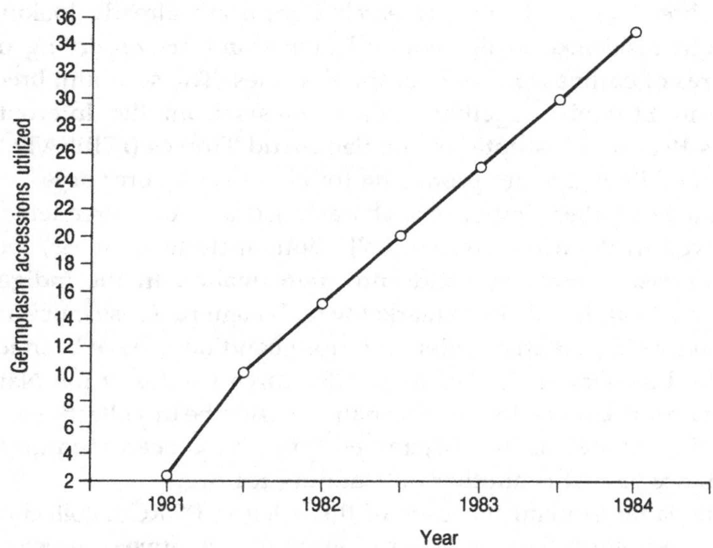

Fig. 1. Use of PGRC/E germplasm in national yield trials.

into the national breeding programmes. At the regional/international level, a three-point-contact approach involving the national breeder in Ethiopia, scientists from the collaborating institution(s) abroad and PGRC/E has often been successful in conducting such activities to mutual benefit.

In the past, cooperative efforts of this nature have often been subject to serious constraints in the national programmes, mainly due to the lack of evaluation data and insufficient interest among qualified plant breeders to deal with the often low-yielding primitive landraces (Worede, 1983b). However, the situation is now changing and as germplasm data accumulate, breeders are becoming increasingly interested in the utility of indigenous landraces. Many PGRC/E accessions are already incorporated into the national crop improvement programmes through the national yield trials and are even utilized in specific areas of breeding, including those related to resistance and adaptation. Figure 1 shows the approximate progress of PGRC/E germplasm utilized in national yield trials since 1981 (Anonymous, 1986).

Much progress in the effective use of the PGRC/E germplasm collection is also attributable to the excellent cooperative relationships formed with the breeders and their active participation in the screening of germplasm material for specific characters of interest to them.

Thus breeders at Holetta Research Centre are already looking for drought resistance in *Brassica* and *Linum* and are observing useful features of earliness in some of the flax lines. The sorghum breeding team in Ethiopia, together with scientists from the International Crops Research Institute for the Semi-Arid Tropics (ICRISAT), have identified lines that are promising for either stalk borer or leaf streak resistance. Other important characteristics these breeders have observed in PGRC/E sorghum collections include earliness, lodging and disease resistance, yield and grain quality. In the indigenous durum wheat, lines with remarkably high stem rust resistance as well as good yield potential under low-input conditions have been identified by breeders at the Debre Zeit Research Centre of the National Agricultural University. In the native castor bean collections, early maturing as well as short-stemmed types have been identified and resistance to rust is another outstanding feature.

The value to plant breeders of the existing PGRC/E collections of indigenous germplasm largely resides in the adaptive complex that is inherent in such stock, apart from such highly prized characters as quality, nutritional value and disease/pest resistance. This must also be viewed from the standpoint of the country's enormous wealth of germplasm resources which have not yet been adequately explored. Much of the diversity which exists among the primitive landraces, their wild relatives or progenitors and the wild and weedy species, is still untapped.

The potential value of these resources, apart from those with characters of resistance and adaptation, is still difficult to assess. Such materials often display crude forms of characters, some desirable and some deleterious. With the development of new and advanced techniques like the new biotechnologies (Hansen *et al.*, 1986) it should be possible to study and efficiently transfer only the useful genes which would, in effect, increase the value of these primitive/wild gene pools. It is imperative, however, that indigenous capabilities be developed for safe and effective employment of such technologies in scientific studies and utilization of germplasm. Pertinent policies must also be formulated to reflect national and international interests.

Germplasm distribution/exchange

Samples of germplasm accessions are distributed to crop improvement programmes in Ethiopia on the basis of specific requests by breeders and other scientists actively utilizing such material. Also, other projects receive germplasm for observation, including adaptive trials, under specific types of farming conditions. In some instances, excess material from multiplication/evaluation activities, bulked on a crop-by-crop basis, has been sent to agricultural extension agents for study purposes, including some screening work, in the various localities.

The Centre also acquires material from external sources for storage and/or distribution as required by the national breeding programmes. This is usually through donations, including repatriation of germplasm of Ethiopian origin or exchange.

Ethiopian germplasm is actively utilized in current breeding work worldwide, perhaps a good deal more than the country is given credit for. Most of these materials were acquired through past explorations but numerous samples representing a wide range of crop types have been collected and utilized by regional projects like the International Livestock Center for Africa (ILCA) (forage grasses and legumes), the International Center for Agricultural Research in the Dry Areas (ICARDA) (grain legumes, ICRISAT (sorghum/chickpea) and the International Development Research Center (IDRC) Oil Crop Network for East Africa and the Indian Region (oilseeds), usually through cooperative arrangements involving the appropriate national projects in Ethiopia. PGRC/E has cooperative links with most of these centres and material is exchanged through specific joint research work involving national breeding programmes and other projects in Ethiopia. A training component is often incorporated.

Among the major constraints on germplasm transfer to and from the genebank are those related to seed health. In the absence of a national quarantine system in Ethiopia, the Centre solicits the assistance of the national crop protection programmes to check the possible introduction of diseases and pests along with incoming material. Seed dispatched abroad is certified by the appropriate agency within the Ministry of Agriculture (MOA). The seed health unit currently being established at PGRC/E is a further development in the Centre's continuing effort to minimize any risk of introducing diseases and pests, thereby facilitating a timely and more efficient flow of germplasm material.

Cooperative links and perspectives

The cooperative role played by PGRC/E at national and international levels has already been mentioned in connection with germplasm utilization and exchange. It should be added that there are cooperative links established bilaterally within Ethiopia with various relevant institutions: ILCA (on forage germplasm resources and exchange of indispensable services); the Pathobiology Institute of the Addis Ababa University (*Phytolacca* spp. for medicinal and industrial use); and the USSR Phytopathology Laboratory at Ambo on disease resistance study. At the international level, collaborative ties exist with IBPGR, the Food and Agriculture Organization of the United Nations (FAO) and various national and international centres working in the same field, namely, ICRISAT, ICARDA, the International Rice Research Institute (IRRI), the International Institute of Tropical Agriculture (IITA), Gatersleben Genebank in the Democratic Republic of Germany, Bari Germplasm Institute in Italy, the USSR Genetic Resource System, and Forschungsanstalt für Landwirtschaft (FAL) in Braunschweig, Federal Republic of Germany, among others.

Genetic resources work on rice, currently being undertaken jointly with IRRI, represents a new collaborative link at the international level. It is directly related to a major PGRC/E objective, i.e. to introduce new germplasm material to Ethiopia. Scientific studies of wild rice found in the country, carried out by an Ethiopian scientist trained at IRRI, are also included in such cooperation. This kind of cooperative development is very significant for Ethiopia as it contributes towards the country's current effort to introduce rice as a staple crop, especially in marginal areas where rice appears to be adapted.

Genetic resources work, as Hawkes (1985a) puts it, consists of a series of distinct activities or stages which follow each other in a logical sequence. Each stage has an effect, good or bad, on the one following it according to the efficiency with which it is carried out. This means that for any genebank to be properly functional, all aspects of genetic resources work must be adequately encompassed in its programme and existing gaps filled.

The Ethiopian genebank has yet to claim a fully fledged status in this regard although most of the activities pertaining to its major objective are taking a definite shape. Many gaps still exist in the strategies and scientific approaches currently employed by the Centre to tackle the enormous qualitative and quantitative dimensions of those conservation problems unique to Ethiopia. An opportunity exists, however, to fill these gaps. New approaches and methodologies for scientific preservation can be developed on the basis of priority objectives through a well defined nationally or internationally integrated network of activities. The Centre will continue to make full use of such an opportunity and whenever possible seek to apply the progressively advancing technologies that the international community is providing.

Gaps also exist in the available base of genetic diversity of the various crops in the collections, as wild gene pools have hardly been included in PGRC/E collecting operations to date. Nor have the distribution and degree of genetic diversity of crops in the different ecological areas of the country been adequately explored to allow a broad enough representation of material. Again based on priority objectives, a relatively wide array of crop types and locations will therefore be covered in future exploration/collecting missions. More comprehensive collections will represent wider crop and plant categories, e.g. food/feed crops, medicinal plants, industrial crops and other less known but potentially useful plant species. Every effort will also be made to salvage germplasm material that is threatened with extinction by the drought prevailing in the country and by factors such as land clearing in forest areas and ploughing under pastures.

Surveys

Much work still remains to be done to define clearly the situation of existing PGRC/E collections which comprise both old and new material. Such a task is significant because it is a prerequisite to any elaborate future exploration, since the loss of genetic diversity of material already in storage must be taken into account.

Survey work, involving an assessment of materials for each crop already in the genebank and for those likely to be still available in the field, is therefore in order and a consultancy report for this complex task is already at hand (Hawkes, 1985b).

In devising strategies and approaches for conservation work, one should take into account emergency situations on the one hand and the need to be selective or rational on the other. Given this situation, it would be only logical for the Centre to capitalize on areas where genetic diversity of a given crop species is concentrated, while emergency operations in response to drought and changes in land use continue. A more comprehensive approach to conservation must also take into account the different requirements of the various plant categories that the programme seeks to encompass.

Need for a diversified approach

Many new advances have been made in standardizing the methodologies and approaches in the conservation of germplasm, the most conventional being *ex situ* conservation, i.e. conservation as seed, tissue or pollen or as plants in field genebanks. The Centre will continue to work on reliable and low-cost, *ex situ* conservation techniques, including storage of seeds in permafrost areas as has been suggested for developing countries (Swaminathan, 1983).

Technology for *in vitro* propagation and preservation of vegetatively propagated materials is developing rapidly and PGRC/E will look into the benefits to be gained in terms of plant health and efficiency of maintenance and distribution of material, not to mention the employment of such a technique in collecting material whose seeds are difficult to conserve (Mix, 1983).

There are other, less conventional approaches to the future conservation of germplasm of various crops species, e.g. pollen storage. Cryogenic (ultra-cold) storage, using liquid nitrogen at a temperature of −196 °C is a further interesting advance and research is in progress in the USA on the possible use of such technology in the preservation of recalcitrant seeds and other plant propagules (National Seed Storage Laboratory, Fort Collins, Colorado, personal communication). At present, little or no information is available regarding the possible routine use of these technologies.

Neither of these technologies nor any single system would, however, represent a mechanism by which safety of genetic diversity is ensured on a long-term basis. It is, therefore, mandatory for a country like Ethiopia with genetic reserves to resort to a more diversified conservation approach with a view to minimize losses of germplasm which are likely to occur where only preservation *ex situ* is employed. New approaches and strategies, as they apply to the conservation problems unique to prevailing situations, should be considered and adopted by the country.

Conservation *in situ*, i.e. conservation of landraces and wild relatives in their natural habitats in areas where genetic diversity exists and where wild/weedy forms are present, often hybridizing with cultivated forms, represents a vital component of preserving diversity. Such an approach is essential because it relates to continuity of the evolutionary systems that are responsible for generation of variability.

As is beginning to happen in Ethiopia, wild relatives of crops could be preserved in natural parks and biosphere reserves in a state of continuing evolution through multi-institutional arrangements in areas where large tracts of land of this nature still exist. Such a conservation approach may provide a less expensive protection of wild gene pools than *ex situ* measures for developing countries which often face practical constraints in providing optimal storage and maintenance facilities for germplasm (Swaminathan, 1983). This type of preservation also represents deliberate protection of remaining habitats and the species they include, as entire preservation of vast tracts with *in situ* conservation of animals and plants is extremely important in slowing the rate of species extinction.

Role of farmers

In its broader sense, *in situ* conservation could include growing out of material without conscious selection on the site where the seed was collected, although strictly speaking this is applied to natural populations regenerated naturally (FAO, 1987).

This aspect of a diversified strategy of germplasm conservation relates to a grass-roots approach involving farmers and community workers. In Ethiopia as well as in many other developing countries, farmers play a central role in the conservation of germplasm, as they hold the bulk of genetic resources. Peasant farmers always retain some seed stock for security unless circumstances dictate otherwise.

With certain horticultural crops grown by small farmers in Ethiopia, not only are crop species maintained in a dynamic state of evolution under conditions that are almost ideal in respect to sustaining original population structures, but also new variations are created. In Arsi region, for example, one often observes new, different forms of fruits in *Brassica carinata* and *B. nigra* on farms where such crops are grown in mixtures. This is attributable to introgression that may have resulted from natural intercrossing between the two closely related species, as presumably also reflected in the relatively high inter-population diversity that is often observed among PGRC/E *Brassica* collections.

With coffee, farmers often plant populations of local types on small areas usually for safety purposes alongside the more uniform coffee berry disease (CBD) resistant lines which are distributed by the Coffee Improvement Project in Ethiopia. This is a tremendous support in the conservation of such a crop which in the first place is difficult to store safely on a long-term basis as seed.

Similarly, as part of the national coffee conservation programme, steps are being taken to conserve the semi-cultivated coffee on peasant farms in areas like Kefa, Ilubabor and part of Welega where the forest coffee occurred spontaneously or, in the case of backyard coffee (Harerge, Sidamo and Welega regions), in nurseries maintained by farmer cooperatives.

Farmers in Ethiopia and other developing countries with similar backgrounds should, therefore, be encouraged to continue to maintain small holdings of seed stock as this would represent some form of *in situ* (on-site) conservation of germplasm across a broad range of agro-ecological conditions. This may even be extended to encourage the farmers or farmer cooperatives in Ethopia to play the curator/farmer's role: grow limited samples of endangered landraces native to the region (Mooney, 1983), exchange material within a network of such activity, etc. This becomes even more significant in the long run as the introduction of modern farming practices progresses in these countries. Such cooperation by the farming community would provide an additional support to *ex situ* measures of landrace preservation, at least in providing a long-term protection against extinction of native cultivars, and would complement *in situ* conservation in natural parks or biosphere reserves.

Some allocation of funds should, however, be made through international support for a wider network of such activity to assist the farmer as he undertakes the curator/farmer's role on account of the cost–benefit implications that are associated with such an exercise. Genebanks in cooperation with locally available agricultural extension agents could assist in the sampling of material and in providing the technical support and monitoring necessary for systematic handling and scientific studies that should accompany such conservation.

Evaluation

Future scientific studies of germplasm will be rather more comprehensive given the diverse possible uses of the country's wide resources of plant material. Further evaluation work will be based on such acute national problems as drought and other stress conditions, as well as disease and pest resistance. Evaluation of germplasm under low-input conditions, already initiated with Brassicas, will be expanded to include various other crop types. New and appropriate methodologies will aim to make the greatest possible use of the adaptive complex that exists within the indigenous stock.

Germplasm utilization/cooperative roles

The precise role of PGRC/E in international cooperation and transfer of material will depend largely on the country's future exchange policy. It is certain, however, that, without prejudice to the principles of international cooperation, mutualism of benefits from collaborative activities will continue to be the focal point. This can be viewed in many different ways depending on the nature of the collaborative work and can be dealt with on a case-by-case basis once pertinent policy guidelines are provided. In general, however, issues that promote national capabilities in plant genetic resources, plant breeding and seed multiplication in Ethiopia would, as has already been happening, provide some basis for such cooperation. For institutions working on common mandate crops, the sharing of responsibilities and cost on the basis of common advantages provides a useful basis for joint exploration activities, evaluation and subsequent utilization of material.

The three-point-contact arrangement mentioned earlier is one such venture which has been successful. Already in some use at PGRC/E, it has proved a useful mechanism in the transfer of technology and sharing of costs that otherwise might have drawn substantially from resources available for the national programme. The joint sorghum/chickpea collecting and characterization work with ICRISAT scientists is one good example of cooperation of this nature (Prasada Rao & Mengesha, 1981).

With regard to coffee, Ethiopia has already been identified as a suitable location for a base collection for Africa on account of its importance as a centre of genetic diversity of the crop. The task of maintaining active collections should be shared among the other African countries concerned, as was decided at the First Regional Workshop on Coffee Berry Disease, held in Addis Ababa in July 1982. Based on prevailing national policy, the Centre will pursue this line of regional cooperation with a view to fostering concerted efforts to salvage and effectively utilize the dwindling germplasm resources of this important commodity crop.

Training

Training of Ethiopian scientists and suitable technicians is another major objective which the Centre has pursued since its inception. As far as possible, training will constitute a vital component of any ongoing and projected activities, especially in areas where collaboration with other institutions exists.

Training of personnel from other developing countries will also be included and will continue to form another line of regional cooperation; indeed, PGRC/E is already playing a vital role in this field. An even better prospect exists with the possibility of a genetic resources training unit being established at the National Agricultural University in Ethiopia.

External support/future role

Most of the activities described so far are resource demanding and are too costly to be handled solely by a developing country like Ethiopia. External support will therefore be needed, especially for new areas or those that represent expansion of ongoing activities. Based on pertinent national policy, such support will be sought through collaborative links with regional/international institutions working in the same field or others that are involved in the promotion of genetic resources activities. The existing bilateral technical cooperative programme with the Federal Republic of Germany provides continued support on a follow-up basis.

It is true that PGRC/E is a national programme, mandated to operate primarily as a nucleus for the conservation of genetic resources in Ethiopia. But by virtue of the significance of the country as a global centre, it should, as is beginning to happen already, play a role as a regional base for genetic conservation in Eastern Africa and neighbouring regions.

References

Anonymous (1986). PGRC/E activity reports. Plant Genetic Resources Centre, Addis Ababa (mimeographed).

Engels, J. M. M. (1983). Documentation and information management at PGRC/E. *PGRC/E–ILCA Germplasm Newsletter*, **9**, 20–7.

Food and Agriculture Organization (1987). *Plant genetic resources, their conservation* in situ *for human use*. Forest Resources Division, Forestry Department, FAO, Rome.

Hansen, M., Busch, L., Burkhardt, J., Lacy, W. B. & Lacy, L. (1986). Plant breeding and biotechnology. *Bioscience*, **36**, 29–39.

Hawkes, J. G. (1985a). Plant genetic resources – the impact of the international agricultural centres. *CGIAR Study Paper no. 3*. World Bank, Washington DC.

Hawkes, J. G. (1985b). Report on a consultancy mission to Ethiopia for GTZ to advise PGRC/E on germplasm exploration, conservation, multiplication and evaluation. Birmingham (mimeographed).

Krauss, A. (1983). Organization of the seed processing and storage section at PGRC/E. *PGRC/E–ILCA Germplasm Newsletter*, **4**, 9–12.

Mengesha, M. H. (1975). Crop germplasm diversity and resources in Ethiopia. *In*: O. H. Frankel and J. G. Hawkes (eds), *Crop Genetic Resources for Today and Tomorrow*. Cambridge University Press, Cambridge, pp. 449–53.

Mix, G. (1983). The importance of *in vitro* techniques in germplasm conservation. *In*: K. J. Neddenriep and D. Wood (eds), *Genetic Resources and the Plant Breeder: the Next Ten Years*. GTZ, Eschborn, pp. 77–88.

Mooney, P. R. (1983). The law of the seed: another development and plant genetic resources. *Development Dialogue*, No. 1–2, pp. 6–173.

Prasada Rao, K. E. & Mengesha, M. H. (1981). A pointed collection of 'Zera-zera' sorghum in the Gambela area of Ethiopia. *Genetic resources progress report 33*, ICRISAT, Patancheru.

Seegeler, C. J. P. (1986). Genetic variability of oilcrops in Ethiopia. Consultancy report by GTZ for PGRC/E. Oosterbeek (mimeographed).

Swaminathan, M. S. (1983). Genetic conservation: microbes to man. Presidential Address, 15th International Congress of Genetics, New Delhi, 12 December 1983. IRRI, Philippines.

Worede, M. (1982). Coffee genetic resources in Ethiopia: conservation and utilization with particular reference to CBD resistance. *Proceedings, First Regional Workshop on Coffee Berry Disease*. Association for the Advancement of Agricultural Sciences in Africa, Addis Ababa, pp. 203–11.

Worede, M. (1983a). Crop genetic resources in Ethiopia. *In*: J. C. Holmes and W. M. Tahir (eds), *More Food from Better Technology*. FAO, Rome, pp. 143–7.

Worede, M. (1983b). Issues relevant to genebank management problems in relation to national and international programmes. *In*: K. J. Neddenriep and D. Wood (eds), *Genetic Resources and the Plant Breeder: the Next Ten Years*. GTZ, Eschborn, pp. 101–6.

Part II

The Ethiopian centre of diversity

2

The Ethiopian gene centre and its genetic diversity

J. M. M. ENGELS AND J. G. HAWKES

Introduction

Based on the concept of gene centres, developed by N. I. Vavilov in the 1920s, Ethiopia represents one of the eight centres in the world where crop plant diversity is strikingly high and where some of the crops concerned became domesticated. The concepts of centres of origin and diversity have evolved considerably since Vavilov's days as shown for instance by Harlan (1975) and by Hawkes (1983). Nevertheless, some of the basic characteristics which apply to the majority of the world's gene centres generally hold also for Ethiopia though varying from one crop to another. The highly dissected highland of Ethiopia includes natural barriers formed by mountains (up to approximately 5000 m above sea level) or ravines (sometimes more than 1300 m deep) where crop plants would have evolved in isolation under primitive agricultural conditions. To this must be added the ancient and very diverse cultural history of its people, thus providing many thousands of years of artificial selection within the landrace populations since the early days of agriculture. However, several crops which possess an extremely high diversity in Ethiopia do not follow all the basic principles required for a centre of origin in a given crop. In the case of barley, for instance, no wild relatives are known within the country nor is there any archaeological evidence to indicate early cultivation or domestication. In such cases the term 'secondary centre of diversity' has been used by Vavilov and would be correctly applicable to various other Ethiopian crop plants with an extremely high diversity (e.g. tetraploid wheats, lentil, faba bean and others). Schiemann (1951) considered Ethiopia to be an 'accumulation centre' for genetic diversity of certain crops which had not originated there (see under barley, below). For many other crops Ethiopia is a main centre of diversity as well as the probable area of domestication.

In the following paragraphs an attempt is made to provide information on the status of the Ethiopian centre for each crop in terms of whether it is a primary or secondary centre of diversity, and whether or not the crop in question is endemic. Relevant facts and figures are presented on the important traits for the major crops as well as for some of the minor crops that are unique to Ethiopia, in order to illustrate the importance of the Ethiopian gene centre for plant breeders throughout the world. As stated by Leppik (1970) it is a generally accepted rule today, though with some exceptions, that the primary and secondary gene centres of cultivated plants are the best places to find genuine resistance to common diseases and insect pests. Furthermore, some factors are mentioned which are causing an alarming threat to genetic diversity and which give full support and justification for the operation of a costly genebank within the Ethiopian gene centre.

Finally, the significance of Ethiopia as a source of important diversity in plants can be illustrated by the germplasm flow in and out of the country since historical times. Not only did the Fertile Crescent have contacts with the Ethiopian highlands in very early times but the Egyptians, the Arabs and the Indians (Mehra, Chapter 11) exchanged plant materials at later periods. From the early days of the European voyages of discovery, European countries began to profit from the plant genetic diversity of Ethiopia: first the Portuguese, then the Italians, Germans, Russians and others.

Cereals

Barley (*Hordeum vulgare*)

Barley ranks only third in terms of acreage in Ethiopia after teff and sorghum, but would seem to rank first or second in terms of phenotypic diversity (Engels, Chapter 9). The crop does not possess any related wild species in Ethiopia and this might support the idea that the species was introduced from the Near East in ancient times, possibly 5000–6000 years ago (Harlan, 1975; Purseglove, 1976). Since then the crop has formed unique morphotype groups, such as the *deficiens* and *irregulare* barleys. This extreme morphological diversity led Schiemann (1951) to propose the concept of 'accumulation centres' for barley in Ethiopia and hexaploid wheats in the Hindu Kush. She argued that both those regions, where the original wild prototypes did not exist, could be considered as areas where highly diverse natural and artificial selection had been responsible for great diversity; similar diversity might have occurred elsewhere in ancient times but had been strongly selected against subsequently. Several mutations unique to Ethiopia have also been found such as barley yellow dwarf virus resistance, high lysine gene, resistance to diseases such as powdery mildew, leaf rust, net blotch, *Septoria*, scald, spot blotch, loose smut and barley stripe mosaic virus (Qualset, 1975). Recently, considerable variation in the degree of drought resistance has been found, related partly to differences in the root phenotypes (Hettinger & Engels, 1986). Another example supporting the remarkable status of the Ethiopian barleys is given by Froest, Holm & Asker (1975) who found a pattern C of flavonoids which was almost completely confined to Ethiopia.

At present, one can still find extensively grown landraces of barley, some of them consisting of 10 or more clearly distinguishable components. Genetic erosion is caused mainly by replacement of barley by crops such as bread wheat, teff and, recently, also oats (J.M.M.E., personal observation). This replacement is clearly shown by the steady decrease in acreage since the early 1970s (from approximately 1.8 million hectares in 1971–2 to 0.75 million hectares in 1983–4).

Sorghum (*Sorghum bicolor*)

The Ethiopian sorghums may well be the most variable of all crops grown in the country. This could be a reflection of the wide variation in environments where the crop is being cultivated, ranging from altitudes below 400 m in a few cases up to almost 3000 m above sea level. The rainfall range over the whole of the country varies from approximately 600 mm to well above 2000 mm in the south-western part.

Of the five morphological sorghum races recognized, i.e. bicolor, guinea, caudatum, durra and kafir, all except kafir are grown in Ethiopia. In addition, many of the intermediate forms as well as several of the wild and weedy forms (e.g. *arundinaceum* and *aethiopicum*), can be found (Snowdon, 1955; Stemler, Harlan & de Wet, 1975; Doggett, Chapter 10; J.M.M.E., personal observation).

The diversity of sorghum encountered by Vavilov in the 1920s led him to believe that Ethiopia was the centre of domestication of the crop, and this hypothesis has been supported by many scientists since then. However, in a detailed account Stemler *et al.* (1975) recognized the overwhelming diversity in Ethiopia but presented arguments that only the race durra might have originated in Ethiopia

Table 1. *Phenotypic diversity expressed in the range, mean and* CV *of Ethiopian sorghum germplasm accessions for some quantitative characters*

Character	Range		Mean	CV (%)	*N*
	Minimum	Maximum			
Plant height (cm)	19	475	233.7	22.7	2599
Ear length (cm)	4	50	21.5	36.6	2511
Ear width (cm)	2	30	9.4	34.1	2599
Peduncle extension (cm)	1	44	14.7	53.8	2254
Number of days to 50% flowering	76	169	116.7	14.3	2603
Crude protein content (%)	5.0	15.3	9.6	13.1	3644
Thousand-grain weight (g)	6.0	61.1	28.5	35.5	200

Source: PGRC/E, unpublished data.

and that most probably the race durra-bicolor has further evolved there subsequently.

Some of the phenotypic diversity for certain quantitative characters is demonstrated in Table 1. Other important traits reported from Ethiopian sorghum are: cold tolerance (Singh, 1985); high lysine and protein content (Singh & Axtell, 1973), glossy seedlings – related to sorghum shootfly resistance (Maiti *et al.*, 1984); grain quality and resistance against grain mould (International Crops Research Institute for the Semi-Arid Tropics, 1985). Other disease and pest resistance has been reported from Ethiopia (Plant Genetic Resources Centre/Ethiopia, unpublished data). Recently, Subramanian *et al.* (1987) have reported high sugar content in the stalks of four Ethiopian accessions of the ICRISAT collection. Drought resistance has been observed in Ethiopia in several areas (J.M.M.E., personal observation).

At present the diversity in the sorghum crop is doomed to be reduced because of the adoption of improved local landraces as well as imported varieties and the replacement of sorghum mainly by maize (J.M.M.E., personal observation).

Wheat (*Triticum* spp.)

Vavilov (1931) was particularly impressed with the diversity of Ethiopian wheats. He collected very widely there in 1926, and recognized five species, though modern breeders prefer to class four of them as one tetraploid species, *T. turgidum*. Vavilov's species were:

1. *T. durum* subsp. *abyssinicum*

2. *T. turgidum* subsp. *abyssinicum*
3. *T. dicoccum*
4. *T. vulgare* (now *T. aestivum* hexaploid)
5. *T. polonicum*

These 'species' can still be seen in farmers' fields today. Vavilov was also impressed by the many endemic characters, such as violet-grained, beardless and half-bearded hard wheats (*durum* types), and similar forms of *turgidum* wheats. He was also struck by the parallel evolution of hard (*durum*) and soft (*turgidum*) forms, and considered very strongly that the divergence of morphological and physiological characters, clearly seen outside Ethiopia, had in this country not yet emerged. He noted that the wild tetraploid ancestor, *T. dicoccoides*, was not present in Ethiopia even though the Ethiopian wheats were considered by him to be very primitive.

Vavilov pointed out that the most widespread species of wheat in Ethiopia was *T. durum* whilst *T. dicoccum* was localized in the Hararge and Addis Ababa regions.

Durum wheat is still one of the major cereals in Ethiopia. It ranks fifth after teff, maize, sorghum and barley in acreage (Central Statistics Office, 1984; Tesemma, Chapter 22). In terms of genetic diversity it might compete well with barley and sorghum. Compared with other major wheat producing countries, the Ethiopian durum wheat accessions in the world germplasm collection, maintained by the United States Department of Agriculture (USDA), showed the highest diversity index (Jain *et al.*, 1975). One of the striking features of the Ethiopian material was the high percentage of polymorphic accessions, especially for glossy leaf sheath and kernel colour. A similar high diversity index was reported by Negassa (1986) and by Porceddu (1976).

Besides high phenotypic diversity, agronomically important genes have also been found in Ethiopian germplasm. Resistance or immunity to *Erysiphe graminis* f.sp. *tritici*, *Puccinia* spp. and *Septoria nodorum* are found and Negassa (1986) reported dwarf genes in several accessions. In other studies very early heading genotypes have been reported (Qualset & Puri, 1974) as well as very late maturing ones (Porceddu, 1976).

Although Vavilov (1951) described the Ethiopian region as a centre of diversity and origin for durum wheat, the absence of wild related species, as well as archaeological findings, strongly suggests that Ethiopia represents a secondary centre of diversity rather than a primary one. The alternative scheme of Hawkes (1983) distinguishing

Nuclear Centres of agricultural origin (i.e. the Near East) and regions of diversity (among others Ethiopia) fits the facts well in the case of durum wheat and also of barley. Genetic erosion does occur, mainly because of replacement of the crop by teff and because of imported bread wheat varieties.

Besides durum wheat, emmer (*T. turgidum* subsp. *dicoccoides*) was introduced into Ethiopia some 5000 years ago and is still fairly widespread (Haile/Mariam & Worede, 1988). *T. turgidum* conv. *aethiopicum* is another convariety of subspecies *turgidum* which can be observed regularly and about 11% of the wheat accessions in the Ethiopian genebank belong to this form. Intermediate forms of *durum* and *aethiopicum* are reported (Habtemariam & Worede, 1988). *T. turgidum* subsp. *turgidum* conv. *polonicum* has been observed to exist in the central highlands of Ethiopia, frequently interplanted with *durum* but also occurring in pure stands. The diversity in the latter is relatively small (J.M.M.E., personal observation).

Finally, it is worth pointing out that diploid einkorn wheats (*T. monococcum*) and hexaploid bread wheats (*T. aestivum*, etc.) do not seem to be native to the Ethiopian gene centre. The bread wheats were in fact all introduced in recent historical times, but einkorn wheats – as far as is known – never penetrated into the Ethiopian region.

Teff (*Eragrostis tef*)

Teff is by far the most important crop in Ethiopia in terms of acreage, thus, in 1983–4 about 1.38 million hectares were grown (Central Statistics Office, 1984). It is cultivated from sea level up to 2800 m on all kinds of soil. However, the waterlogged soils of the central Ethiopian highlands seem to be the 'cradle' of teff, a unique environment with limited agricultural use. Although *E. tef* has a wide distribution in Africa, it is cultivated as a food crop only in Ethiopia and North and South Yemen (D. Wood & L. Guarino, personal communication). No detailed information is available as to when the species was brought into cultivation, but it might be several thousands of years. Shaw (1976) argues that teff must have been domesticated before the introduction of wheat and barley to Ethiopia 'or else the teff, sorghum and finger millet never would have been cultivated'. *E. pilosa* is believed to be the progenitor of the cereal (Harlan, 1976) though, according to Lester & Bekele (1981), this species does not show any greater similarities to teff in its amino acid composition than several other wild *Eragrostis* species. In a later study

(Bekele & Lester, 1981) in which data from leaf phenolic chromatography and seed protein electrophoresis were compared, it was found that different teff cultivars showed similarities to several different wild *Eragrostis* species, though *E. pilosa* seemed closest.

The phenotypic diversity in teff is clearly visible but less conspicuous compared with some of the other cereals. The height of the plant, size and compactness of the culm and colour of the culm and seeds are the most variable traits. Ebba (1975) described 35 landrace varieties but undoubtedly more exist. Since diseases and storage pests do not play an important role in teff cultivation, little research has been done to find resistance (Ketema, 1986, Chapter 26).

The threat of genetic erosion to the diversity in teff is almost nonexistent. The crop is still expanding its acreage and almost no replacement of the landraces by improved varieties is taking place.

Miscellaneous cereal species

Finger millet (*Eleusine coracana*) has been mentioned frequently in the literature as being very likely of Ethiopian origin (Doggett, 1965; Harlan, 1969; de Wet *et al.*, 1984). Purseglove (1976) considered it to have originated in Uganda or a neighbouring country. It is grown mainly in the western part of the country and does not show such extreme diversity as some of the other cereals. Seed size, straw length and flowering date do, however, vary considerably (J.M.M.E., personal observation). The wild species *E. africana* occurs as a weed in the finger millet fields and is regarded as the progenitor of the cultivated species. The crop is facing some genetic erosion, due mainly to replacement by other crops.

Pearl or bulrush millet (*Pennisetum americanum*) is another minor millet which Doggett (1965) also believes to have originated in Ethiopia; others, however, such as Purseglove (1976) see the centre of domestication in West Africa. Indeed, in Ethiopia it is not an important crop, since it is only grown in the north-western part of the country in the marginal lower areas of Eritrea. At present the crop is expanding quickly in these marginal environments, mainly at the expense of sorghum and, to some extent, of finger millet (Teferie Michael, personal communication).

Maize (*Zea mays*) is a relatively recent introduction into Ethiopia. It is one of the fastest expanding crops and is causing a real threat to some of the sorghum diversity (J.M.M.E., personal observation). In the few hundred years since its introduction it has already built up quite a remarkable diversity.

In Ethiopia two wild weedy tetraploid oat species (*Avena abyssinica* and *A. vaviloviana*) occur in the 'cereal belt' between 2200 and 2800 m (Vavilov, 1957; Ladizinsky, 1975). Both are endemic to Ethiopia and the Yemen, and are related to the Mediterranean *A. barbata*. Besides being sometimes noxious weeds and adjusting themselves well to changing conditions, the Ethiopian oats are of no commercial value.

Finally, a mention should be made of the wild rice species which have been reported from the western lowlands. Both *Oryza barthii* and *O. longistaminata* are collected from the wild by the local inhabitants in times of food scarcity (Dadi & Engels, 1986).

Oil crops

Ethiopian mustard (*Brassica carinata*)

Ethiopian mustard is an important oil crop as well as a leaf vegetable. It is extensively grown throughout the highlands and shows a considerable diversity for several vegetative characters, mainly regarding leaves and growth habit (Engels, 1984). This diversity may be due to the introgression of genes from *B. oleracea*, since both species are often found growing in close proximity (Tcacenco, Ford-Lloyd & Astley, 1985). During the evaluation of the *Brassica* germplasm collection many intermediate forms between these two species have been observed (J.M.M.E., personal observation). No wild relative of Ethiopian mustard is known, a fact which further supports the hypothesis that it is an (allotetraploid) hybrid between *B. nigra* and *B. oleracea*. No clear indications for genetic erosion exist. *B. nigra* is used mainly for medicinal purposes but also for greasing the 'injera' pan. The species is predominantly grown in backyards but can regularly be observed in the same field as *B. carinata*. *B. oleracea* is cultivated mainly for its leaves and is perrenial. Tcacenco *et al*. (1985) conducted chromosomal examinations and found that *B. carinata* and *B. oleracea* make up a separate group of 'cabbage' types.

Niger seed, noog (*Guizotia abyssinica*)

This crop is treated as one of the 'classical Ethiopian crops' in terms of its origin (Seegeler, 1983). The highlands of central Ethiopia are the home of noog and because of the high demand for its oil the crop is expanding its acreage yearly. Regarding its phenotypic diversity, the crop does not show such striking features as some of the other crops, which might be partly due to its strong outbreeding behaviour. The highest diversity exists for characters such as number of days to flower and number of days to maturity as well as for head size (J.M.M.E., personal observation). No systematic research on oil content and oil quality has been conducted so far.

The frequent presence of the wild/weedy related species *G. scabra* around the fields of noog indicates the possible presence of a crop/weed complex which would be worth further study. Because of the steadily expanding acreage of the crop and the absence of released high-yielding varieties genetic erosion for noog will be mild, but because of the further improvement of agricultural practices it could be expected for the wild related species *G. scabra*.

Linseed (*Linum usitatissimum*)

Linseed is at present the second most important oil crop in Ethiopia; only noog is more popular. Linseed is grown only for oil production and its use as flax is hardly known. Vavilov (1957), however, observed that the seeds were used mainly for the preparation of fodder meals, by grinding after drying. The phenotypic diversity is not very striking: some variation in flower colour, plant height, number of days to flower and maturity as well as in capsule size has been noted (J.M.M.E., personal observation). Despite this, Ethiopia was regarded as a centre of flax diversity, though not of origin, by Vavilov (1951). There are strong indications that the existing diversity is seriously threatened by genetic erosion.

Sesame (*Sesamum indicum*)

The origin of sesame is still under dispute. However, since all the wild species but one are native to Africa, Ethiopia would seem very probably to be its centre of origin (Vavilov, 1951; Purseglove, 1968; Harlan, 1969; Nayar, 1976). It is certainly a very ancient cultigen in Africa. At present sesame is the third most important oil crop in economic terms in Ethiopia. The phenotypic diversity encountered in the country is considerable for pod shape and size, for seed size and colour (Seegeler, 1983) and for number of days to maturity and plant habit (J.M.M.E., personal observation). The genetic erosion can be considered as critical and more collecting remains to be done.

Castor bean (*Ricinus communis*)

Castor bean is widespread as a wild plant through East and North Africa, the Yemen and the Middle East. According to Purseglove it was cultivated in Egypt from 4000 BC onwards. Although no commercial production exists in Ethiopia, the plant is widely distributed almost from sea level up to the highlands, generally as a wild plant, e.g. Eritrea (Bruecher, 1977), or as a weed in disturbed habitats. 'Weedy' plants with dehiscent pods and cultivated types with indehiscent pods can be observed (personal observations). The diversity for many plant, fruit and seed characters is enormous and this would fully justify the assumption of Vavilov (1951) and Zeven & Zhukovsky (1975) that the cultivated castor bean might be of Ethiopian origin. Although no threat of genetic erosion exists at the moment it would be an exciting study to collect and evaluate castor bean more extensively. At present, the plant is used mainly for home consumption, as a medicinal plant and as a producer of oil for lighting.

Safflower (*Carthamus tinctorius*)

Purseglove (1968) mentions three primary centres: Afghanistan, the Nile Valley and Ethiopia. Since it was probably derived from the wild *C. oxycantha* which occurs as a weed from India to Turkey it is very difficult to be certain of its exact point of origin, and indeed it may have had several. Vavilov (1951) and Knowles (1976) considered Ethiopia to be its probable centre of domestication. However, this assumption was not supported by a diversity study conducted on a world germplasm collection where the Ethiopian diversity index was fairly low (Wu & Jain, 1977). In addition, the observed diversity in the field was relatively small (J.M.M.E., personal observation) and the crop is grown only on a small scale, frequently in the borders around cereal fields. Since safflower can be treated as a minor crop its diversity might well be endangered by genetic erosion.

Crambe (*Crambe abyssinica*)

This 'new' crop has its home in the Ethiopian highlands (Leppik & White, 1975). Wild populations as well as 'cultivated' fields of crambe can be found, but relatively rarely. Because of ever expanding agricultural lands many natural habitats are disturbed and the existing diversity undergoes considerable genetic erosion.

Pulses

Faba bean (*Vicia faba*)

The faba bean is the most important pulse in Ethiopia, occurring mainly as the small-seeded type, a character also typical for germplasm from Afghanistan, probably one of the primary centres of diversification (Lafiandra *et al.*, 1981). In the same study a high protein content in Ethiopian material was reported and, recently, considerable resistance against chocolate spot has been found in Ethiopian germplasm (PGRC/E, unpublished data). As in the other pulses, at present no real threat to the diversity exists.

Field pea (*Pisum sativum*)

The field pea is an old crop in Ethiopia and is still one of the dominant pulses in the country. During the several thousand years of its presence, a unique subspecies developed in Ethiopia (subsp. *abyssinicum*) and this earlier led Vavilov to seek the origin of the pea partly in Ethiopia. However, in a later publication (Vavilov, 1957) he admitted that this pea had probably come from the Yemen. From our own observations it should be noted that the phenotypic diversity was found to be rather limited. No information about the degree of genetic erosion exists; however, it is expected to be low.

Chickpea (*Cicer arietinum*)

After faba beans and field peas, chickpeas are the most important pulse in terms of acreage. They are eaten immature as a snack or as matured seeds in a roasted, boiled or ground form. Chickpea is an ancient crop in Ethiopia and archaeological evidence in the caves of Lalibela has shown an age of 500 BC (Dombrowski, 1969). Although some authors have mentioned Ethiopia as a centre of origin and a related wild species has been found in northern Ethiopia (e.g. *C. cuneatum*), there are strong indications that the origin lies in south-west Asia (Harlan, 1969). Nevertheless, the diversity encountered in the Ethiopian fields is considerable and Pundir *et al.* (1985) reported unique flower colours, high anthocyanin in the leaves and non-beige seed colours. Furthermore, some disease resistance and drought tolerance has been found in initial testing. Genetic erosion is of little importance and is no real threat to the chickpea diversity.

Lentil (*Lens culinaris*)

The lentil is one of the crops which was introduced into Ethiopia in the early days from west Asia, probably via the Yemen, and which belongs to the 'South-Western Asian Complex' (Harlan, 1969). Another study suggests that lentils might have originated in India as a selection from a wild form (Williams, Sanchez & Jackson, 1974) but this will require further investigation. Because of the early introduction a considerable diversity in the crop has built up. Erskine & Witcombe (1984) reported the following interesting characteristics found in Ethiopian germplasm: earliness, high seed yield, high harvest index, high number of seeds per pod and good cold tolerance. On the other hand, the Ethiopian material showed a low seed protein content. Some genetic erosion can be expected since the acreage of lentils is steadily declining.

Miscellaneous pulse species

Of the minor pulses, cowpea (*Vigna unguiculata*) should be mentioned first. This African species may have been domesticated in Ethiopia (Vavilov, 1951; Steele, 1976) and several wild (related) species exist in the country. Not much is known about the diversity in this crop with its limited cultivation, mainly in Konso and the Gambela region (Engels & Dadi, 1986). The wild species might be endangered because of drastic changes in land use.

Fenugreek (*Trigonella foenum-graecum*) is a crop with a long history in Ethiopia. Although it is considered more as a medicinal plant it is also used as a pulse. The genetic diversity is considerable (J.M.M.E., personal observation).

Grass pea (*Lathyrus sativus*) is a fairly common crop in the Ethiopian highlands, frequently grown on fallow lands. Its diversity is not yet systematically studied but considerable drought resistance can be expected (J.M.M.E., personal observation).

The hyacinth bean (*Lablab purpureus*) and pigeon pea (*Cajanus cajan*) were reported by Vavilov (1951) and Zeven & Zhukovsky (1975) to be of Ethiopian origin but there is considerable doubt about this. A more likely hypothesis is that they were domesticated in India (Purseglove, 1968; Royes, 1976; Smith, 1976). Both crops are frequently cultivated in southern Ethiopia (Konso) where they form an important component in the highly developed and sustainable agricultural system.

Miscellaneous crops of Ethiopian origin

Coffee (*Coffea arabica*)

From an economic point of view this may be the most important Ethiopian crop. Almost the entire diversity originated in Ethiopia, mainly in the south-western rainforest area, where almost undisturbed patches of forest with coffee as undergrowth can still be found. The phenotypic diversity is overwhelming for qualitative as well as for quantitative characters (Tadesse & Engels, 1986), not to mention the variation in disease and pest resistance (Wondimu, 1987), quality characteristics and others. Because of extensive deforestation and replacement of primitive coffee populations by maize, chat and other crops, together with changing patterns in land use, the diversity is highly threatened by erosion (Hawkes *et al.*, 1986; Bellachew, 1987).

Ensete or false banana (*Ensete ventricosum*)

In the more humid highlands, especially in southern Ethiopia, this unique Ethiopian crop species is an important staple. The starch of the pseudostem is used after fermentation treatment for several days. Furthermore, the plant or parts of it are used as fodder, for fuel and packing material, etc. (Olmstead, 1974). Although the plant is propagated only vegetatively, an astonishing variation in several characters can be found. Demeke *et al.* (1986) described 76 named varieties which vary mainly in colour of pseudostem and leaf midribs, their earliness and quality of the final product as well as for disease resistance, e.g. bacterial wilt. The most serious factors causing genetic erosion in ensete are bacterial wilt and, to a lesser extent, drought.

Chat (*Catha edulis*)

This evergreen bush is the source of the fresh young leaves which are chewed in Eastern Africa and the Arabian Peninsula for their mild narcotic properties. The plant, which grows wild in East Africa, was first domesticated in Ethiopia and is now expanding in the eastern part of the country at the expense of coffee. Some striking leaf colour variants have been observed, especially in south-east Ethiopia (personal observation). No genetic erosion takes place at present.

Okra (*Abelmoschus esculentus*)

Okra is grown only in the western lowlands. It was also in this part of the country that, beside the domesticated species, two more related species were encountered, one being 'half domesticated' (*A. manihot*) and the other, *A. moschatus*) (?) being collected from the wild (Engels & Dadi, 1986). These findings support the assumption of Harlan (1969) and Vavilov (1951) that okra could be of Ethiopian origin even though Joshi & Hardas (1976) cast some doubt on this.

Roots and tubers

Several species are cultivated or semi-cultivated in restricted parts of Ethiopia but only for their roots or tubers. Especially in southern Ethiopia (e.g. Konso), these crops play an important role.

The better known species are *Plectranthus edulis* (galla potato), *Coccinia abyssinica* (anchote), *Amorphophallus abyssinicus* (bagana) and *Sauromatum nubicum* (Dadi & Engels, 1982). Furthermore, several yam species (*Dioscorea* spp.) might have their origin in Ethiopia as well (Harlan, 1969; Zeven & Zhukovsky, 1975). Leon (1978) reports *Sphenostylis stenocarpa* to be a 'yam bean' of possible Ethiopian origin.

Spices and medicinal plants

Several spices and medicinal plants are of Ethiopian origin, or expected to be so. The most important species are *Aframomum korarima* (false cardamom), *Carum copticum* (nech azmud), *Coriandrum sativum* (coriander), *Embelia schimperi* (enkoko), *Hygenia abyssinica* (koso), *Lepidium sativum* (garden cress, feto), *Nigella sativa* (black cumin) and *Rhamnus prinoides* (buckthorn, gesho). The bark of the latter is used in place of hops in beer. Wilson & Gebre-Mariam (1979) have published a list of 37 species which are used commonly as medicinal plants on the Ethiopian plateau. Details of a wide variety of medicinal plants are presented by Abebe & Hagos (Chapter 6).

Fibre plants

Nicholson (1960) presents arguments that cotton (*Gossypium herbaceum* L. var. *acerifolium*) might have been domesticated in Ethiopia. Also kenaf (*Hibiscus cannabinus*) is reported to be of Ethiopian origin (Purseglove, 1968; Zeven & Zhukovsky, 1975). In south-west Ethiopia some wild *Hibiscus* species have been observed which were reported to be used for their fibres (personal observation).

Moringa stenopetala (cabbage tree) is an important vegetable tree in south Ethiopia (Konso). Some other trees are used extensively for fodder (e.g. *Balanites aegyptica* and *Terminalia brownii*), a practice unique to Ethiopia.

Conclusions

From the foregoing paragraphs it is obvious that Ethiopia is an important centre of genetic diversity for a wide range of crops. This is partly due to its very dissected terrain, with consequently an extremely wide range of agro-ecological conditions. It is also due to its geographical position at the crossroads between the Near-Eastern centre of diversity on the one hand and the Indian centre on the other. It has also benefited from its connections with the mountain chain and Rift Valley, following southwards in East Africa, and its connections with West Africa via the Sudan and Sahel regions.

Many of Ethiopia's most important crops such as barley and the tetraploid wheats have come from elsewhere several thousand years ago. Under Ethiopian conditions they have developed much diversity, often of a unique character. Hexaploid or bread wheats came much later, perhaps not more than 100 years ago. Diploid wheat and rye do not appear at all, and oats only as weeds.

Sorghum, with its tremendous diversity in Ethiopia, is something of an enigma. Whilst several authorities such as Vavilov and Doggett believe that Ethiopia may be the centre of origin of sorghum, others, such as Harlan and de Wet, are rather doubtful.

However, there is no doubt about teff. This certainly was domesticated in Ethiopia and is still its most important cereal.

Concerning the minor millets (*Eleusine coracana* and *Pennisetum americanum*), doubts exist as to their having originated in Ethiopia.

Of the oil crops, *Guizotia abyssinica*, *Brassica carinata* and *Sesamum indicum* appear to be native, whilst others are of uncertain origin. Flax was almost certainly introduced.

Pulses such as faba bean, field pea, chickpea and lentil seem all to have been introduced, though they probably have been in Ethiopia for several millennia. Castor bean and safflower may be indigenous crops but there is insufficient evidence to be really certain. The same problem is seen with okra.

At least we are on firmer ground with coffee (*Coffea arabica*), ensete (*Ensete ventricosum*) and chat (*Catha edulis*). These are all Ethiopian in origin, though from their very nature it is difficult to point to characters in them that are 'domesticated'. In other words, they are still at the stage of wild species that have been taken into cultivation but could probably still exist quite well in the wild. Nevertheless, they *were* taken into cultivation, even though the boundary between cultivation and domestication in these cases is not very clear.

The main problem in all the arguments and discussions, as to the place and time of origin of Ethiopian crops, is that we have very few hard facts to lean on in terms of reliable carbon-dated archaeological finds. When and if these appear, as they already have in and near the Middle East, India and the Americas, we shall be able to back up our often tentative ideas with greater certainty. However, whether or not Ethiopia is a centre of origin for cereal crops such as wheat, barley, sorghum and certain millets, as well as the oil crops and pulses about which there often is some doubt, one thing is certain: Ethiopia is undoubtedly a world centre of crop plant diversity of great importance for a number of important domesticates. Agriculture must be very ancient in Ethiopia, and its position at the crossroads of ancient crop migrations has resulted in the creation of a centre of genetic diversity which all crop scientists from the time of Vavilov onwards have considered of the greatest theoretical importance, and one of very great practical value to plant breeders.

References

Bekele, E. & Lester, R. N. (1981). Biochemical assessment of the relationships of *Eragrostis tef* (Zucc.) Trotter with some wild *Eragrostis* species (Gramineae). *Annals of Botany*, **42**, 717–25.

Bellachew, B. (1987). Coffee (*Coffea arabica* L.) genetic erosion and germplasm collection in Harerge region. Institute of Agricultural Research, Jima Research Centre, Addis Ababa (mimeographed).

Bruecher, H. (1977). *Tropische Nutzpflanzen, Ursprung, Evolution and Domestikation*. Springer-Verlag, Berlin.

Central Statistics Office (1984). *Time series data on area, production and yield of principal crops by regions, 1979/80–1983/84*. Central Statistics Office, Addis Ababa, 219 pp.

Dadi, T. & Engels, J. M. M. (1982). Crop genetic resources of Ethiopia. *PGRC/E–ILCA Germplasm Newsletter*, **1**, 5–9.

Dadi, T. & Engels, J. M. M. (1986). Exploration and collection of wild rice in Ethiopia. *PGRC/E–ILCA Germplasm Newsletter*, **12**, 9–11.

Demeke, T., Zelleke, A., Ghirmay, A. & Shikur, H. G. (1986). A preliminary note on the variation of *Ensete ventricosum*. Alemaya University of Agriculture, Debre Zeit, 9 pp. (mimeographed).

de Wet, J. M. J., Prasada Rao, K. E., Brink, D. W. & Mengesha, M. H. (1984). Systematics and evolution of *Eleusine coracana* (Gramineae). *American Journal of Botany*, **71**, 550–7.

Doggett, H. (1965). The development of the cultivated sorghums. *In*: Sir J. B. Hutchinson (ed.), *Essays on Crop Plant Evolution*. Cambridge University Press, Cambridge, pp. 50–69.

Dombrowski, J. (1969). Preliminary report on excavation in Lalibela and Natchabiet caves, Begemeder. *Annales d'Ethiopie*, **8**, 21–9.

Ebba, T. (1975). Teff (*Eragrostis tef*) cultivars: morphology and classification. Part II. *Debre Zeit Agricultural Research Centre Bulletin no. 66*, pp. 1–73. Alemaya University of Agriculture, Dire Dawa, Ethiopia.

Engels, J. M. M. (1984). Genetic variation in Ethiopian *Brassica* spp. *Cruciferae Newsletter (Eucarpia)*, **9**, 59–60.

Engels, J. M. M. & Dadi, T. (1986). Germplasm exploration in Gambela: local okra and cowpea. *PGRC/E–ILCA Germplasm Newsletter*, **11**, 15–19.

Erskine, W. & Whitcombe, J. R. (1984). *Lentil Germplasm Catalog*. ICARDA, Aleppo.

Froest, S., Holm, G. & Asker, S. (1975). Flavonoid patterns and the phylogeny of barley. *Hereditas*, **79**, 133–42.

Habtemariam, G. & Worede, M. (1988). A preliminary classification of Ethiopian wheat germplasm accessions. *In*: J. M. M. Engels (ed.), The conservation and utilization of Ethiopian germplasm. Proceedings of an international symposium, Addis Ababa, 13–16 October 1986, pp. 133–9 (mimeographed).

Harlan, J. R. (1969). Ethiopia: a centre of diversity. *Economic Botany*, **23**, 309–14.

Harlan, J. R. (1975). *Crops and Man*. American Society of Agronomy, Madison, Wisconsin.

Harlan, J. R. (1976). The origins of cereal agriculture in the old world. *In*: C. A. Reed (ed.), *Origins of Agriculture*. Mouton Publishers, The Hague, pp. 357–83.

Hawkes, J. G. (1983). *The Diversity of Crop Plants*. Harvard University Press, Cambridge, Massachusetts.

Hawkes, J. G., Engels, J. M. M. & Tadesse, D. (1986). Suggested sampling and conservation system for coffee in Ethiopia. *PGRC/E–ILCA Germplasm Newsletter*, **11**, 25–8.

Hettinger, B. & Engels, J. M. M. (1986). Screening methods for drought resistance in indigenous Ethiopian barley. *PGRC/E–ILCA Germplasm Newsletter*, **13**, 26–30.

International Crops Research Institute for the Semi-Arid Tropics (1985). *ICRISAT research highlights 1984*. ICRISAT, Patancheru.

Jain, S. K., Qualset, C. O., Bhatt, G. M. & Wu, K. K. (1975). Geographical patterns of phenotypic diversity in a world collection of durum wheats. *Crop Science*, **15**, 700–4.

Joshi, A. B. & Hardas, M. W. (1976). Okra. *In*: N. W. Simmonds (ed.), *Evolution of Crop Plants*. Longman, London, pp. 194–5.

Ketema, S. (1986). Food self sufficiency and some roles of teff (*Eragrostis tef*) in the Ethiopian agriculture. Institute of Agricultural Research, Addis Ababa (mimeographed).

Knowles, P. F. (1976). Safflower. *In*: N. W. Simmonds (ed.), *Evolution of Crop Plants*. Longman, London, pp. 31–3.

Ladizinsky, G. (1975). Oats in Ethiopia. *Economic Botany*, **29**, 238–41.

Lafiandra, D., Polignano, G. B., Filippetti, A. & Porceddu, E. (1981). Genetic variability for protein content and s-aminoacids in broad-beans (*Vicia faba* L.). *Kulturpflanze*, **29**, 115–27.

Leon, J. (1978). Origin, evolution and early dispersal of root and tuber crops. Tropical root and tuber crops symposium. CATIE, Turrialba, Costa Rica (mimeographed).

Leppik, E. E. (1970). Gene centers of plants as sources of disease resistance. *Annual Review of Phytopathology*, **8**, 324–44.

Leppik, E. E. & White, F. A. (1975). Preliminary assessment of *Crambe* germplasm resources. *Euphytica*, **24**, 681–9.

Lester, R. N. & Bekele, E. (1981). Amino acid composition of the cereal teff and related species of *Eragrostis*. *Cereal Chemistry*, **58**, 113–15.

Maiti, R. K., Prasada Rao, K. E., Raju, P. S. & House, L. R. (1984). The glossy trait in sorghum: its characteristics and significance in crop improvement. *Field Crops Research*, **9**, 279–89.

Nayar, N. M. (1976). Sesame. *In*: N. W. Simmonds (ed.), *Evolution of Crop Plants*. Longman, London, pp. 231–3.

Negassa, M. (1986). Estimates of phenotypic diversity and breeding potential of Ethiopian wheats. *Hereditas*, **104**, 41–8.

Nicholson, G. E. (1960). The production, history, uses and relationships of cotton (*Gossypium* spp.) in Ethiopia. *Economic Botany*, **14**, 3–36.

Olmstead, J. (1974). The versatile ensete plant: its use in the Gamu Highland. *Journal of Ethiopian Studies*, **12**, 147–53.

Plant Genetic Resources Centre/Ethiopia (1986). *Ten years of collection, conservation and utilization, 1976–1986*. PGRC/E, Addis Ababa, 40 pp.

Porceddu, E. (1976). Variation for agronomical traits in a world collection of durum wheats. *Zeitschrift für Pflanzenzüchtung*, **77**, 314–29.

Pundir, R. P. S., Rao, N. K. & van der Maesen, L. J. G. (1985). Distribution of qualitative traits in the world germplasm of chickpea (*Cicer arietinum* L.). *Euphytica*, **34**, 697–703.

Purseglove, J. W. (1968). *Tropical Crops: Dicotyledons, vols 1 and* 2. Longman, London.

Purseglove, J. W. (1976). The origins and migrations of crops in tropical Africa. *In*: J. R. Harlan, J. M. J. de Wet and A. B. L. Stemler (eds), *Origins of African Plant Domestication*. Mouton Publishers, The Hague, pp. 291–309.

Qualset, C. O. (1975). Sampling germplasm in a centre of diversity: an example of disease resistance in Ethiopian barley. *In*: O. H. Frankel and J. G. Hawkes (eds), *Crop Genetic Resources for Today and Tomorrow*. Cambridge University Press, Cambridge, pp. 81–96.

Qualset, C. O. & Puri, Y. P. (1974). Sources of earliness and winter habit in durum wheat. *Wheat Information Service*, **38**, 13–15.

Royes, W. V. (1976). Pigeon pea. *In*: N. W. Simmonds (ed.), *Evolution of Crop Plants*. Longman, London, pp. 154–6.

Schiemann, E. (1951). New results on the history of cultivated cereals. *Heredity*, **5**, 305–20.

Seegeler, C. J. P. (1983). *Oil Plants in Ethiopia, their Taxonomy and Agricultural Significance*. PUDOC, Wageningen.

Shaw, T. (1976). Early crops in Africa: a review of the evidence. *In*: J. H. Harlan, J. M. J. de Wet and A. B. L. Stemler (eds), *Origins of African Plant Domestication*. Mouton Publishers, The Hague, pp. 107–53.

Singh, R. & Axtell, J. D. (1973). High lysine mutant gene (hl) that improves protein quality and biological value of grain sorghum. *Crop Science*, **13**, 535–9.

Singh, S. P. (1985). Sources of cold tolerance in grain sorghum. *Canadian Journal of Plant Science*, **65**, 251–7.

Smith, P. M. (1976). Minor crops. *In*: N. W. Simmonds (ed.), *Evolution of Plants*. Longman, London, pp. 301–24.

Snowdon, J. D. (1955). The wild fodder sorghums of the section Eusorghum. *Journal of the Linnean Society, London*, **55**, 191–260.

Steele, W. M. (1976). Cowpeas. *In*: N. W. Simmonds (ed.), *Evolution of Crop Plants*. Longman, London, pp. 183–5.

Stemler, A. B. L., Harlan, J. R. & de Wet, J. M. J. (1975). Evolutionary history of cultivated sorghums (*Sorghum bicolor* [Linn.] Moench) of Ethiopia. *Bulletin of the Torrey Botanical Club*, **102**, 325–33.

Subramanian, V., Prasada Rao, K. E., Mengesha, M. H. & Jambunathan, R. (1987). Total sugar content in sorghum stalks and grains of selected cultivars from the world germplasm collection. *Journal of the Science of Food and Agriculture*, **39**, 289–95.

Tadesse, D. & Engels, J. M. M. (1986). Phenotypic variation in some fruit characters in coffee collected from Chora wereda. *PGRC/E–ILCA Germplasm Newsletter*, **12**, 2–8.

Tcacenco, F. A., Ford-Lloyd, B. V. & Astley, D. (1985). A numerical study of variation in a germplasm collection of *Brassica carinata* and allied species from Ethiopia. *Zeitschrift für Pflanzenzüchtung*, **94**, 192–200.

Vavilov, N. I. (1931). The wheats of Abyssinia and their place in the general system of wheats. *Bulletin of Applied Botany, Genetics and Plant Breeding, Supplement* **51**, p. 233.

Vavilov, N. I. (1951). The origin, variation, immunity and breeding of cultivated plants. *Chronica Botanica*, **13**, 1–366.

Vavilov, N. I. (1957, published in English in 1960). *World Resources of Cereals, Leguminous Seed Crops and Flax, and their Utilization in Plant Breeding. Jerusalem: Israel Program for Scientific Translations.*

Williams, J. T., Sanchez, A. M. C. & Jackson, M. T. (1974). Studies on lentils and their variation, I. The taxonomy of the species. *SABRAO Journal*, **6**, 133–45.

Wilson, R. T. & Gebre-Mariam, W. (1979). Medicine and magic in central Tigre: a contribution to the ethnobotany of the Ethiopian plateau. *Economic Botany*, **33**, 29–34.

Wondimu, M. (1987). The variability in physiologic groups of *Coffea arabica* L. to rust (*Hemileia vastatrix* B. & Br.) and their pattern of distribution in Ethiopia from the 1981 germplasm collections. Coffee Plantation Development Corporation, Addis Ababa, 8 pp. (mimeographed).

Wu, K. K. & Jain, S. K. (1977). A note on germplasm diversity in the world collections of safflower. *Economic Botany*, **31**, 72–5.

Zeven, A. C. & Zhukovsky, P. M. (1975). *Dictionary of Cultivated Plants and their Centres of Diversity*. PUDOC, Wageningen.

3

Crops with wild relatives found in Ethiopia

SUE B. EDWARDS

Introduction

All our modern crops have been developed from wild plants. The domestication of a plant passes through stages from intensified usage of the wild plant to the development of a domesticate so dependent on Man that it cannot survive in the wild. All stages are seen in the crop complement of Ethiopia. There are many wild plants which are used for food, particularly in times of food shortage such as the period between seed sowing and harvest. It is hardly surprising that the majority of such plants are those used as leafy vegetables, followed by those with edible fruits, tubers or roots. Another example is the grass, *Snowdenia polystachya* (Fresen.) Pilg., whose seeds are collected and used in a similar way to teff. The following account includes only those that are related to domesticates. Examples of semi-domesticated plants are *Avena abyssinica* and *Coccinia abyssinica*, both of which are discussed further below. There are also wild plants now attracting attention as potential crops, for example *Vernonia galamensis* (Cass.) Less. (Perdue, 1988) and *Cordeauxia edulis* (Hemsl.) (Polhill & Thulin, 1989). Ethiopia also has fully domesticated endemic crops, the best known being teff, *Eragrostis tef* and ensete, *Ensete ventricosum*. For fully domesticated plants the wild species from which the crop developed has in some cases been identified; in others it seems to have disappeared after the plant was domesticated.

Both environmental degradation and modern agriculture are putting traditional crops and their wild relatives at risk. The now inadequate traditional agriculture must change if Ethiopia is to feed itself and this is one of the major tasks being faced by the Government. However, it is hoped that the following account gives some idea of the size of the task facing conservationists who are working to preserve the traditional varieties and their wild relatives for use in developing modern and appropriate cropping systems. There is no part of the country where some crop or other and/or its wild relatives do not occur: for example, *Thymus* spp. in the Afro-alpine regions; *Ensete ventricosum* in the medium to higher altitudes and *Gossypium* spp. in the lowlands.

But crops are not the only plants used by Man. Any consideration of the Ethiopian wealth of plants must take into account those other plants used by Man although it is not possible to cover them in detail in the present account. An important group of useful plants is those used in traditional medicine. There is no precise, modern account of all these plants. However, the Ministry of Health now has a department responsible for studying traditional medicine and assisting its practitioners. Although some of the better known plants, such as *Brucea antidysenterica* J. F. Mill. (Simaroubaceae) and *Hagenia abyssinica* (Bruce) J. F. Gmel. (Rosaceae), are left when surrounding vegetation is cut down and naturally occurring seedlings may be moved and/or protected, the major medicinal plants in Ethiopia are on the whole not cultivated. However, some cultivated herbs are also used medicinally and these are mentioned below.

The Ethiopian region, including the Sudan and Somalia, is well known for its resins and gums. These come mainly from three genera, *Acacia* in the Fabaceae (Leguminosae), and *Boswellia* and *Commiphora* in the Burseraceae. True gum arabic is extracted from *Acacia senegal* (L.) Willd. which grows most abundantly in the lowlands of Ethiopia, Somalia and the Sudan (Asfaw Hunde & Thulin, 1989). *Boswellia* and *Commiphora* have a centre of diversity in the Somalia–Massai regional centre (White, 1983) which gives the Ogaden region of Bale, Gamo Gofa, Harerge and Sidamo its distinctive vegetation. There are 52 species of *Commiphora* recorded for Ethiopia, of which 35 (67 per cent) belong to the Somalia–Masai region and are found only in south and south-eastern Ethiopia (Vollesen, 1989b).

Ethiopian agriculture is heavily dependent on animals for power and the forage and browse for these all has to come from natural vegetation and crop residues. Although germplasm taken from eastern Africa has been developed elsewhere into important forage plants, for example *Chloris gayana* Kunth. (Bermuda grass), work on indigenous forage plants is only beginning. This has already shown Ethiopia to be a centre of diversity for *Trifolium*, with 10 of its 26 indigenous species being endemic (Thulin, 1989).

Then there are the many plants used as cosmetics, producing perfume and colouring, as fumigants and cleansers, and as dyes and inks. There are also the many plants used to construct houses and furniture, make agricultural tools and provide fuelwood. The majority of these species are wild plants. They will not be replaced by plastic and other artificial materials as has happened in developed countries, because the source for these artificial materials is the fossil fuels which have a limited world supply. Thus conservationists must consider the whole Ethiopian environment if these many useful plants are to be available for future generations.

However, the following account gives only the known occurrences in Ethiopia of the wild relatives of crops. The crops include both those grown in Ethiopia and those not grown in this country but which have some importance in international trade, for example pistachio nut and the drug senna, and which have wild relatives in the country. The common name and usage for each of the different crops is given in Table 1.

Dicotyledons

Amaranthaceae

Amaranthus spp. The account for the Flora of Ethiopia gives 11 common weeds in fields and in open disturbed habitats. In many places the young plants are eaten cooked and there are records of the seeds of *A. caudatus* being used in making 'tala' (Townsend, in preparation).

Anacardiaceae

Pistacia vera L. There are two species in Ethiopia. *P. aethiopica* Kokwaro is recorded from Eritrea, Gamo Gofa, Sidamo and Bale from a variety of habitats. *P. falcata* Mart. grows in Shewa where it is the dominant tree on recent lava flows. It is also recorded from Eritrea (Gilbert, 1989b).

Apiaceae (Umbelliferae)

Anethum graveolens L. is semi-cultivated, being also found growing wild (Heywood *et al.*, in preparation; personal observation).
Apium graveolens L. is said to occur as an escape from cultivation. There are two wild species, *A. nodiflorum* (L.) Lag. which grows in wet places in Eritrea, Tigray, Gondar and Harerge and *A. leptophyllum* (Pers.) Muell. ex Benth., which is a weed (Heywood *et al.*, in preparation).
Carum carvi L., *Coriandrum sativum* L., *Cuminum cyminum* L., and

Table 1. *Crops with important gene pools in Ethiopia with their common names and uses*

Scientific name (Family)	Common names in English and Amharic	Part used in Ethiopia
Dicotyledons		
Abelmoschus esculentus (Malvaceae)	Okra, Ladies Fingers	Young fruit used as cooked vegetable.
Amaranthus caudatus *A. hybridus* (Amaranthaceae)	Grain Amaranths, 'alma'	Seed for brewing; young plants as cooked vegetable.
Anethum graveolens (Apiaceae)	Dill, 'insilal'	Whole plant used in traditional medicine and to flavour alcoholic drinks.
Apium graveolens (Apiaceae)	Celery	Leaves and stems as herb: fruits as a spice.
Brassica campestris *B. carinata* *B. integrifolia* (Brassicaceae)	Field mustard Ethiopian Kale, 'gomen' for young plants and leaves, 'gomenzer' for seeds used for oil	Young plants and leaves used as leafy vegetable; seeds used for crushing and oiling the baking plate for cooking 'injera'.
Brassica juncea *B. nigra* (Brassicaceae)	Indian mustard Mustard 'senafetch'	Seeds used to prepare a special fasting dish 'siljo'.
Brassica oleracea (Brassicaceae)	Cabbage, etc. 'tikil gomen', etc.	Used as cooked vegetable and salad.
Canavalia ensiformis (Fabaceae)	Jack Bean or Sword Bean	Young pods and seeds may be used as vegetable.
Cannabis sativa (Cannabaceae)	Hemp (fibre) Cannabis (drug) 'itse faris'	Leaves used medicinally; elsewhere grown for fibre.

Table 1 (*cont.*)

Scientific name (Family)	Common names in English and Amharic	Part used in Ethiopia
Capsicum spp. (Solanaceae)	Chili pepper, 'karya' for fresh fruit; 'berbere' for red mature fruit	Fruits eaten fresh and dried fruits are the main spice in 'berbere'.
Carthamus tinctorius (Asteraceae)	Safflower, 'suf'	Fruits for edible oil.
Carum carvi (Apiaceae)	Caraway	Fruits used to flavour drinks, cakes and bread.
Catha edulis (Celastraceae)	Khat, 'chat'	Leaves used as stimulant.
Cicer arietinum (Fabaceae)	Chickpea, 'shimbira'	Ripe seed as a pulse: green seeds also eaten fresh.
Citrullus lanatus (Cucurbitaceae)	Watermelon, 'birchik', also 'habhab'	Fruit eaten fresh.
Citrus aurantifolia (Rutaceae)	Lime, 'lomi'	Fruits used fresh for sucking, cleaning meat and treating skin problems.
Coccinia abyssinica (Cucurbitaceae)	'anchote'	Tuber eaten as a cooked vegetable.
Coffea arabica (Rubiaceae)	Arabica coffee, 'buna'	Roasted seeds, as well as dried fruit walls and leaves, used to make hot drinks.
Corchorus oligatorius (Tiliaceae)	Jute	Young plants used as leafy vegetable. (Elsewhere a source of fibre.)
Coriandrum sativum (Apiaceae)	Coriander, 'dimbilal'	Fresh leaves sometimes as a herb: fruits used in spicing 'berbere'.

Crambe abyssinica (Brassicaceae)	Crambe	Only use recorded is medicinal.
Cucumis melo (Cucurbitaceae)	Melon	Fruit eaten fresh.
Cucurbita spp. (Cucurbitaceae)	Pumpkin, 'duba'	Fruit eaten as cooked vegetable, both fresh and after drying.
Cuminum cyminum (Apiaceae)	Cumin, 'cumin'	Fruits as a spice, particularly with 'berbere'.
Cyamopsis tetragonoloba (Fabaceae)	Cluster Bean	Young pods may be used as a vegetable.
Daucus carota (Apiaceae)	Carrot, 'carrot'	Root as food.
Diplolophium abyssinicum (Apiaceae)	'dog'	Used to flavour alcoholic drinks; also medicinal, usually growing wild.
Eruca sativa (Brassicaceae)	Garden rocket, 'jirjir' (Tigrinya)	Young plants as salad.
Ficus carica (Moraceae)	Fig, 'beles'	Fruits eaten fresh.
Foeniculum vulgare (Apiaceae)	Fennel	Leaves and stems as salad and herb; fruits as spice and flavouring, also medicinal.
Gossypium spp. (Malvaceae)	Cotton, 'tit'	The lint covering the seeds spun to make cloth; oil extracted from the seeds.
Guizotia abyssinica (Asteraceae)	Niger seed, 'noug'	Fruits for edible oil.
Hibiscus cannabinus (Malvaceae)	Kenaf, Jute	Elsewhere stems a source of fibre, in Ethiopia young leaves sometimes eaten as cooked vegetable.

Table 1 (*cont.*)

Scientific name (Family)	Common names in English and Amharic	Part used in Ethiopia
Indigofera arrecta	Indigo	Leafy branches for dye.
I. tinctoria (Fabaceae)	Indigo	
Lablab purpureus (Fabaceae)	Hyacinth Bean, Lablab	Whole plant as forage: elsewhere white seeded varieties as cooked pulse.
Lagenaria siceraria (Cucurbitaceae)	Bottle gourd, 'kil'	Dried fruit as container.
Lathyrus sativus (Fabaceae)	Grass Pea, 'guaya'	Seed as a pulse.
Lens culinaris (Fabaceae)	Lentil, 'miser'	Seeds as a pulse.
Lepidium sativum (Brassicaceae)	Cress, 'feto'	Seeds used medicinally; before introduction of *Capsicum* used as a spice.
Linum usitatissimum (Linaceae)	Linseed, 'telba'	Seeds used as source of edible oil; also used medicinally.
Luffa cylindrica (Cucurbitaceae)	Luffa	Young fruits may be eaten as cooked vegetable; dried fruit used as a cleaning tool.
Lupinus albus (Fabaceae)	White lupin, 'gibto'	Seed eaten whole after soaking and boiling; also used to produce high quality 'araki'.
Meriandra bengalensis (Lamiaceae)	Sage, 'nihba' (Tigrinya)	Leaves used as a herb.
Momordica charantia (Cucurbitaceae)	Bitter gourd	Young fruits may be eaten

Moringa oleifera *M. stenopetala* (Moringaceae)	Horse-Radish Tree	Seeds used as a source of oil; can also be used to purify water. Leaves used as a cooked vegetable.
Mucuna pruriens (Fabaceae)	Bengal Bean, Velvet Bean	Seeds can be eaten after repeated boiling.
Nasturtium officinale (Brassicaceae)	Watercress	Leafy branches as salad.
Nigella sativa (Ranunculaceae)	'tikur azmud'	Seed used as a spice in bread and in spicing 'berbere'.
Ocimum basilicum *O. gratissimum* (Lamiaceae)	Sweet Basil, 'bessobila' Holy Basil	Flowering shoots used as herb in cooking and clarifying butter; same parts used medicinally. Leaves used medicinally and as fumigant.
Olea europea subsp. *africana* (Oleaceae)	Olive, 'weira'	Fruits eaten by children: leaves and twigs used as a fumigant and tooth brush.
Phaseolus vulgaris (Fabaceae)	Haricot Beans, 'fasolya' for green pods; 'bolokie' for dried beans	Young pods used as cooked vegetable; dried seeds infrequently used.
Piper nigrum (Piperaceae)	Black pepper, 'kundo-berbere'	Fruit a high quality spice.
Pistacia aethiopica (Anacardiaceae)	Ethiopian mastic	Resin for mastic.
Pisum sativum var. *abyssinicum* (Fabaceae)	Field Pea, 'ater'	Seeds used as a pulse; and unripe seeds eaten fresh.
Plantago afra (Plantaginaceae)	Psyllium	Seed a source of mucilage, but not used in Ethiopia.
Plectranthus edulis (Lamiaceae)	Hausa Potato, 'Oromo dinich'	Tubers eaten as cooked vegetable.

Table 1 (*cont.*)

Scientific name (Family)	Common names in English and Amharic	Part used in Ethiopia
Prunus persica (Rosaceae)	Peach, 'kok'	Ripe fruits eaten fresh.
Punica granatum (Lythraceae)	Pomegranate, 'roman'	Fruit eaten fresh; root, bark, fruit, rind of the fruit and flowers all used medicinally.
Rhamnus prinoides (Rhamnaceae)	'gesho'	Leaves and smaller branches used to flavour home-made beer 'tala' and honey wine 'tej'.
Ricinus communis (Euphorbiaceae)	Castor, 'gulo'	Used medicinally. Mostly seeds crushed and used to oil the baking plate for cooking 'injera'.
Ruta chalapensis (Rutaceae)	Rue, 'tenaddam'	Leafy and flowering shoots used as a spice and medicinally.
Salvia nilotica	'antate-welakha' (Tigrinya)	Seeds used medicinally.
S. schimperi (Lamiaceae)	'mai-sendedo' (Tigrinya)	Seeds used medicinally.
Satureja spp. (Lamiaceae)	Savory or Bean herb	Occasionally leafy shoots used as herbs.
Senna alexandrina (Fabaceae)	Senna, 'sono'	Leaves and fruits used medicinally.
Sesamum indicum (Pedaliaceae)	Sesame, 'selit'	Seed a source of high quality edible oil.
Tamarindus indica (Fabaceae)	Tamarind, 'roka' and 'humer'	Pulp from pods used in cooking and preparing non-alcoholic drinks.
Thymus spp. (Lamiaceae)	Thyme, 'tossin'	Whole plant used as a herb.

Trachyspermum ammi (Apiaceae)	'netch azmud'	Fruits used as a spice, important in spicing 'berbere'.
Vicia faba (Fabaceae)	Horse Bean, 'bekela'	Seeds used as a pulse; young seeds eaten fresh.
Vigna radiata (Fabaceae)	Mung Bean, Green Gram	Seeds used as a pulse but only found in a few areas.
Vinga unguiculata (Fabaceae)	Cowpea, 'adenguare'	Young leaves and pods eaten as a vegetable; ripe seeds used as a pulse.
Ziziphus spina-christi (Rhamnaceae)	'geba' and 'qwrqwra'	Ripe fruits eaten both fresh and after drying.
Monocotyledons		
Aframomum korarima (Zingiberaceae)	False Cardamom, 'korarima'	Seeds used as a spice important in 'berbere'.
Allium cepa (Alliaceae)	Onion, 'kei shinkurt' Shallot, 'kei shinkurt'	Fleshy bulb used as both a vegetable and a spice; leaves also used sometimes as spice.
Allium sativum (Alliaceae)	Garlic, 'netch shinkurt'	Fleshy bulbils used as a spice and also medicinally.
Amorphophallus abyssinica (Araceae)	'hamba guita' (Tigrinya)	Tuber is said to be edible.
Asparagus spp. (Asparagaceae)	Asparagus, 'kestenitcha'	Young shoots eaten fresh or as a cooked vegetable.
Avena abyssinica (Poaceae)	Ethiopian Oats, 'senar'	Seed used in admixture with barley as food and for brewing.
Colocasia esculenta (Araceae)	Taro, 'godere'	Tubers are eaten as a cooked vegetable.
Cymbopogon citratus (Poaceae)	Lemon Grass, 'tej sar'	Leaves used as a spice and fumigant.

Table 1 (*cont.*)

Scientific name (Family)	Common names in English and Amharic	Part used in Ethiopia
Dioscorea bulbifera	Aerial Yam, 'kota hari' in SW Ethiopia	Aerial tubers eaten as cooked vegetable.
Dioscorea alata *D. cayenensis–D. rotundata* complex (Dioscoreaceae)	Yam, 'boyye'	Underground tubers can be eaten.
Eleusine coracana (Poaceae)	African Finger Millet, 'dagusa'	Grain used mainly for brewing but also used to make food.
Eragrostis tef (Poaceae)	'teff'	Grain used mainly to make a flat fermented bread called 'injera', also used for other types of bread.
Ensete ventricosum (Musaceae)	'enset' for food types; 'koba' for types where leaves are used	Pseudocorm is processed to form a starchy food. Fibre from the leaves used for rope. Leaf lamina for wrapping bread during cooking, for eating out and for wrapping many other materials.
Hordeum vulgare (Poaceae)	Barley, 'gebs'	Grain used for both food and brewing.
Hyphaene thebaica (Arecaceae)	Dum Palm, 'dum' and 'arkokobay' for plant; 'akat' for fruit; 'lakha' for the leaves (all Tigrinya names)	Outer covering of fruit edible; leaves used to weave mats and baskets; stems for fuel.
Musa spp. (Musaceae)	Banana, 'mooz'	Ripe fruits eaten fresh; leaves used in same way as 'enset'.
Oryza sativa (Poaceae)	Rice, 'rooz'	Grain used as boiled food.

Pennisetum glaucum (Poaceae)	Pearl Millet, 'bultug'	Grain used to make 'injera', usually mixed with other cereals, and other types of food.
Phoenix dactylifera	Date Palm, 'temer'	Fruits edible fresh and dried.
P. reclinata (Arecaceae)	'zembaba'	Fruits are edible; leaves used for making mats.
Sorghum bicolor (Poaceae)	Sorghum, 'mashila' for 'injera' types; 'zengada' for brewing types	White grain used to make 'injera', either alone or mixed with 'teff' and other types of food. Dark grain used in brewing.
Triticum aestivum (Poaceae)	Bread Wheat, 'sindi' or 'dabo sindi'	Grain used to make raised bread. Whole grain eaten boiled or roasted.
Triticum durum (Poaceae)	Durum Wheat, 'sindi' or 'habesha sindi'	Used in same way as bread wheat plus as chipped grains and in making pasta.
Triticum polonicum *T. spelta*, *T. turgidum* (Poaceae)	'adja'	Used to make special food for nursing mothers and invalids.
Zingiber officinale (Zingiberaceae)	Ginger, 'zinjib'	Rhizome used as spice; also important medicinally.

Sources: personal notes and Cufodontis (1953–72); Purseglove (1968); Grieve (1976); FAO (1984).

Trachyspermum ammi (L.) Sprague ex Turrill (synonyms *Ammi copticum* L. and *Carum copticum* (L.) Benth. & Hook. ex Hiern) are widely cultivated throughout the highlands and are also found as escapes. There are no wild relatives of these species in Ethiopia (Heywood *et al.*, in preparation).
Daucus carota L. A wild form, often named as var. *abyssinica* A. Braun, is found in grassland and bushland on better drained soil in the highlands of Eritrea, Tigray, Gondar, Gojam, Shewa and Harerge. There is a second species, *D. hochstetteri* Engl. which is endemic to Ethiopia. It occurs in similar habitats to *D. carota* and is recorded from Eritrea, Tigray, Gondar, Shewa and Sidamo (Heywood *et al.*, in preparation).
Foeniculum vulgare Miller is a recent introduction. It is often confused with *Diplolophium africanum* Turcz. because the plants have similar leaf types and smells. *Diplolophium* occurs growing wild and can be a conspicuous member of the natural vegetation in open meadows. The soft young stems are eaten by children (personal observation).

Asteraceae (*Compositae*)

Carthamus tinctorius L. According to Cufodontis (1953–72, pp. 1177–8), *C. flavescens* Willd. (given as *C. persicus* Desf. ex Willd.) occurs only in northern Somalia and north-eastern Sudan. The wild species in Ethiopia is *C. lanatus* L. which is widespread throughout the highlands. It is a very aggressive weed in vertisols where *C. tinctorius* is normally cultivated and this could be the result of introgression with *C. tinctorius* (student research project, unpublished).
Guizotia abyssinica (L. f.) Cass. is most likely derived from *G. scabra* (Vis.) Chiov. This is a widespread weed often growing in the same fields as *G. abyssinica*. However, naturally occurring hybrids are not common. Both the crop and weedy species show a great deal of phenotypic variation (Seegeler, 1983, pp. 87–110).

Brassicaceae (*Cruciferae*)

Brassica spp. According to Seegeler (1983) there are six species of cultivated *Brassica* in Ethiopia. These are *B. campestris* L., *B. carinata* A. Braun, *B. integrifolia* (West) Rupr., *B. juncea* (L.) Czern., *B. nigra* (L.) Koch and *B. oleracea* L. There are no completely wild species given in Cufodontis (1953–72, pp. 148–50) but throughout the highlands of Ethiopia there are weedy forms of *Brassica* which are gathered to be eaten as a leafy vegetable. Mature plants of these weedy forms may also be collected for their seeds which are crushed and used to oil the earthenware plate on which 'injera' is baked. These weedy forms have not been studied intensively. Seegeler (1983, pp. 85–7) has recorded species of *Erucastrum*, which mostly occur wild, being used in a similar way to those of *Brassica*.

Crambe abyssinica Rich. Cufodontis (1953–72, p. 151) gives two other species, *C. kilimandscharica* Schulz, which occurs throughout East Africa and *C. sinuato-dentata* Petri. The latter is endemic and could well be conspecific with *C. abyssinica* as both *C. abyssinica* and *C. kilimandscharica* have been included in *C. hispanica* L. by Jonsell. However, Jonsell considered the Ethiopian material to form a distinct group within *C. hispanica* and this has been confirmed in chromosome studies. *Crambe* seems to be a rather difficult crop to find either in markets or growing in fields (Seegeler, 1983, pp. 82–5).

Eruca sativa Hill. occurs in northern Ethiopia and is cultivated and eaten as a salad before the inflorescences develop. It also occurs as a weed (Cufodontis, 1953–72, p. 145; personal observation).

Lepidium sativum L. Cufodontis (1953–72, pp. 140–2) gives four more species, *L. alpigenum* Rich. found in Eritrea and Arabia, *L. armoracia* Fisch. & Mey. found in Eritrea, Kenya and Arabia, *L. divaricatum* Soland. subsp. *subdentatum* (Burch.) Engl. found in north and probably also central Ethiopia and *L. intermedium* Rich. found only in Eritrea.

Nasturtium officinale R. Br. occurs naturally throughout the highlands and is sometimes gathered and sold to expatriates (Cufodontis, 1953–72, p. 152).

Cannabaceae

Cannabis sativa L. is most likely found in all regions as a weed and also sometimes cultivated. It is not used as a source of fibre in Ethiopia but in traditional medicine to treat epilepsy and similar emotional disorders (Verdcourt, 1989).

Celastraceae

Catha edulis (Vahl.) Forssk. ex Engl. is an important crop but it also occurs naturally in evergreen montane and medium altitude forest, usually near the margins or along valleys with rocky slopes (Robson, 1989).

Cucurbitaceae (Jeffrey, in preparation)

Citrullus lanatus (Thunb.) Matsum. & Nakai. The second species in Ethiopia is the wild *C. colocynthis* (L.) Schrad. which has small fruits

Table 2. *Distribution of* Coccinia *spp. in Ethiopia*

Species name and altitude range	Region according to *Flora of Ethiopia* (1989)[a]															
	EE	AF	EW	TU	GD	GJ	WU	SU	AR	WG	IL	KF	GG	SD	BA	HA
C. *schliebenni* Harms. 1220–2000 m										×	×	×				
C. *adoensis* (Hochst. ex A. Rich.) Cogn. 550–1850 m				×		×				×	×	×	×			
C. sp. A. 350–760 m																×
C. *abyssinica* 1300–2360 m				×		×				×	×	×				
C. sp. B 1220–1350 m																×
C. *megarrhiza* C. Jeffrey 1600–1800 m														×		
C. sp. C 1250–1300 m.														×		
C. *grandis* (L.) Voigt 300–1900 m	×	×	×		×			×	×		×	×		×		×

[a] Regions used to describe the distribution of plants in the *Flora of Ethiopia*:
EW – Eritrea West, west and above 1000 m contour.
EE – Eritrea East, east and below 1000 m contour.
TU – Tigray region, west and above 1000 m contour.
AF – Afar region, east and below 1000 m contour to Eritrean border in the east and Harerge border in the south.
WU – Welo region, west and above 1000 m contour.
SU – Shewa region, west and above 1000 m contour.
GD – Gondar region.
GJ – Gojam region.
WG – Welega region.
IL – Ilubabor region.
KF – Kefa region.
GG – Gamo Gofa region.
AR – Arsi region.
SD – Sidamo region.
BA – Bale region.
HA – Harerge region.

with bitter flesh. It is found below 1300 m in Eritrea, the Afar and Harerge.

Coccinia abyssinica (Lam.) Cogn. is an endemic plant found both cultivated and wild. The genus in Ethiopia needs further study as the most recent account (Jeffrey, in preparation) has a further seven taxa, three of them unnamed. The distribution of these taxa is given in Table 2.

Cucumis melo L. and *C. sativa* L. There are no wild relatives of *C. sativa* in Ethiopia. The wild *C. melo* subsp. *agrestis* (Naud.) Grebensc. is found in open woodland, especially on river margins and also in cultivation in eastern Eritrea, the Afar, Shewa and Kefa below 1100 m. The fruits are recorded as being used as food. The cultivated subsp. *melo*, is also grown. There are 10 other wild species of *Cucumis* found in Ethiopia. *C. humifructus* Stent has a subterranean fruit and is recorded from woodland and wooded grassland in southern Shewa and Sidamo. *C. metuliferus* E. Mey. ex Nadu. has been collected only once from Gondar; elsewhere it is sometimes cultivated. The following eight species – *C. figarei* Del. ex Naud., *C. ficifolius* A. Rich., *C. aculeatus* Cogn., *C. prophetarum* L., *C. insignis* C. Jeffrey, *C. dipsaceus* Ehrenb. ex Spach and the species of two unnamed taxa – are genetically compatible to varying degrees and can form hybrids. Many of the wild species are highly poisonous but are used in traditional medicine.

Cucurbita. There are no wild species found in Ethiopia but four species are recorded as cultivated. *C. ficifolia* Bouché is a perennial recorded from Asbe Teferi in Harerge. *C. moschata* (Duchesne ex Lam.) Duchesne ex Poir. is an annual recorded from the Lower Omo Valley. *C. pepo* L. and *C. maxima* Duchesne ex Lam. are annuals which are grown in many parts of the country but herbarium collections do not exist to give an accurate record of the distribution of these two species. It is highly likely that at least one of these species has been cultivated for a long time in Ethiopia (Tewolde Berhan Gebre Egziabher, 1984).

Lagenaria siceraria (Molina) Standl. is found both cultivated and wild or escaped in bushland and grassland throughout the country up to 1850 m. A completely wild species, *L. abyssinica* (Hook. f.) C. Jeffrey, is found in forest and scrub and is recorded from Gondar, Gojam, Shewa, Arsi, Kefa and Sidamo between 1600 and 2750 m.

Luffa cylindrica (L.) M. J. Roem. grows wild on riverbanks and in cultivated areas in the western lowlands where it is recorded from Eritrea and Ilubabor and also the Webi Shebelli Valley. It may be cultivated in some areas. The wild species *L. echinata* Roxb. is found

on riverbanks and along irrigation ditches in the Afar and Kefa.
Momordica charantia L. is found in cultivation at lower altitudes; it is widely cultivated in some other countries such as India. The closely related wild species *M. balsamina* L. is found in deciduous bushland on banks and in dry beds of rivers on sandy soil in Eritrea and Harerge. There are 11 other species of *Momordica* found in Ethiopia.

Euphorbiaceae

Ricinus communis L. is the only species in the genus and is generally recognized as having originated in Africa. It is widespread throughout Ethiopia, being grown as a house garden plant where the ripe seeds are crushed and used mainly to oil the baking plate for 'injera'. The seeds are also collected from wild stands, which can range from small shrubs to fairly robust trees (Seeleger, 1983, pp. 204–38). Some plants have dehiscent (wild type) fruits and others indehiscent (cultivated type) fruits.

Fabaceae (Leguminosae), subfamily *Caesalpinioideae* (Polhill & Thulin, 1989)

Senna alexandrina Mill, previously *Cassia senna* L., is the source of the drug senna. Both the commercially exploited var. *alexandrina* and the wild var. *obtusata* (Brenan) Lock occur in Ethiopia. Var. *alexandrina* occurs in Eritrea and the Afar in semi-desert scrub and grassland; var. *obtusata* is found in *Acacia–Commiphora* bushland and semi-desert in Eritrea and Harerge.
Tamarindus indica L. is not cultivated in Ethiopia. It is found throughout the country most frequently in river valleys, but also in *Combretum* woodlands where there is adequate ground water.

Fabaceae (Leguminosae), subfamily *Papilionoideae* (Thulin, 1989)

Canavalia ensiformis (L.) DC. Its wild relative, *C. africana* Dunn (synonym *C. virosa* (Roxb.) Wight and Arn.), occurs in Eritrea, Harerge, Ilubabor and Gamo Gofa. In Eritrea it is cultivated as a cover crop or to give shade.
Cicer arietinum L. The wild species, *C. cuneatum* A. Rich., is found in grassland and as a weed in cultivations in Eritrea, Tigray and Shewa.
Cyamopsis tetragonoloba (L.) Taub. is recorded as cultivated in western Eritrea as a house garden crop. Its wild relative *C. senegalensis* Guil and Perr. reaches the north-eastern limit of its distribution in western Eritrea.

Indigofera. There are four species of *Indigofera* in Ethiopia which were once important as the source of the internationally traded dye, indigo. These are *I. articulata* Gouan found in dry grassland and bushland in Eritrea, the Afar, Shewa and Harerge, *I. arrecta* Hochst. ex A. Rich. found also in dry grassland and bushland and recorded from all parts of the country except eastern Eritrea and the Afar, *I. coerulea* Roxb. with two varieties (var. *coerulea* from the arid coastal plains and var. *occidentalis* Gillett & Ali from all the drier parts of the country), and *I. tinctoria* L. found only in the Lower Omo Valley. Only *I. tinctoria* has been found in cultivation in Ethiopia. *Indigofera* is a large genus in Ethiopia with 78 species in the most recent account of the family (Thulin, 1989).

Lablab purpureus (L.) Sweet (synonym *Dolichos lablab*. L.), has both a wild subspecies, subsp. *uncinatus* Verdc., which can also be cultivated and a cultivated subspecies, subsp. *purpureus*; subsp. *uncinatus* is widespread. *Dolichos* has six species, all wild, in the present treatment of the family.

Lathyrus sativus L. can also be found growing as an escape. The wild species found in Ethiopia are *L. pratensis* L. and *L. sphaericus* Retz. growing in upland grassland. There are two more introduced cultivated species, the ornamental *L. odoratus* L. and the forage *L. aphaca* L. which is found as an escape in Eritrea.

Lens culinaris Medik. The wild species, *L. ervoides* (Brign.) Grande, grows in montane grassland and is found in Tigray, Gondar and Shewa.

Lupinus has six species, *L. albus* L. (synonym *L. termis* Forssk.) which is cultivated, particularly in Gojam and Gondar, and four more introduced and being grown by Soil and Water Conservation Projects. The sixth, *L. princei* Harms., is found in grassland in southern Sidamo.

Mucuna pruriens (L.) DC. var. *utilis* (Wall. ex Wight) Bak. ex Burck. The endemic *M. melanocarpa* Hochst. ex A. Rich. (sometimes misidentified as *M. pruriens* var. *pruriens*) is found in woodland and forest margins in western Eritrea, Tigray, Welega, Arsi, Harerge, Kefa, Gamo Gofa and Sidamo.

Pisum sativum L. var. *abyssinicum* (A. Br.) Alef. is endemic and only known as a cultivated plant.

Vicia faba L. has no close relative in Ethiopia. *V. villosa* Roth. is cultivated for forage and has escaped in some areas of Shewa and Arsi. *V. sativa* L. var. *sativa* is reported to be cultivated for fodder while var. *angustifolia* L. is wild and widespread in upland grassland and scrub. *V. paucifolia* Bak. and *V. hirsuta* (L.) S. F. Gray are wild species from montane grassland.

Vigna unguiculata (L.) Walp. is divided into five subspecies – three cultivated and two wild. Subsp. *unguiculata* and subsp. *cylindrica* occur as unimproved landraces with good drought resistance in Eritrea and Harerge; there are no recent records of subsp. *sesquipedalis* (L.) Verdc. Subsp. *dekindtiana* (Harms.) Verdc. occurs wild in Eritrea, Tigray, Gondar and Ilubabor and subsp. *mensensis* (Schweinf.) Verdc. in Eritrea, Kefa and Gamo Gofa.

V. radiata (L.) Wilczek (synonym *Phaseolus radiatus* L.) has a wild variety, var. *sublobata* (Roxb.) Verdc., recorded from Tigray and Gondar. Material of this variety was collected in the early 19th century. There are 15 other wild species of *Vigna* recorded from Ethiopia.

Lamiaceae (Labiatae)

Ocimum basilicum L. is an important cultivated spice and herb occurring in two varieties, var. *basilicum* and var. *thyrsiflorum* (L.) Benth. It is part of a complex of four species. *O. canum* Sims is smaller than *O. basilicum* and grows both wild and cultivated. *O. forskolei* Benth. is a wild species close to *O. basilicum* with white or light blue flowers which is found in many drier parts of the country. *O. stirbeyi* Schweinf. & Volkens is confined to the Ogaden region of Sidamo, Harerge, Bale, northern Kenya and southern Somalia (Ryding & Sebald, in preparation).

O. gratissimum L. has probably been introduced and is found only in cultivation. However, this species belongs to a group of species which are all used in a similar way and which are found both wild and cultivated. The most widespread are *O. urticifolium* Roth. (synonym *O. suave* Willd.) and *O. trichodon* Baker ex Gurke which can be difficult to distinguish from *O. gratissimum*. *O. lamiifolium* Hochst. ex Benth. is a more distinctive plant which is important in traditional medicine. It is usually found in forests and abandoned in fields. However, it has also been seen in gardens. The remaining two species in this group, *O. spicatum* Deflers and *O. jamesii* Sebald, are both wild and found in drier parts of the country (Ryding & Sebald, in preparation).

Plectranthus edulis (Vatke) Agnew, synonym *Coleus edulis* Vatke, occurs both wild and cultivated for its small irregularly shaped edible tubers. The crop is found in the wetter south and south-west, but wild forms are found throughout the country. *Plectranthus* is a large genus in Ethiopia with about 30 wild species recorded. The nearest to *P. edulis* is *P. punctatus* L'Hérit. which is sometimes considered conspecific with it. However, *P. punctatus* never forms tubers. The other species which forms tubers is *P. esculentus* N.E. Br. but this has not been recorded as growing in Ethiopia. *Solenostemon* is closely related to *Plectranthus* and *S. rotundifolius* (Poir.) Morton has tuberous roots which are edible, but it has not been confirmed as grown in Ethiopia. There are three other species of *Solenostemon* found in Ethiopia (Ryding & Morton, in preparation).

Salvia spp. Although none of the indigenous species are cultivated, *S. nilotica* Juss. ex Jacq. and *S. schimperi* Benth. have seeds with a good oil content (Seegeler, 1983). The ornamental species have all been introduced. The shrub *Meriandra bengalensis* (König & Roxb.) Benth. has often been mistaken as belonging to the genus *Salvia*. It is found both wild and cultivated in Eritrea and possibly also Gondar (Ryding, in preparation).

Satureja spp. and *Thymus* spp. The two species of *Thymus*, *T. schimperi* Ronniger and *T. serrulatus* Hochst. ex Benth. are extensively collected for the local market from their natural habitat above 2500 m. They are never cultivated and the wild populations show considerable phenotypic variation. There are 10 species of *Satureja* recorded in Cufodontis (1953–72, pp. 821–5), the most widespread being *S. biflora* (Ham. ex Don) Briquet. None are used as extensively as *Thymus*. Other species widely grown in house gardens for their aromatic foliage are *Origanum majorana* L. and *Rosmarinus officinalis* L.

Linaceae

Linum usitatissimum L. Cufondontis (1953–72, pp. 354–6) gives four other species: *L. holstii* Engl., found in southern Ethiopia and eastern Africa; *L. keniense* Fries, a rare plant found only in southern Ethiopia and northern Kenya; *L. strictum* L., found in northern Ethiopia; and *L. trigynum* L. var. *sieberi* (Planch.) Cuf., recorded from Eritrea, Shewa and Harerge. These are plants of grassland and edges of woodland. The closest relative of *L. usitatissimum*, *L. bienne* Miller, does not occur in Ethiopia (Seegeler, 1983).

Lythraceae

Punica granatum L. has been grown in the northern and central highlands for a long time but it is not found wild. It is now found cultivated in most of the larger towns above 1500 m. The only other species in this genus is *P. protopunica* Balf. f. which is endemic to Socotra (Gilbert, in preparation, a).

Malvaceae

Abelmoschus esculentus (L.) Moench. *A. ficulneus* (L.) Wright & Arn.

occurs in lowland Eritrea and Gamo Gofa in grassland on seasonally waterlogged black cotton soil. It closely resembles *A. esculentus* and is probably more widespread than existing records suggest (Vollesen, in preparation, a).

Gossypium has eight species recognized in the treatment for the Flora of Ethiopia (Vollesen, in preparation, a). These fall into three groups:

- indigenous wild species whose seeds are glabrous or covered with short brown hairs;
- indigenous cultivated species which are diploid and have seeds covered in a cottony lint which does not separate cleanly from the seed; and
- introduced cultivated species which are tetraploid and have seeds covered in a cottony lint which separates cleanly from the seed.

Wild species

Gossypium anomalum Wawra & Peyr. subsp. *senarense* (Fenzl. ex Wawra & Peyr.) Vollesen is found in the western Eritrean lowlands at the eastern end of its range in Africa. It grows on alluvial soil in *Acacia* bushland and grassland. This is a B genome species (Saunders, 1961).

G. somalense (Gürke) Hutch. is recorded from the Awash Valley, Sidamo and Harerge on gravelly granitic, volcanic or limestone soils. This is an E genome species (Saunders, 1961) also found outside Ethiopia. *G. bricchettii* (Ulbr.) Vollesen, known only from Bale administrative region and southern Somalia, is found in open *Acacia–Commiphora* bushland on gypsum hills. *G. benadirense* Mattei has been collected from the Dolo area of Sidamo, north-east Kenya and southern Somalia in similar habitats to *G. bricchettii*. These two species were included in *G. somalense* by Hutchinson (1947) and Fryxell (1980) but Vollesen (in preparation, a) considers them to be distinct although closely related to *G. somalense*. The latter two species probably have the same genome type as *G. somalense* and are an example of the important species-rich flora of the Somalia–Masai region which has a high level of endemism (White, 1983).

Cultivated species

The use of cotton has a long history in Ethiopia as seen in the clothing of the woman in the 'Statue of Haoulti' from the Pre-Axumite period (de Cotenson, 1981, p. 360). The indigenous cultivated species are *G. arboreum* L. and *G. herbaceum* L. According to Seegeler (1983) *G. arboreum* seems to have disappeared completely from cultivation although it is sometimes found in a feral state. Ramanathan (1947) is doubtful if this species is *G. arboreum*. He suggests that it may well have been perennial *G. herbaceum*. It was last recorded in cultivation from the lower parts of the Webi Shebele Valley around 1960. *G. herbaceum* is still cultivated but to a reduced extent in the Konso area together with *G. hirsutum*. It used to be widespread in northern Ethiopia but there are no recent records. It does not seem to have become established as an escape.

The introduced species, of American origin, are *G. hirsutum* L. and *G. barbadense* L. Both were introduced before the 1830s as there are records from the time of W. G. Schimper (1837–78). Formerly *G. hirsutum* was the species most commonly cultivated by peasant farmers but *G. barbadense* has become increasingly popular since it was introduced on a larger scale by the Italians around 1910.

Hibiscus is a large genus with 48 species in 10 sections. Both *H. cannabinus* L. and *H. sabdariffa* L. belong to section Furcaria which has six other wild species in Ethiopia (*H. diversifolius* Jacq., *H. berberidifolius* A. Rich., *H. sparseaculeatus* Bak. f., *H. surattensis* L., *H. rostellatus* Guill. & Perr. and *H. noldeae* Bak. f.), all of which are widespread, and *H. acetosella* Welw. ex Hiern. which is an introduced ornamental. *Hibiscus cannabinus* L. occurs wild in a variety of habitats (*Acacia* woodland and wooded grassland on grey to black alluvial soil, swamps, seepages, etc.) where there is sufficient ground moisture. There are no records of its cultivation except at the research level. *Hibiscus sabdariffa* L. is the only cultivated species (Vollesen, in preparation, a).

Moraceae

Ficus carica L. The closely related *F. palmata* Forssk. is a common shrub or small tree in hedgerows, secondary scrub, forest edges and riverine forest and scrub in Eritrea, Tigray, Gondar, Gojam, Welo, Shewa, Harerge, Arsi and Kefa (Friis, 1989).

Moringaceae

Moringa oleifera Lam. Of the 14 species recognized in this genus, nine are more or less endemic to north-east Africa. *M. oleifera* has only been recorded from Harerge and Eritrea but could occur more widely. *M. peregrina* (Forssk.) Fiori is recorded from eastern Eritrea and the Afar below 700 m. The other four species are all part of the Somali–Masai regional flora (White, 1983) and include *M. stenopetala* (Bak. f.) Cuf. which is an important and conspicuous part of the agricultural system practised in the Konso area of Gamo Gofa (Verdcourt, in preparation).

Oleaceae

Olea europea L. Subspecies *africana* (previously *O. africana* Mill.) is found throughout the drier parts of the highlands. There have been two expeditions by American entomologists to Eritrea to find natural enemies of the Olive Black Scale, *Saissetia oleae* (Bernard), which occurs naturally on *O. europea* subsp. *africana*. The scale was causing devastation to citrus orchards in California but after the introduction of the parasite, *Metaphycus helvolus*, the Black Olive Scale was reduced to a minor pest (Andemeskel, 1987).

Pedaliaceae

Sesamum indicum L. is found both cultivated and escaped or wild below 1800 m. Cufodontis (1953–72, p. 918) records two other species, *S. alatum* Thonn. from Eritrea, and *S. latifolium* Gillett from western Gojam and Gondar.

Piperaceae

Piper nigrum L. *Piper* has three species in Ethiopia of which *P. guineense* is sometimes eaten and has a fruit similar to *P. nigrum* (Gilbert, in preparation, b).

Plantaginaceae

Plantago afra L. (synonym *P. psyllium* L.) grows wild in northern Ethiopia and can be abundant on poorer, well drained soil. There are five other wild species found in Ethiopia (Cufodontis, 1953–72, pp. 980–2).

Ranunculaceae

Nigella sativa L. is reported to occur only in cultivation but it has been found growing abundantly as an ephemeral on a steep bank beside the road in the Bale mountains. There are no other species recorded for Ethiopia (Cufodontis, 1953–72, p. 106; personal observation).

Rhamnaceae

Rhamnus prinoides L'Hérit. grows as both a cultivated plant and a natural component of montane and riverine forest, usually on the edges or in clearings, in all parts of the country from 1400 to 3200 m. *R. staddo* A. Rich., which is used in a similar way to *R. prinoides*, occurs only wild, usually at the edges of montane forest, in wooded and scrub grassland from 1400 to 2900 m in Eritrea, Tigray, Gondar, Shewa, Arsi, Kefa, Gamo Gofa, Sidamo, Bale and Harerge (Vollesen, 1989a).

Ziziphus jujuba Mill. The preferred fruit, *Z. spina-christi* (L.) Desf., is found in wooded grassland on limestone slopes, *Acacia* bushland, in and along dry riverbeds, as well as edges of cultivation and gardens up to 2400 m in Eritrea, the Afar, Tigray, Gondar, Welo, Shewa, Gamo Gofa, Bale and Harerge. Other species with edible fruits are: *Z. abyssinica* Hochst. ex Rich. (sometimes considered a subspecies of *Z. jujuba*), with a similar distribution to *Z. spina-christi* but growing in woodland, wooded grassland and bushland; *Z. mucronata* Willd., growing in a wide range of dry woodland in all areas except Tigray, the Afar and Welega; *Z. mauritiana* Lam., recorded from riverine thickets and riverbanks in southern Ethiopia; and *Z. hamur* Engl., confined to soils derived from limestone and gypsum in Sidamo, Bale and Harerge. None of these have as good-tasting a fruit as *Z. spina-christi* (Vollesen, 1989a; personal observation).

Rosaceae

Prunus persica (L.) Batsch. has been grown in house gardens throughout the highlands of Ethiopia for a long time and was widespread in the early 16th century (Alvares, 1961). Although suffering severely from peach curl, with adequate moisture, the trees produce large crops of small green to yellow fruits.

Rubiaceae

Coffea arabica L. is found throughout the country, mostly between 1500 and 1900 m. It can occur as low as 1000 m in the very wet extreme south-west and as high as 2500 m in gardens and backyards. It grows as a genuinely wild, moist montane forest shrub or small tree, as a semi-wild crop in moist montane forests, as a properly cultivated crop in shade under rainfed conditions in the moist montane forests, as an irrigated crop without shade in some drier areas, and as a garden plant often mixed with fruit trees and herbs in the backyard, or simply watered from water jars (Tewolde Berhan Gebre Egziabher, 1990).

Rutaceae

Ruta chalepensis L. (often misnamed as *R. graveolens* L.) is very widely cultivated but is not known as an escape (Gilbert, 1989a).

Citrus aurantifolia (Christm.) Swingle is the most widely cultivated citrus and is sometimes found naturalized (Gilbert, 1989a). Like *Prunus persica* it has been grown in Ethiopia for a long time (Alvares, 1961).

Solanaceae

Capsicum spp. Introduced some time in the 16th or early 17th century (Tewolde Berhan Gebre Egziabher, 1984), *Capsicum* now shows a wide range of types cultivated for different purposes throughout the country. Cufodontis (1953–72, pp. 859–61) records three species, *C. annum* L., *C. frutescens* L. and *C. abyssinicum* Rich. *C. abyssinicum* is sometimes considered conspecific with *C. frutescens*.

Solanum is a large genus in Ethiopia with over 50 species recorded by Cufodontis (1953–72, pp. 861–80). Neither *S. tuberosum* L. nor *S. melongena* L. has wild relatives in the country. The wild species are of interest because a number of them are used in traditional medicine, including *S. marginatum* L., and in removing the hair prior to tanning animal skins.

Tiliaceae

Corchorus olitorius L. grows in grassland on black cotton soil, along riverbeds and as a weed of irrigated fields. It is not cultivated in Ethiopia. There are nine other species found in Ethiopia; none of them are cultivated but some are collected at a young stage and eaten as a cooked vegetable (Vollesen, in preparation, b).

Monocotyledons

Alliaceae

Allium cepa L. (including both shallot and onion) and *A. sativum* L. The native landraces of *A. cepa* are all shallots, onion being a recent introduction. Shallot is variable, ranging from landraces cultivated for their well developed bulbs to those with virtually no bulbs, the whole plant being chopped up as a vegetable. There are two wild species, *A. alibile* Rich., which is sometimes included in *A. ampeloprasum* L. from northern Ethiopia, and *A. subhirsutum* L. subsp. *spathaceum* (Steud. ex Rich.) Duyfjes from Eritrea, Gondar and Harerge (Tewolde Berhan Gebre Egziabher, in preparation).

Araceae

Amorphophallus abyssinicus (Rich.) N. E. Brown. Cufodontis (1953–72, p. 1501) gives two more species, both endemic, *A. gallaensis* (Engl.) N. E. Brown from Sidamo and *A. gomboczianus* Pichi-Sermolli from Gondar, Gojam, south-western Shewa and Kefa.

Colocasia esculenta (L.) Schott is widely grown in the wetter south and south-west. It is also recorded from Eritrea and Gojam. It easily escapes and can appear to occur spontaneously (Cufondontis, 1953–72, pp. 1501–2).

Arecaceae (Palmae)

The importance of palms as multi-purpose plants in traditional economies is being increasingly realized. The following are found in Ethiopia.

Borassus aethiopum Mart. is found in the lowlands and river valleys of western Ethiopia and south-western Sidamo (Cufodontis, 1953–72, p. 1499).

Hyphaene thebaica (L.) Mart. is cultivated in the lowlands of Eritrea, Tigray, Gondar and Harerge including the Afar. According to Cufodontis (1953–72, pp. 1496–9) there are two other species in Ethiopia: *H. dankaliensis* Beccari, found only in eastern Eritrea and Djibouti; and *H. nodularia* Beccari, recorded only from western Eritrea and southern Gamo Gofa. A further 10 species are recorded from Somalia, nine of them endemic.

Phoenix dactylifera L. is recorded from Eritrea, Tigray and Gamo Gofa. *P. reclinata* Jacquin is found throughout the country and also outside Ethiopia, but *P. abyssinica* Drude is an endemic found in Eritrea, Tigray, Gondar, Gojam, Sidamo, Gamo Gofa, Kefa and Ilubabor.

Asparagaceae

Asparagus spp. Cufodontis (1953–72, pp. 1562–6) records 10 species for Ethiopia. They grow from the dry hot lowlands to the cold, frost-prone mountains just below 3000 m. The most widespread species in the highlands are *A. africanus* Lam. and *A. asiaticus* L. The young shoots are eaten by children and are sometimes found being sold in Addis Ababa (personal observation).

Dioscoreaceae

Dioscorea. Yams are not a staple crop in Ethiopia although both root and aerial tubers, with aerial tubers being more common, can be found in local markets of the wetter, western half of Ethiopia (personal observation). Tubers of wild species are said to be eaten in times of food shortage.

Miège (1986) has identified seven species and two species groups from herbarium collections. *Dioscorea bulbifera* L. has both cultivated and wild forms, the latter are said to be violently poisonous. The other wild species with aerial tubers is *D. schimperana* Kunth. It is recorded as thriving on terraces between 800 and 2100 m where annual rainfall is between 900 and 1400 mm. It grows wild in *Acacia* and Combretaceous woodlands and at forest edges. Most cultivated yams belong to the section Enantiophyllum and material of both *D. alata* L. and the *D. cayenensis* Lamk.–*D. rotundata* Poir. complex have been identified. Some cultivars of *D. alata* produce both root and aerial tubers. Enantiophyllum also includes the *D. abyssinica* Hochst., *D. lecardii* de Wild. and *D. odoratissima* Pax complex, all of which only occur wild. *D. abyssinica* grows in hilly areas covered in wooded grassland and Combretaceous woodland between 1000 and 1800 m. Of the remaining four species, the most peculiar is *D. gilletti* Milne-Red. with its nearest relatives in the Pyrenees of south-west Europe. It is the most drought-resistant of all the Ethiopian species being found in areas with less than 700 mm of rain a year. The distribution of the various species according to Miège (1986) is given in Table 3.

Musaceae

Ensete ventricosum (Welw.) Cheesman occurs throughout the country both wild and cultivated wherever there is adequate moisture. Its optimal altitudinal range is between 1600 and 2400 m but it can be found up to 3000 m in the Gurage highlands and below 1000 m in the wet south-west. Wild stands grow in forests and are found as far north as Tigray in isolated moist pockets of forest. This crop shows a very wide range of variation which has not been systematically studied throughout its range (Food and Agriculture Organization, 1984; personal observation).

Musa spp. are grown both on a large scale and by peasant farmers. The latter show quite a range of types which do not get into the larger markets because they do not travel well. There are no wild species of *Musa* in Ethiopia, although stands may be found apparently separated from any habitation. These are either left after a household has moved or have been deliberately planted in a place, such as a sheltered valley, which is more favourable for the production of the crop (FAO, 1984; personal observation).

Poaceae (Gramineae)

Avena abyssinica Hochst. This is the oat often grown in a mixture with barley. It is closely related to the weedy *A. vaviloviana* (Malz.) Mordv. and hybrids are formed where these two species meet. Specimens from these two species, as well as the hybrid between them have also

Table 3. *Distribution of* Dioscorea *spp. in Ethiopia*

Species name and altitude range	Region according to *Flora of Ethiopia* (1989)[a]															
	EE	AF	EW	TU	GD	GJ	WU	SU	AR	WG	IL	KF	GG	SD	BA	HA
D. quartiniana	×		×	×	×			×	×	×			×	×		×
D. dumetorum			×	×	×					×						
D. cochleari–apiculata				×												
D. gillettii															×	×
D. bulbifera	×		×		×							×				
D. schimperana				×	×	×	×	×	×	×		×	×	×		
D. alata														×		
D. cayenensis–rotundata								×				×		×		
D. abyssinica	×		×	×	×					×	×	×	×	×		
D. odoratissima																×

[a] Abbreviations as for Table 2.

been referred to the species *A. barbata* Pott. from which the Ethiopian species may have been derived. *A. barbata* Pott., in the narrow sense, does not occur in Ethiopia (Phillips, in preparation).

Cymbopogon citratus (DC. ex Nees) Stapf is cultivated in house gardens throughout the country. According to Cufodontis (1953–72, pp. 1392–5) there are seven other species, *C. commutatus* (Steud.) Stapf., *C. excavatus* (Hochst.) Stapf., *C. floccosus* (Schwfth.) Stapf., *C. giganteus* (Hochst.) Chiov., *C. nervatus* (Hochst.) Chiov., *C. proximus* (Hochst. ex Rich.) Stapf. and *C. schoenanthus* (L.) Sprengel, all of which contain aromatic essential oils. All these wild species grow below 1700 m.

Eleusine coracana (L.) Gaertn. can grow as an escape. It also forms hybrids with both subspecies of *E. indica* (L.) Gaertn., subsp. *indica* and subsp. *africana* (Kennedy-O'Byrne) Phillips. Both subspecies are found as weeds and in open disturbed habitats from sea level to 2400 m. Hybrids of subsp. *africana* with the crop species usually have longer and narrower spikes than true *E. coracana*; also the spikelets are not so closely packed on the rachis, the grain is intermediate in size and the spikelets usually shatter (Phillips, 1974 and in preparation).

Eragrostis tef (Zucc.) Trotter. The closest wild relative is generally considered to be *E. pilosa* (L.) P. Beauv. This species is recorded from Eritrea, Tigray, Gondar and Shewa where it grows as an annual in open places and as a weed in cultivated fields, often near ditches (Jones, 1988; Phillips, in preparation).

Oryza sativa L. *O. barthii* A. Chev. is found in the Gambella plains of western Ilubabor and *O. longistaminata* A. Chev. & Roehr. occurs in the swamps and marshes up to 2500 m around Lake Tana, where it sometimes forms pure stands (Phillips, in preparation).

Pennisetum glaucum (L.) R. Br. The supposed wild source of pearl millet is *P. violaceum* (Lam.) A. Rich. with the eastern extension of its distribution reaching the lowlands of western Eritrea (Phillips, in preparation).

Sorghum bicolor (L.) Moench. This species forms hybrids with *S. arundinaceum* (Willd.) Stapf., which occurs in Ethiopia, and possibly other species are found in Ethiopia.

There are no wild relatives in Ethiopia of either *Hordeum vulgare* L. or the several species of *Triticum*.

Zingiberaceae

Aframomum korarima (Per.) Engl. Cufodontis (1953–72, pp. 1594–5) records two more species, *A. polyanthum* (K. Schum.) K. Schum. and *A. sanguineum* (K. Schum.) K. Schum., both from Kefa and extending from southern Sudan to north-eastern Zaïre.

References

Alvares, Francisco, ed. C. F. Beckingham and G. W. B. Huntingford (1961). *The Prester John of the Indies, vols 1 and 2*. Cambridge University Press, Cambridge.

Andemeskel Woldehaimanot (1987). Handbook of insect pests of major crops in Eritrea Administrative Region, Ethiopia, and their control. Asmara University, p. 59 (mimeographed).

Asfaw Hunde & Thulin, M. (1989). 95. Fabaceae (Leguminosae), subfamily Mimosoideae. *In*: I. Hedberg and S. Edwards (eds), *Flora of Ethiopia, vol. 3*. The National Herbarium, Addis Ababa University, Ethiopia, and the Department of Systematic Botany, Uppsala University, Sweden, pp. 71–96.

Cufodontis, G. (1953–72). *Enumeratio Plantarum Aethiopiae, Spermatophyta*. Facsimile 1974 from Bulletin du Jardin Botanique de l'Etat, Bruxelles (**23**, 1953 – **36**, 1966) and Bulletin du Jardin Botanique National de Belgique (**37**, 1967 – **42**, 1972). 1657 pp.

de Cotenson, H. (1981). Pre-Aksumite culture. *In*: G. Mokhtar (ed.), *General History of Africa. II. Ancient Civilizations of Africa*, Heinemann, California, UNESCO p. 360.

Food and Agriculture Organization (1984). *Assistance to Land Use Planning, Ethiopia. Land Evaluation. Part Two. Land utilization types*. Based on the work of Sue B. Edwards, Abebe Mengesha, J. K. W. Niemeyer and R. B. Ridgway. AGOA: ETH/78/003, Technical Report 5, Part Two. UNDP/FAO, pp. 137–40.

Friis, I. (1989). 104. Moraceae. *In*: I. Hedberg and S. Edwards (eds), *Flora of Ethiopia, vol. 3*. The National Herbarium, Addis Ababa University, Ethiopia, and the Department of Systematic Botany, Uppsala University, Sweden, pp. 280–1.

Fryxell, P. A. (1969). A Classification of *Gossypium* L. (Malvaceae). *Taxon*, **18**, 585–91.

Gilbert, M. G. (1989a). 120. Rutaceae. *In*: I. Hedberg and S. Edwards (eds), *Flora of Ethiopia, vol. 3*. The National Herbarium, Addis Ababa University, Ethiopia, and the Department of Systematic Botany, Uppsala University, Sweden, pp. 419–20, 431–2.

Gilbert, M. G. (1989b). 127. Anacardiaceae. *In*: I. Hedberg and S. Edwards (eds), *Flora of Ethiopia, vol. 3*. The National Herbarium, Addis Ababa University, Ethiopia, and the Department of Systematic Botany, Uppsala University, Sweden, pp. 419–20, 530–2.

Gilbert, M. G. (in preparation, a). Lythraceae. To be published in Flora of Ethiopia, vol. 2.

Gilbert, M. G. (in preparation, b). Piperaceae. To be published in Flora of Ethiopia, vol. 2.

Grieve, M. (1976). *A Modern Herbal*. Penguin Books, London.

Heywood, V., Jury, S. & Abebe, D. (in preparation). Apiaceae (Umbelliferae). To be published in Flora of Ethiopia, vol. 4.

Hutchinson, J. B., Silow, R. A. & Stephens, S. G. (1947). *The Evolution of*

Gossypium *and the differentiation of the Cultivated Cottons*. Empire Cotton Growing Corporation, London.

Jeffrey, C. (in preparation). Cucurbitaceae. To be published in Flora of Ethiopia, vol. 2.

Jones G. (1988). Endemic crop plants of Ethiopia. I. T'ef (*Eragrostis tef*). *Walia* **11**, 37–43.

Miège, J. (1986). Dioscoreaceae of Ethiopia. Acta University Uppsala, *Symb. Bot. Ups. xxvi:2*, 157–68.

Onwueme, I. C. (1978). *The Tropical Tuber Crops: yams, cassava, sweet potato, cocoyams*. John Wiley and Sons, Chichester, pp. 3–16.

Perdue, R. E., Jr (1988). *Vernonia galamensis* update, cyclostyled newsletter, 2 pp.

Phillips, S. M. (1974). 62. Eleusine. *In*: W. D. Clayton, S. M. Phillips and S. A. Renvoize, Gramineae (Part 2): R. M. Polhill (ed.), *Flora of Tropical East Africa*, Crown Agents for Oversea Governments and Administration, pp. 262–4.

Phillips, S. M. (in preparation). Poaceae (Gramineae). To be published in Flora of Ethiopia, vol. 7.

Polhill, R. M. & Thulin, M. (1989). 95. Fabaceae (Leguminosae), subfamily Caesalpinioideae. *In*: I. Hedberg and S. Edwards (eds), *Flora of Ethiopia, vol. 3*. The National Herbarium, Addis Ababa University, Ethiopia, and the Department of Systematic Botany, Uppsala University, Sweden, pp. 49–70.

Purseglove, J. W. (1968). *Tropical Crops: Dicotyledons, vols 1 and 2*. Longmans, London, pp. 658–90.

Ramanathan, V. (1947). The evolution of *Gossypium*. *Indian Journal of Genetics and Plant Breeding*, **7**, 55.

Robson, N. K. B. (1989). 108. Celastraceae. *In*: I. Hedberg and S. Edwards (eds), *Flora of Ethiopia, vol. 3*. The National Herbarium, Addis Ababa University, Ethiopia, and the Department of Systematic Botany, Uppsala University, Sweden, pp. 340–1.

Ryding, O. (in preparation). Lamiaceae (Labiatae). To be published in Flora of Ethiopia, vol. 5.

Ryding, O. & Morton, J. (in preparation). Lamiaceae (Labiatae), genus *Plectranthus*. To be published in Flora of Ethiopia, vol. 5.

Ryding, O. & Sebald, O. (in preparation). Lamiaceae (Labiatae), genus *Ocimum*. To be published in Flora of Ethiopia, vol. 5.

Saunders, J. H. (1961). *The Wild Species of* Gossypium *and their Evolutionary History*. Empire Cotton Growing Corporation, London.

Seegeler, C. J. P. (1983). *Oil Plants in Ethiopia, their Taxonomy and Agricultural Significance*. PUDOC, Wageningen.

Stannard, B. (1989). 122. Simaroubaceae. *In*: I. Hedberg and S. Edwards (eds), *Flora of Ethiopia, vol. 3*. The National Herbarium, Addis Ababa University, Ethiopia, and the Department of Systematic Botany, Uppsala University, Sweden, pp. 440–1.

Tewolde Berhan Gebre Egziabher (1984). Some important New World plants in Ethiopia. *In*: S. Rubenson (ed.), *Proceedings of the 7th International Conference of Ethiopian Studies*, University of Lund, 26–29 April 1982. Institute of Ethiopian Studies, Addis Ababa University, Addis Ababa, Scandinavian Institute of African Studies, Uppsala, and African Studies Center, Michigan State University, East Lansing, Michigan, pp. 187–94.

Tewolde Berhan Gebre Egziabher (1990). The importance of Ethiopian forests in the conservation of *Arabica* coffee gene pools. Paper presented at AET-

FAT (Association pour l'Etude Taxonomique de la Flore d'Afrique Tropicale) 12th Plenary Meeting, 4–10 September 1988. Institut für Allgemeine Botanik und Botanischer Garden, Hamburg (in press).

Tewolde Berhan Gebre Egziabher (in preparation). Alliaceae. To be published in Flora of Ethiopia, vol. 6.

Thulin, M. (1989). 95. Fabaceae (Leguminosae), subfamily Papilionoideae (Faboideae). *In*: I. Hedberg and S. Edwards (eds), *Flora of Ethiopia, vol. 3*. The National Herbarium, Addis Ababa University, Ethiopia, and the Department of Systematic Botany, Uppsala University, Sweden, pp. 97–251.

Townsend, C. C. (in preparation). Amaranthaceae. To be published in Flora of Ethiopia, vol. 2.

Verdcourt, B. (1989). 106. Cannabaceae. *In*: I. Hedberg and S. Edwards (eds), *Flora of Ethiopia, vol. 3*. The National Herbarium, Addis Ababa University, Ethiopia, and the Department of Systematic Botany, Uppsala University, Sweden, pp. 327–8, 419–20.

Verdcourt, B. (in preparation). Moringaceae. To be published in Flora of Ethiopia, vol. 2.

Vollesen, K. (1989a). 118. Rhamnaceae. *In*: I. Hedberg and S. Edwards (eds), *Flora of Ethiopia, vol. 3*. The National Herbarium, Addis Ababa University, Ethiopia, and the Department of Systematic Botany, Uppsala University, Sweden, pp. 390–7.

Vollesen, K. (1989b). 123. Burseraceae. *In*: I. Hedberg and S. Edwards (eds), *Flora of Ethiopia, vol. 3*. The National Herbarium, Addis Ababa University, Ethiopia, and the Department of Systematic Botany, Uppsala University, Sweden, pp. 442–78.

Vollesen, K. (in preparation, a). Malvaceae. To be published in Flora of Ethiopia, vol. 2.

Vollesen, K. (in preparation, b). Tiliaceae. To be published in Flora of Ethiopia, vol. 2.

White, F. (1983). The vegetation of Africa – a descriptive memoir to accompany the UNESCO/AETFAT/UNSO vegetation map of Africa. *Natural Resources Research*, **XX**, 110–30.

4

Diversity of the Ethiopian flora

TEWOLDE BERHAN GEBRE EGZIABHER

Introduction

It is generally accepted that Ethiopia is an important domestication and genetic diversification centre of crop species (Purseglove, 1968; Mooney, 1979). Likewise, it is instinctively felt that it must have a rich flora. But this is not known quantitatively, partly because efforts at documenting Ethiopian plants have been sporadic (Friis, 1982) and as a result, many plants remain unrecorded. On the other hand, hasty recording has often meant that a species goes by different names, causing double counting. The situation is made more confusing because the plant specimens collected from Ethiopia are scattered in various herbaria, mostly in Europe. The information published on them is equally scattered and in numerous European languages (Cufodontis, 1953–72). Compiling information on Ethiopian plants is thus a daunting task.

The Ethiopian Flora Project, supported financially by both the Ethiopian Government and the Swedish Agency for Research Cooperation with Developing Countries (SAREC), was launched to meet this challenge. The project is therefore building a reference herbarium and library so that, in some years' time, information on Ethiopian plants will be found organized at one reference point in Ethiopia. The project is training young Ethiopian taxonomic botanists so that this information can be continually augmented, managed and updated. It is also writing a Flora of Ethiopia, covering the whole country, so that plant identification, both in the field and in the herbarium, becomes possible. When the Flora has been completed, it will be possible to compile reliable data on the number of species and their distribution as well as on the interesting question of endemism, thus quantifying the diversity of the Ethiopian flora.

One volume of the Flora of Ethiopia has now been published (Hedberg & Edwards, 1989). It should, therefore, be possible to extrapolate from it and obtain impressions of the whole flora. Another very important source of information is Cufodontis' list of plants of the Horn of Africa (Cufodontis, 1953–72). He collected written information on the plants of Ethiopia, Somalia and Djibouti. It is a rather uncritical compilation, which is understandable since it is the work of one man and was based on literature. Nevertheless, we will try to juxtapose the information in this work with the more critically compiled information in the completed volume of the Flora of Ethiopia, and extrapolate. This will give only a rough estimate of both the size of the Ethiopian flora and its endemism; given the present state of knowledge, however, this will be the best that can be obtained. But, before extrapolating, we must look at the affinities of the Ethiopian flora as these affinities have a direct bearing on the number of species involved and on the likelihood of a species being endemic: in short, on the diversity of the flora.

Affinities of the Ethiopian flora

Thulin (1983), in the introduction to his Leguminosae of Ethiopia, has summarized the information on this issue. The highlands of Ethiopia, together with the highlands of East Africa, Cameroun and the Sudan, constitute the Afro-montane floristic region (White, 1978). The flora is uniform and endemism in a given country is therefore low as many of the Afro-montane taxa of one country will also be found in another country. Though the Ethiopian highlands are the most extensive of the African mountainous regions, the number of species in them is lower than in the less extensive East African mountains. This is probably because the Ethiopian highlands are, on the whole, drier than their East African counterparts.

To the west of the Ethiopian plateau, encroaching into it along river valleys, is the adjoining Sudanian floristic region (White, 1979). This region extends westwards from Ethiopia all the way to the Atlantic Ocean. Endemism in countries in this region is, therefore, understandably low as the region is divided into many countries.

To the east of the Ethiopian plateau, and thence south to northern Tanzania, is the relatively small (in terms of area) but very distinct Somalia–Masai floristic region. This is a region of high endemism. It has similarities with the Madagascan and Arabian floristic regions since the bodies of water separating it from them are narrow. It also bears a similarity with the Kalahari floristic region in spite of the intervening wide Zambesian floristic region.

Since Ethiopia contains these regions, its flora has affinities with all but Equatorial Africa. It also has affinities with Arabia, both because of its Somalia–Masai component and because of its Afro-montane component. In floristic and geological terms, the plateau of southern Arabia is, in fact, African (Mohr, 1971).

The higher parts of the Ethiopian plateau, with their medium to low temperatures, have enabled primarily Mediterranean and temperate Eurasian taxa to occur. With increased human movement, new taxa are being added, establishing the similarity at even specific and infraspecific levels. There is a similarity between the floras of Ethiopia and the Canary Islands, which is difficult to explain. Furthermore, human movement is also introducing Australian and American taxa into Ethiopia. Who can imagine Addis Ababa without *Eucalyptus globulus*, an Australian species?

In the following text some examples are given to illustrate the affinities summarized above. The genera *Cadia* and *Delonix* are Madagascan, but *C. purpurea* and *D. elata* occur in Ethiopia. Of the many taxa found in Arabia and Ethiopia, *Rosa abyssinica* and *Polygala aethiopica* can be mentioned, and of course teff (*Eragrostis tef*) from among the montane species; *Barbeya oleoides, Commicarpus pedunculosus, Erythrococca abyssinica* and a number of *Commiphora* and *Boswellia* species from among the Somalia–Masai species.

It is surprising that the Kalahari link exists even at the species level: *Indigofera trigonelloides* is found only in Namibia and Ethiopia, and nowhere in between (Thulin, 1983). *Commiphora* and *Boswellia* are two characteristic genera of the Somalia–Masai floristic region, the centre of diversity for these genera. Forty-eight *Commiphora* species and six *Boswellia* species occur in Ethiopia (Vollesen, 1989), firmly establishing the Somalia–Masai character of its flora.

The dominant montane forest trees, perhaps Ethiopia's most conspicuous species, occur in other montane areas of Africa; *Juniperus procera, Podocarpus gracilior, Olea africana, Aningeria adolfi-friedericii* and *Celtis africana* can be mentioned as examples (Eggeling, 1952; Dale & Greenway, 1961).

The Sudanian floristic character is shown in the savannahs of the lowlands and river valleys of western Ethiopia, e.g. *Panicum maximum* (elephant grass), *Anogeissus leiocarpus* and many savannah woodland species of *Combretum* and *Terminalia*.

The Mediterranean and temperate Eurasian traits of the Ethiopian flora can be illustrated with a number of genera, e.g. *Festuca, Dianthus, Silene* and *Trifolium* illustrating the temperate connection, *Scorpiurus, Medicago* and *Satureja* illustrating the Mediterranean connection and a number of families, e.g. Cruciferae (Brassicaceae), Umbelliferae (Apiaceae) and Caryophyllaceae illustrating both connections. Recent European introductions have been deliberate, e.g. varieties of *Brassica oleracea* (cabbage), *Beta vulgaris* (beetroot) and *Daucus carota* (carrot). Note that wild forms of *D. carota* are native to Ethiopia but the cultivated form is an introduction. Introductions have been accidental and sometimes harmful, e.g. *Galinsoga parviflora*, which became a serious weed; and sometimes not harmful – would it be better to say 'not yet harmful'? – e.g. *Dactylis glomerata*, *Silybum marianum*. Some introductions are ancient, e.g. *Arundo donax* which still cannot flower in Ethiopia but keeps growing vigorously, vegetatively.

The puzzling Canarian connection can be seen from the genera *Canarina*, with only two species occurring in Ethiopia and the other eastern African highlands; *Hypagophytum*, with one species in Ethiopia only (*H. abyssinicum*); and *Aeonium*, with two species in Ethiopia and southern Arabia, but each with numerous species in the Canary Islands and none in the vast intervening African hinterland.

Size and endemism of the Ethiopian flora

Cufodontis' (1953–72) list includes about 6370 species. The qualifying term 'about' is needed because Cufodontis has included uncertain records, even only plant names without taxonomic descriptions or voucher specimens. Dealing with such records without going to the source material involves personal judgement. Going to source material is a long process and cannot be contemplated when making only an estimate of species numbers; it forms a major part of the ongoing work on the Flora of Ethiopia.

Of the 6370 species Cufodontis records, 4865 are Ethiopian, the other 1505 being endemic to the Somalia–Djibouti region (Cufodontis, 1953–72). The number of endemics to Ethiopia is only 1182, while the number of endemics to the whole of the Horn of Africa is 2291. These figures can be used as the basis for estimating the size of the Ethiopian flora.

The method of estimating species numbers to be adopted here is that of comparing the figures derived from Cufodontis' work for the whole flora with the more precise estimate obtained from the work of the Ethiopian Flora Project. When doing this, the families in the completed volume of the Ethiopian Flora cannot be taken as a random sample for simple extrapolation because the choice of the families involved is systematic in the taxonomic sense, and hence is not a random sample in the statistical sense. The choice had to be systematic because related families must come close together in a volume. A simple extrapolation from the numbers in the completed volume to the whole, by calibrating Cufodontis' records against the records of the completed Ethiopian Flora volume, would thus not be justified. Instead, a method of arriving at an informed guess is being adopted as more realistic. Two families from the completed volume of the Flora are being used to arrive at what intuitively feels correct. The two families chosen are Leguminosae (Fabaceae) and Burseraceae.

The family Leguminosae is represented all over Ethiopia, and its estimate of endemism using Cufodontis' records is about average (31 per cent for the Horn of Africa and 22 per cent for Ethiopia for Leguminosae, compared with 36 per cent for the Horn of Africa and 24 per cent for Ethiopia for the whole flora. Endemism in the family Burseraceae, on the other hand, is average according to Cufodontis' records for Ethiopia (24 per cent) but very high for the Horn of Africa (87 per cent). This suggests a much more complete recording of species for Somalia and Djibouti than for Ethiopia. More new records for Ethiopia would, therefore, be expected in the Burseraceae than in the Leguminosae. The work of the Ethiopian Flora has shown this to be the case; Leguminosae has risen from Cufodontis' records of 492 to 607, an increase of only 23 per cent, while Burseraceae has risen from Cufodontis' records of 29 to 54, an increase of 86 per cent. In either case, the increase in the species is proportional to the number of regional (Horn of Africa) species not found in Ethiopia as it is mostly from among these species that new Ethiopian records can be expected. Using this formula:

$$(\text{Regional species} - \text{Ethiopian species}) \times \frac{\text{Ethiopian species}}{\text{Regional species}} + \text{Ethiopian species}$$

an estimate of the likely number of Ethiopian species can be obtained from the numbers in Cufodontis' records. This will be an overestimate because it assumes that the less known regional endemics and the better recognized widespread species equally have failed to be recorded in Ethiopia in the past and will be equally represented in future new records. The opposite view is to assume that only regional endemics will continue to be discovered in Ethiopia. The reality is probably somewhere between the two. An average of both figures is, therefore, being used to estimate the size of the Ethiopian flora. Predicted and actual values for the two big, and in terms of distribu-

Table 1. *Estimates of the Ethiopian flora*

	Number of Ethiopian species				
	A	B	C	D	E
Leguminosae	492	604	562	583	607
Burseraceae	29	50	52	51	54
All seed plants in Ethiopia	4865	6014	5712	5863	?

A = Counted from Cufodontis (1953–72).

B = (Regional spp. − Ethiopian spp.) × $\frac{\text{Ethiopian spp.}}{\text{Regional spp.}}$ + Ethiopian spp.

C = (Regional endemic − Ethiopian endemic) × $\frac{\text{Ethiopian spp.}}{\text{Regional spp.}}$ + Ethiopian spp.

D = Average value of B and C.

E = Counted from the completed volume of *Flora of Ethiopia*.

tion interesting, families in the completed volume of the Flora of Ethiopia, Leguminosae and Burseraceae, are used to check this out (Table 1).

Using the adopted formulae, the actual numbers of species for Leguminosae and Burseraceae are shown to be greater than the estimated numbers. This is because, while Ethiopia adjoins the Sudan and Kenya, Cufodontis' records exclude these two countries; yet plants already recorded in these countries, but not yet in Ethiopia, will also be expected to appear as new records in Ethiopia. The figure of 5863, or roughly 6000, is therefore a low estimate for the Ethiopian species. Since endemism in the Sudanian zone is low, the rise owing to contribution from it will not be as large as that from the Somalia–Masai region. However, it could be estimated that it will be about half as much, thus giving us an estimate for the seed plants of Ethiopia of approximately 6360, rounding off to a more realistic figure, about 6500. The ferns of Ethiopia will consist of a few hundred species, thus giving us an estimate of the higher plants of Ethiopia of about 6700–6900, or, more realistically, between 6500 and 7000.

Endemism in Ethiopia has been estimated by a few authors. Brenan (1978), counting a sample portion of Cufodontis' list, arrived at the conclusion that about 21 per cent of the species of Ethiopia are endemic. Using the whole of Cufodontis' list, it was calculated that 24 per cent of the species are endemic. The endemism according to Cufodontis' list is 22 per cent for Leguminosae and 24 per cent for Burseraceae. Endemism in these two families, according to the newly completed Ethiopian Flora volume, is 11 per cent (Thulin, 1983) and

15 per cent (Vollesen, 1989), respectively. These figures are not very different from each other in spite of the very high regional endemism of Burseraceae. This is because even the small Somalia–Masai region is divided among several countries (Ethiopia, Djibouti, Somalia, Kenya and Tanzania), thus reducing endemism within national boundaries. The regional (i.e. Ethiopia–Somalia–Djibouti) endemism, considering only those species found in Ethiopia, gives a different picture: 15 per cent for Leguminosae and 42 per cent for Burseraceae. Endemism in the Horn of Africa is thus high, in the order of 20 per cent, but it is reduced in Ethiopia to about 12 per cent, or about half for the region.

The implicit belief that has existed hitherto, that the Ethiopian flora is rich both in species numbers and in endemics is, therefore, valid.

References

Brenan, J. P. M. (1978). Some aspects of the phytogeography of tropical Africa. *Annals of the Missouri Botanical Garden*, **65**, 437–78.

Cufodontis, G. (1953–72). *Enumeratio Plantarum Aethiopiae, Spermatophyta.* Bulletin du Jardin Botanique National de Belgique, Bruxelles.

Dale, I. R. & Greenway, P. J. (1961). *Kenya Trees and Shrubs.* Buchanans, Kenya Estates Ltd, Nairobi, in association with Hatchards, London.

Eggeling, W. J. (1952). *Indigenous Trees of the Uganda Protectorate, 2nd edn revised by I. R. Dale.* Crown Agents, London.

Friis, I. (1982). A list of botanical collectors in Ethiopia. University of Copenhagen (unpublished).

Hedberg, I. & Edwards, S. (eds) (1989). *Flora of Ethiopia, vol. 3.* The National Herbarium, Addis Ababa University, Ethiopia, and the Department of Systematic Botany, Uppsala University, Sweden. (Further volumes in preparation.)

Mohr, P. (1971). *The Geology of Ethiopia.* Haile Selassie I University Press, Addis Ababa, pp. 7–8.

Mooney, P. R. (1979). *Seeds of the Earth.* Canadian Council for International Cooperation, Ottawa.

Purseglove, J. W. (1968). *Tropical Crops, vols 1, 2, & 3.* Longmans, London.

Thulin, M. (1983). Leguminosae of Ethiopia. *Opera Botanica*, **68**, pp. 7–13.

Vollesen, K. (1989). 123. Burseraceae. *In*: I. Hedberg and S. Edwards (eds), *Flora of Ethiopia, vol. 3.* The National Herbarium, Addis Ababa University, Ethiopia, and the Department of Systematic Botany, Uppsala University, Sweden, pp. 442–78.

White, F. (1978). The Afromontane region. *In*: M. J. A. Werger (ed.), *Biogeography and Ecology of Southern Africa, vol. 1.* Dr W. Junk, The Hague, pp. 463–513.

White, F. (1979). The Guineo–Congolian region and its relationships to other phytochoria. *Bulletin du Jardin Botanique National de Belgique*, **49**, 11–55.

5

Forest genetic resources of Ethiopia*

J. DE VLETTER

Introduction

The Ethiopian Government attaches a high priority to plantation forestry. Through institutions like the State Forests Conservation and Development Department, the Soil Conservation and Community Forestry Development Department and the revolutionary mass organizations, more than 40 000 ha of degraded lands annually are put under some kind of forest-like vegetation cover, partly as pure stands, partly in combination with other land uses such as agriculture or grazing.

This is an effort that requires the introduction of suitable trees, growing from seeds of known origin, if the forests (or agroforests) are to fulfil their objectives: erosion control and production of fodder, fuelwood, construction wood and timber. Tree species suitable for plantation forestry should have a maximum adaptability to a wide spectrum of prevailing (and sometimes rapidly changing) environmental conditions. In order to meet this requirement, the forester must be able to select from natural or planted tree populations, which must have a certain degree of genetic diversity.

Unfortunately, the Ethiopian natural tree populations have been, and still are, subject to indiscriminate destruction. Shifting cultivation and traditional grazing have been practised for centuries in Ethiopia. This, and the relentless cutting for fuel and building needs by a dense and rapidly growing population, have led to an almost complete deforestation of the Ethiopian highlands today.

The remaining forests are very unequally distributed. The northern

* The content of this paper does not necessarily represent the ideas of the State Forests Conservation and Development Department (SFCDD).

and central parts of the country are almost bare. Most of the forests are found in the south-western and southern parts of the country, mainly as closed broadleaved forests ('rainforests'). Elsewhere, in areas of lower altitude, woodlands, open bush and shrublands occur. These still cover more or less extensive areas, but overgrazing, charcoal production and man-made fires pose a serious threat.

The closed broadleaved forests form Ethiopia's best developed forests, representing the most eastern extension of the African equatorial rainforest belt. These forests are confined to the administrative regions of Kefa, Ilubabor, Sidamo and Bale. They are extremely important for the supply of raw material for the saw-milling industry. Moreover, the area decreases year after year, even before 'domestication' of the commercial tree species could start. Whereas Ethiopia's agriculture can already benefit from more than 10 years of collection, conservation and evaluation of agricultural crop germplasm resources, carried out by the Plant Genetic Resources Centre (PGRC/E) in Addis Ababa, similar work in the field of forestry is still in its initial stages.

It is true that species elimination trials have a fairly long tradition and that today, from a moderately wide range, suitable species can be chosen for the numerous bio-climatic zones of Ethiopia. However, systematic exploration and conservation of indigenous tree species is a neglected field of activity which should receive highest priority. All forest-like vegetation types should be included in the programme, not only the coniferous forests and the broadleaved forests (where most of the timber trees are found), but also the woodlands and riparian forests. Woodlands are composed mainly of drought-resistant *Acacia* species, which are widely used as a source of fodder and fuel. They are also used as multi-purpose trees in agroforestry systems and in plantation forestry in Ethiopia as well as in other African countries.

Agro-climatic belts in Ethiopia

The characteristics of Ethiopia's natural vegetation are to a large extent determined by two main factors:

- elevation (and temperature);
- rainfall.

On the basis of elevation, Ethiopia has been divided traditionally into five agro-climatic belts:

1. Berha the dry and hot belt below 500 m above sea level

2. K'olla the dry to moist, warm belt between 500 and 1500 m
3. Weyna Dega the dry to wet, moderately warm belt between 1500 and 2300 m
4. Dega the moist to wet, temperate belt between 2300 and 3200 m
5. Wurch the moist to wet, cold belt above 3200 m

'Dry' is defined as having less than 900 mm annual rainfall, 'moist' as having between 900 and 1400 mm and 'wet' as having more than 1400 mm annual rainfall.

Forests are found in the moist and wet Weyna Dega and to a lesser extent also in the moist K'olla and the moist and wet Dega. The other agro-climatic zones carry mainly woodlands, bushlands, savannahs, steppes or alpine formations.

Vegetation classification

The classification of the Ethiopian vegetation is still in a preliminary stage. In the introduction to his manual of indigenous trees of Ethiopia, von Breitenbach (1963) gives a scheme of plant associations, listing more than 50 different plant communities. Knapp (1968) has developed a complicated system of vegetation units on the basis of dominant species.

Steppe

Nearly treeless grasslands; here and there widely scattered shrubs occur. This vegetation type covers extensive areas in west Ethiopia, the Danakil and Ogaden plains and the coastal parts of Eritrea.

Savannah

Lands covered with perennial grasses, with scattered trees and shrubs. The trees shed their leaves during the extended dry season. Fire occurs frequently.

Lowland savannahs occupy vast areas in the Rift Valley, on the plains surrounding Lake Tana, on the Sudan plain, on the lower plateau escarpments and in west Kefa, south Gamo Gofa and Welega.

In the hot and dry lowland areas (Berha and K'olla), lowland steppes and lowland savannahs are found.

Steppes and savannahs are also found at the other end of the altitudinal range, in the moist and wet Wurch zones. Their upper limits are situated around 4000 m. Their lower limits are not clear on account of the extensive land clearings and grazing in the area of mountain woodlands and forests. On abandoned cultivations and pastures they spread as secondary vegetation. Thus this formation, originally confined to comparatively small surfaces at high altitudes, now occupies a major part of the Ethiopian plateau. It is easily recognized that this vegetation is secondary because small remnants of former forest or woodland are often found.

Woodlands

These are lands dominated by trees, which are heavily branched and which have a height of up to 20 m. The flat crowns do not form a closed canopy, but cover more than 20 per cent of the ground and are leafless for some part of the year. The ground is covered with grasses, herbs and shrubs. Fires are frequent.

Vegetation types with intermediate characteristics between savannahs and woodlands are shrublands and bushlands.

Shrublands

Lands supporting a stand of shrubs, usually not exceeding 6 m in height, with a canopy cover greater than 20 per cent. Trees are rare. The ground cover is often poor. Fires are usually infrequent.

Bushlands

Lands supporting an assemblage of trees and shrubs, often dominated by plants with a shrubby habit but with trees always conspicuous, with a single or layered canopy, usually not exceeding 10 m in height and total canopy cover greater than 20 per cent. Ground cover is poor and fires infrequent. Thickets are an extreme form of bushlands where the woody plants form an impenetrable closed stand.

Lowland woodlands, bushlands and shrublands include a wide variety of woody vegetation types, often difficult to separate clearly, mainly occurring in the upper dry K'olla and lower dry Weyna Dega zones. They are distributed over large areas in Eritrea, the Awash region, east and south Harerge, the Rift Valley, south Sidamo, west Ilubabor, Welega and the slopes of the eastern and central highlands. The main genera are various Acacias, *Boswellia, Commiphora, Balanites, Euphorbia, Combretum, Croton* and many others.

Mountain woodlands are found at the other end of the altitudinal range (upper moist and wet Dega, lower moist and wet Wurch). Their physiognomy is similar to the one of lowland woodlands: an upper (more or less) open canopy formed by 5–12 m high trees. Poor specimens of *Juniperus procera* occur. Other genera and species are *Acacia abyssinica, Protea, Cussonia, Hagenia abyssinica, Erica arborea, Hypericum* and *Arundinaria alpina* (bamboo forest).

Forests

These can be defined as a vegetation type which is dominated by trees, forming a closed, deep and complex, often multi-storeyed, canopy. The height of the largest trees may exceed 45 m. Most trees are columnar in shape, having a straight and clear bole. Many species are evergreen. The forest floor has a micro-climate which is clearly moderated by the tree cover and carries a wide variety of herbs, shrubs, seedlings and saplings (regeneration of the trees which form the upper storeys).

Forests are found in the moist and wet Weyna Dega, the moist and wet Dega and (upper) moist K'olla zones (Fig. 1).

Closed broadleaved moist forests are found in south and southwest Ethiopia (mainly Ilubabor and Kefa, partly also in Sidamo and Bale), at elevations between 1200 and 2200 m. Rainfall is more than 1400 mm (locally even up to 2000 mm) annually and the dry period is restricted to 2–3 months. These forests are sometimes referred to as upland rainforests. They form the best developed forests in Ethiopia, but are less impressive than, for instance, the equatorial rainforests of West and Central Africa. The upper storey is discontinuous and consists of scattered 40–60 m high 'emergents' dominating the dense and close canopy of the intermediate storey. *Aningeria adolfi-friedericii* is generally the only emergent species. It forms the highest, non-continuous stratum of the forest; therefore, the term Aningeria-forest is sometimes used. A continuous stratum about 30 m above the forest floor consists of 10–20 species of trees, all with a comparatively similar appearance. The following species can be found here: *Albizia schimperiana, Celtis africana, Cordia abyssinica, Croton macrostachys, Ekebergia* spp., *Ficus* spp., *Olea hochstetteri, O. welwitschii, Ocotea kenyensis, Polyscias* spp., *Sapium ellipticum, Syzygium* spp., *Schefflera abyssinica, Trichilia* spp. and some others.

An 8–10 m high lower storey of small trees consists of: *Allophyllus, Apodytes, Bersama abyssinica, Brucea, Teclea nobilis, Coffea arabica, Millettia ferruginea, Galiniera* and many others.

Climbers are common along forest edges (where more light can

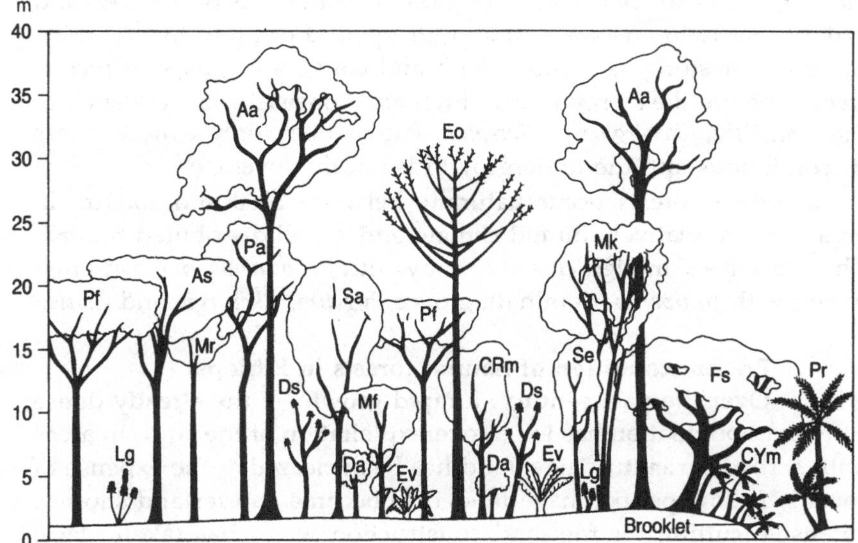

Fig. 1. Transect through a closed broadleaved forest showing storeys and different size classes. Aa, *Aningeria adolfi-friedericii*; As, *Albizia schimperana*; CRm, *Croton macrostachys*; CYm, *Cyathea manniana*; Da, *Dracaena afromontana*; Ds, *Dracaena steudneri*; Eo, *Euphorbia obovalifolia*; Ev, *Ensete ventricosum*; Fs, *Ficus sur*; Lg, *Lobelia giberroa*; Mf, *Millettia ferruginea*; Mk, *Macaranga kilimandscharica*; Mr, *Mitragyna rubrostipulata*; Pa, *Prunus africanus*; Pf, *Polyscias fulva*; Pr, *Phoenix reclinata*; Sa, *Schefflera abyssinica*; Se, *Sapium ellipticum* (from Friis *et al.*, 1982).

penetrate), but are not dominant in high forest. The herbaceous stratum on the forest floor is rich in species, but mostly discontinuous in mature forest. Several species of ferns, seed plants and broad-leaved grasses occur. Epiphytes are frequent, particularly ferns.

Other types of closed broadleaved forests include: broadleaved semi-humid forests (less developed than the broadleaved moist forests, forming the transition towards the coniferous forests of higher elevations); the dense *Acacia* forests (which occur in dry and windy areas between 1800 and 2400 m, and are characterized by the absence of an upper storey of big trees and a closed crown cover dominated by *Acacia xiphocarpa* trees); and riparian forests (which occur at several different altitudes along creeks and rivers).

A second main group of forest types consists of the coniferous forests, which fall into two groups, the *Juniperus* forests and the

Podocarpus forests. *Juniperus* forests occur naturally between 2500 and 3200 m, partly in areas with a long dry period of up to five months. Their upper storey is 30–45 m high and consists of *Juniperus procera* trees. The middle storey is 20 m high and consists of genera such as *Pygeum, Olea, Ekebergia* and *Bersama*. The middle storey is more or less discontinuous and the undergrowth is poorly developed.

Podocarpus forests occur naturally between 2000 and 2500 m, in areas with a relatively humid climate and a well distributed rainfall. There is a 40–45 m high closed canopy with *Podocarpus gracilior*, often mixed with *Juniperus*, dominating over *Pygeum, Ekebergia* and *Celtis*.

Present condition of natural forests in Ethiopia

Over the past century, a rapid growth of the already dense Ethiopian population has led to overexploitation of the land. In areas with settled agriculture new land has been cleared at the expense of forests. Fallow periods have tended to become shorter and shorter. Intensive cutting for fuel and construction wood has taken place. Pasture grounds suffer from overgrazing. All this has led to deforestation, erosion, land degradation and, connected with this, declining agricultural productivity.

It is often assumed, though not established as a fact, that 40 per cent of the Ethiopian highlands were still covered with closed high forests (mainly mixed and coniferous forests) not more than a century ago. By 1950 the forest cover had decreased to about 8 per cent. Today, Ethiopia's forest estate is reduced to 3.6 per cent of the total land area. Table 1 shows the various environments covering Ethiopia and their percentage of the total land area. The figures given are estimates based on Landsat image photo interpretation (Anonymous, 1986).

It can be seen that the total forest area now covers only 3.6 per cent of Ethiopia's land surface, an alarmingly low figure. Also degraded and overexploited forests are included here, and it can be safely assumed that only a minor part of the remaining forest lands consists of undisturbed, virgin forest.

The distribution of the remaining forests is very unequal. They are concentrated in the less densely populated southern and south-western parts of the country. The central and northern parts of Ethiopia are almost completely deforested (Fig. 2).

About 80 per cent of the forests are broadleaved forests ('rainforests') and mixed broadleaved/coniferous forests. The rest consists of more or less homogeneous coniferous forests (*Podocarpus* and

Table 1. *Present types of land cover in Ethiopia*

Environment	Area (km^2)	Percentage
Forest lands	45 120	3.6
Afro-alpine vegetation	1080	0.1
Woodlands	59 195	4.7
Bushlands	207 380	16.6
Xerophilous bushlands	234 905	18.8
Shrublands	264 155	21.2
Grass/range lands	166 990	9.4
Riparian vegetation	19 565	1.6
Wetlands	11 305	0.9
State farms	4730	0.4
Agricultural lands[a]	263 300	21.1
Bare lands	20 615	1.6
Total	1 298 340	100.0

[a] Agricultural lands also include lands partly in combination with other cover types, such as forests, wood-, shrub- or grassland.

Juniperus forests). All these forests are important for the raw material supply of the country's saw-milling industry. Although for the larger part the forests enjoy official status as State Forest or National Priority Area and have been or are being demarcated as such, effective management aiming at sustained yield still has to be implemented. The area and quality of these forests decrease year after year, mainly due to:

- timber exploitation practices, which destroy the (potentially sustainable) production capacity of the forest;
- agricultural encroachment, cutting for fuelwood and man-made fire;
- extensions of coffee and tea production areas;
- other changes in land use, including settlement.

The other woody vegetation type of interest in a discussion on forest genetic resources consists of the woodlands, which cover about 4.7 per cent of the total land area. Although their area probably has not shown such a spectacular decrease as in the case of forests, charcoal burning, fuelwood exploitation, overgrazing and man-made fire form a serious threat to their diversity.

Elsewhere in Ethiopia, in areas of lower altitude and less rainfall, bushlands and shrublands still occur in sizeable areas, together covering more than 50 per cent of the total land area. These vegetation types are not considered here.

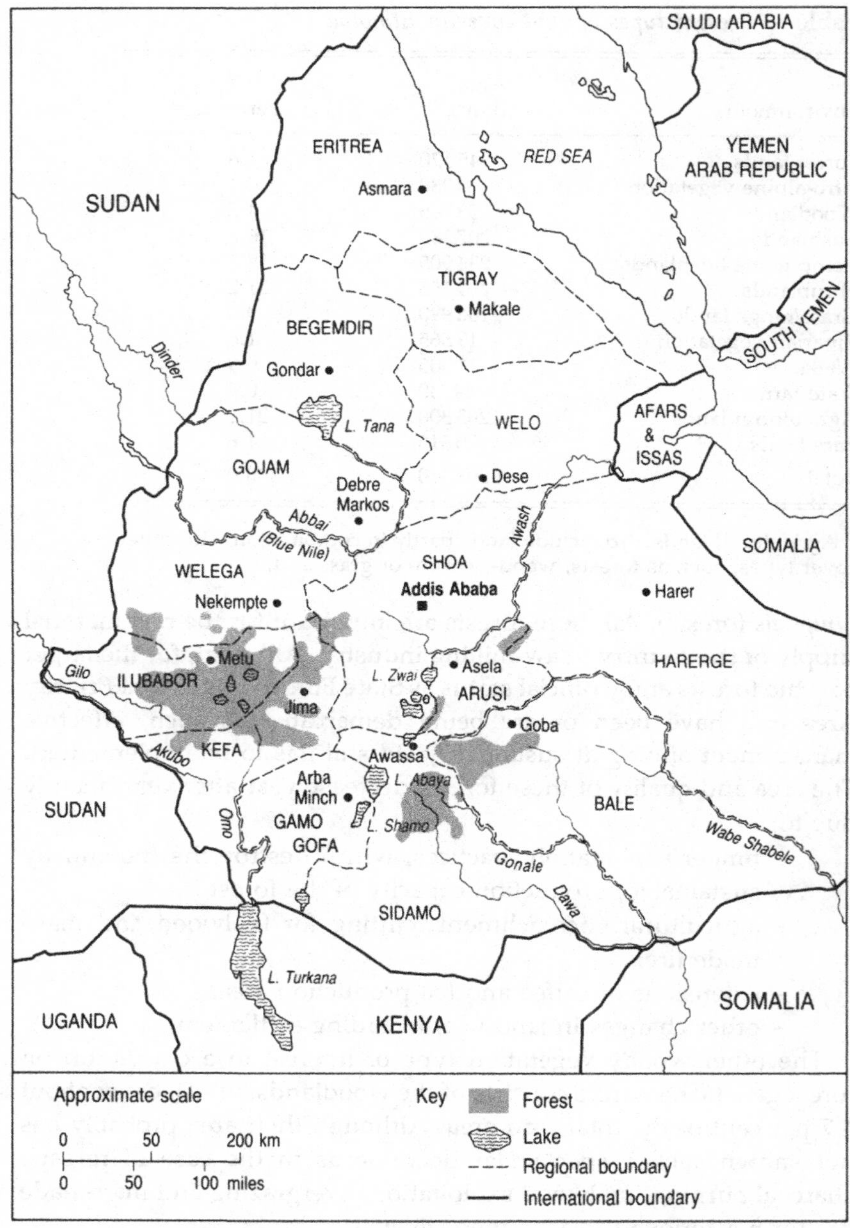

Fig. 2. Main forest areas in south-west Ethiopia (from Chaffey, 1980).

For comparison purposes, the present area of man-made forests can be subdivided as follows:

Urban plantations	80 000 ha
State forest plantations	120 000 ha
Community forests	200 000 ha
Total plantations	400 000 ha

The total plantation area is only 9 per cent of the total natural forest area and represents only 0.3 per cent of Ethiopia's land area. Plans exist to extend considerably the plantation area but it will be clear that even if the present rate of afforestation of 40 000 ha annually could be increased significantly, it will hardly be possible to balance the losses which result from destruction of natural forests.

Regarding the figures for the areas occupied by the various vegetation types (Table 1) and the changes occurring at present, the following should be noted:

1. It is very difficult to give exact information on the rates of deforestation which have taken place, since comparable data are hardly available. Sources are scarce and these use different vegetation classification systems or are based on very tentative area estimations. In order to close this gap a project with foreign assistance is planned, the Woody Biomass Inventory. The negotiations are still not concluded.
2. The estimates of plantation areas are based on numbers of seedlings produced and are therefore only tentative.
3. Most plantations are young, in the seedling or sapling stage. Many have only a low stocking rate. On the Landsat images they probably do not appear as forests, but as woodlands or even bushlands.

Despite these possible inaccuracies the conservation of the remaining natural forests as well as the introduction of sound management practices should receive highest priority.

Conservation and evaluation of forest genetic resources in Ethiopia

The Forestry Research Centre of Ethiopia has made a start with selection and improvement of a number of exotic and indigenous tree species suitable for various plantation objectives. Species elimination trials have been established on some scale since the mid-1950s and today, for a variety of environmental conditions, a choice can be made from a modest range of suitable species.

In view of the strong and continuing human pressure on the remaining natural forests, and directly connected with this the danger of irrevocable disappearance of whole populations of indigenous trees, the ongoing selection programme must be supported and supplemented by a programme of systematic exploration and conservation of natural forest genetic resources. This broadening of the scale of activities should receive highest priority.

Conservation and evaluation of forest genetic resources should be considered as part of one consistent programme, of a long-term nature, and follow the recognized steps of:

- exploration;
- collection of seed for evaluation;
- evaluation;
- conservation;
- utilization.

Exploration and collection of seed for evaluation

Forest inventories and systematic botanical explorations of remaining natural forests will lead to a better knowledge of the distribution of the species, will identify the useful provenances and will indicate the populations of trees which have become endangered in their natural habitat. Since these populations are often already depleted as a result of man-made destruction, the number of areas where successful seed collections can be made will be limited. One strategy could be simply to sample what is left in a systematic way. Under optimal growing conditions, a given species will generally show the largest genetic variation. Populations under marginal conditions often contain valuable genes for adaptability to extreme conditions but are generally less diverse. It is very important to include such populations as well in order to cover the complete range of genetic diversity of a species.

Evaluation

Evaluation methods of forest genetic material by means of comparative, replicated field trials on 'representative' sites are well developed in Ethiopia. General aspects of species and provenance evaluation, including experimental design, layout, assessment, statistical analysis and interpretation have been outlined by Burley & Wood (1976).

The first species trials in Ethiopia were established about a century ago. Some *Eucalyptus* were introduced in order to find suitable species to counter the severe shortage of fuelwood around Addis Ababa. In the mid-1950s and the early 1960s more trials were planted in a number of regions. From 1975 onward, and after the establishment of the Silvicultural Section of the Forestry Research Centre, more systematic trials followed. About 130 species, indigenous as well as exotic, have been tested and evaluated. Today, a range of suitable species for various environmental conditions is known. In the coming years, emphasis will not be so much on the introduction of new species as on new provenances. It is anticipated that high gains in productivity can be achieved by selection of the best adapted provenances for prevailing environmental conditions.

Conservation

Conservation of forest genetic resources can be achieved in the following ways:

- *in situ* conservation of representative parts of forest ecosystems;
- *ex situ* conservation of seeds in specially equipped genebanks or seed centres;
- *ex situ* conservation of species in plantations or field genebanks.

In view of the circumstances prevailing in Ethiopia, *in situ* conservation of provenances and populations or even samples of whole ecosystems, will probably be the main approach.

In principle, *in situ* conservation could well be integrated with the existing system of national parks, managed by the Wildlife Development and Conservation Organization. A disadvantage of this possibility is that the present nine National Parks incorporate only a few forest types, such as *Podocarpus gracilior* and *Juniperus procera* forest. The other parks include only several types of *Acacia* woodlands. Not a single park represents closed broadleaved forest (rainforest), the natural forest type which still covers the widest area in Ethiopia.

A better possibility would be to integrate *in situ* conservation of forest genetic resources with the existing system of officially designated State Forests and National Priority Areas. For these forests at least the intention exists to apply some kind of systematic management, and although this management still has to be implemented, this intention might provide the guarantee for protection in the long run. For the time being, additional management provision must be made in order to achieve real protection and prevent unwanted human disturbance or cattle grazing.

In situ conservation requires the reservation of larger areas than *ex situ* conservation, whereby research has to find an answer to the question of whether many relatively small areas are to be selected or few relatively large areas. For management reasons, the second option seems to be the more realistic.

Conservation of forest tree seeds, for instance within the facilities of the Plant Genetic Resources Centre in Addis Ababa, and in the long run in a proper equipped Seed Centre of the Forestry Research Centre, is another necessary approach to conserve forest genetic resources. For many forest species, long-term storage under low-temperature conditions can be used. However, much research is still needed in order to ascertain the best storage conditions for less orthodox forest tree seeds. Whether the *ex situ* conservation of tree pollen or tissue material in a genebank is a practical proposition at the moment, is questionable. Again, much research will be needed in order to reveal the appropriate techniques.

Ex situ conservation by species provenance plantations is justified only for a rather limited number of important commercial or socially valuable tree species suitable for growing in plantation monocultures or in agricultural schemes.

Utilization

It is extremely important that the genetic reserves, once established, are not only preserved, but also utilized. The management must make possible controlled seed collections in bulk and allow for scientific research or extensive activities. Sometimes the management should also allow active human interference in the forest ecosystem. From a silvicultural point of view light-demanding, fast growing species ('colonizers') are often more interesting than the extremely slow growing, shade-tolerant climax species. Some genetic reserves should be kept deliberately at a point somewhere in between the 'pioneer' and climax stages of succession.

Priorities

The implementation of a consistent, long-term programme of exploration, evaluation, conservation and utilization of forest genetic resources is a huge task, which requires careful planning, including the determination of priorities. Priorities must be allocated to:

- the different *operations* involved;
- different *species* or groups of species;
- different *regions*.

Establishing priorities means that for each species or group of species, on the basis of a social and commercial importance rating, a decision is to be taken concerning the urgency of each possible operation. This urgency is then expressed in the form of a phasing-in time of the operation. This process can be repeated for each of the different zones of Ethiopia; thereby it is conceivable to start the implementation of the programme first in those zones where the pressure on the remaining vegetation is most serious, e.g. the central highlands and north Ethiopia.

Concerning the priorities of species or the groups of species it is important to take the following into consideration. Von Breitenbach (1963) describes 68 families, 158 genera and 326 indigenous species. It is clear that priorities have to be set since not all these species or all ecosystems can be conserved at once.

Under the prevailing conditions, the highest priority should be attached to the group of multi-purpose tree species. Plantation efforts will be largely directed towards rural communities through the establishment of small village woodlots or agroforestry systems.

Ethiopia has a large number of indigenous multi-purpose tree species. Tentatively, the following species can be considered to be of most importance:

Acacia abyssínica	*Acacia tortilis*
Acacia albida	*Balanites aegyptiaca*
Acacia mellifera	*Cordia africana*
Acacia nilotica	*Croton macrostachys*
Acacia senegal	*Erythrina abyssinica*
Acacia seyel	*Erythrina brucei*
Acacia sieberiana	*Terminalia brownii*

A study is urgently needed in order to make this list more complete and to decide which species should receive highest priority.

Once the natural distribution ranges of these species have been determined, systematic seed collections must be made. At the same time more information has to be collected about their phenology, biology and reproduction systems. The next step is to identify the threatened populations and to implement conservation measures. *In situ* conservation of the species in their natural ecosystems should have highest priority, but complementary *ex situ* conservation activities in areas under less pressure should also be initiated.

In the case of species for industrial use, construction and fuel purposes as well as for conservation activities, the priorities should be set according to their threat by genetic erosion.

Looking at the merchantability for timber processing wood, the following species have commercial value:

Juniperus procera
Podocarpus gracilior
Aningeria adolfi-friedericii
Apodytes dimidiata
Albizia schimperiana/gummifera
Celtis africana (=*kraussiana*)
Chlorophora excelsa
Cordia abyssinica
Croton macrostachys
Dalbergia melanoxylon
Ekebergia capensis (=*rueppelliaria*)
Hagenia abyssinica
Linociera giordanii
Olea africana/hochstetteri/welwitschii
Polyscias fulva
Prunus africana (=*Pygeum africanum*)
Syzygium guineense

All these species can be conserved *in situ* by establishing reserves within the coniferous forests, the broadleaved forests and the mixed forests. Since the (climax) rainforest species probably have a poor potential for plantation forestry, special *ex situ* conservation stands or provenance trials should not receive a very high priority. For the important species *Podocarpus gracilior* and *Juniperus procera,* and possibly also *Chlorophora excelsa,* additional *ex situ* conservation plantations and provenance trials are necessary.

In situ conservation of rainforest species, and species of the mixed coniferous/broadleaved forests, would conserve at the same time a number of potentially commercial species, i.e. those species which do not at present have a high market value but might do so in the future. Some examples of potentially usable hardwoods are:

Allophyllus abyssinicus
Diospyros spp.
Dombeya spp.
Lepidotrichilia volkensii
Mimusops kummel
Ocotea kenyensis
Schefflera volkensii
Warburgia ugandensis

There is also a number of exotic species which are already more or less extensively used in plantations, and which belong to the groups of principal use mentioned earlier. A tentative list comprises the following species:

Acacia decurrens (construction wood)
Acacia mearnsii (construction wood)
Acacia saligna (=*cyanophylla*) (B)*
Azadirachta indica (multi-purpose tree)
Casuarina equisetifolia (construction wood)
Cupressus lusitanica A
Eucalyptus camaldulensis (construction wood)
Eucalyptus cladocalyx (construction wood)
Eucalyptus globulus (construction wood)
Eucalyptus grandis (construction wood)
Eucalyptus saligna (construction wood)
Grevillea robusta (A/B)
Melia azedarach (multi-purpose tree)
Leucaena leucocephala (multi-purpose tree)
Pinus patula (A)
Pinus radiata (A)
Parkinsonia aculeata (B)
Prosopis juliflora (B)
Prosopis tamarugo (B)
Sesbania spp. (multi-purpose tree)
Schinus molle (B)

The species with an asterisk are especially suited for industrial processing (A), for conservation purposes (B) or for both (A/B).

Since these species have been used for a reasonable period of time under different conditions, it can be expected that locally adapted ecotypes have developed. Therefore, systematic collections should be made and provenance trials need to be set up.

A considerable amount of research with some of these species has been carried out on a global scale; much can be gained, therefore, if participation is sought in ongoing selection and improvement programmes which are carried out abroad.

Present organizational structure

The State Forests Conservation and Development Department and the Soil Conservation and Community Forestry Development Department are the two governmental institutions responsible for the bulk of the country's reforestation programmes. In this work, both institutions receive support from a specialized Forestry Research

Centre. This Centre has a Seed Section, established in 1957, which is responsible for the collection, storage, testing and distribution of high quality seed of known origin. There is also a Silvicultural Section, which carries out general silvicultural research related to forest establishment and management, such as spacing/increment/thinning trials and species elimination/testing trials. This last category of research has been and still is the main component of the programme.

Provenance research has received much less attention. Recently, 'wide range' provenance trials with a few indigenous species were started (*Croton, Cordia*). For the future, it will be necessary that systematic provenance trials with all socially and commercially valuable indigenous species are established, by using certain priorities, for instance, as indicated in the previous paragraph.

The Forestry Research Centre has for practical reasons put emphasis on applied research, at the evaluation and utilization end of the range of the operations which are necessary in the selection and genetic improvement work. Exploration and conservation have been neglected areas and a considerable increase in efforts in these fields will be of highest importance.

A possible outline of a set-up in the coordination of exploration and conservation activities at the national level could be as follows:

- The Forestry Research Centre would be responsible for the botanical and phenological explorations and collections. Cooperation with the Inventory Section of the State Forests Conservation and Development Department and the University of Addis Ababa would be intensified.
- The Forestry Research Centre would indicate which genetic reserves have to be established and would develop the management guidelines for these reserves.
- The State Forests Conservation and Development Department would be made responsible for the management of these *in situ* reserves; a special unit within this department might have to be founded for this purpose.
- The Forestry Research Centre would establish and manage *ex situ* conservation plantations.
- The Plant Genetic Resources Centre/Ethiopia would take care of the *ex situ* conservation of forest tree seeds. In the future an upgraded Seed Centre might take over this task.

International cooperation

Long-term conservation and evaluation of forest genetic resources is very expensive and needs specialized skills. Financial and skilled manpower resources are both scarce in Ethiopia. Because of this, and since many of the valuable indigenous tree species mentioned also occur in other African countries, international cooperation is essential.

IUFRO with its specific working parties is one of the main platforms for international cooperation. Ethiopia participated in the IUFRO Research Planning Workshop for Africa, Sahelian and North Sudanian Zones held in Nairobi in January 1986. This workshop was centred around the topic 'selection and genetic improvement of indigenous and exotic multipurpose woody species; including seed collection, handling, storage and exchange'. Other international cooperation programmes have been established for a range of exotic species. Coordinating agencies are, for example, the Commonwealth Forestry Institute, Oxford, for Central American tropical pines (not yet very important for Ethiopia); the Commonwealth Scientific and Industrial Research Organization (CSIRO), Canberra, for Australian Acacias, Casuarinas and *Eucalyptus* (highly important for Ethiopia); the Food and Agriculture Organization (FAO), for drought-resistant Acacias and *Prosopis* species (important for Ethiopia); the Centre Technique Forestier Tropical, Nogent-sur-Marne, France, for African hardwoods and Pacific insular *Eucalyptus* (less important for Ethiopia); and NAS, Washington, for nitrogen fixing trees and species for the Sahel zone (relevant for Ethiopia).

The Forestry Research Centre already receives some outside assistance. In the near future there will be an extension of outside assistance by inputs through agencies like the World Bank, Finnida, FAO and SIDA in connection with projects in the field of forestry which are presently carried out by these agencies. It is strongly recommended that the project proposals which have to be prepared contain components in the field of exploration and conservation of forest genetic resources.

References

Anonymous (1986). *Biomass Energy Sources.* Final report of a cooperation in the energy sector between the Ministry of Mines and Energy of the Provisional Military Government of Socialist Ethiopia, ENEC and CESEN–Ansaldo/Finmeccanica group, Addis Ababa.

Burley, J. & Wood, P. J. (1976). A manual on species and provenance research with particular reference to the tropics. *Tropical Forests Paper No. 10.* CFI.

Chaffey, D. R. (1980). *South West Ethiopia Forest Inventory Project. A glossary of vernacular plant names and a field key to the trees.* Land Resources Development Centre, UK.

Friis, I., Rasmussen, N. & Vollesen, K. (1982). Studies of the flora and vegetation of southwest Ethiopia. *Opera Botanica*, **63**, 8–28.

Knapp, R. (1968). Hoehere Vegetationseinheiten von Aethiopien, Somalia, Natal, Transvaal, Kapland und einige Nachbargebieten. *Geobotanische Mitteilungen (Giessen)*, **56**, 1–36.

von Breitenbach, F. (1963). *The Indigenous Trees of Ethiopia, 2nd edn*. Addis Ababa: Ethiopian Forestry Association.

6

Plants as a primary source of drugs in the traditional health practices of Ethiopia

DAWIT ABEBE AND ESTIFANOS HAGOS

Introduction

The maintenance of health by means of various techniques and substances is almost as old as the history of human evolution itself. Although the resources available were easily drawn from the natural environment, their efficacy for solving problems which reduce life expectancy was established only through rigorous trials over a considerable period of time, often involving Man himself as subject of the experiment.

Our ancestors, and millions of people in modern Africa, have relied heavily on plants, animals and minerals to ward off pathogens and to maintain the functional balance of each organ. Many species of plants had to be tested and retested in the endless search for drugs that could prolong life or, it was believed, even confer immortality. Many plants have been found to possess the desired effects as a result of well planned experiments, while some discoveries were just a product of serendipity. Many sacrifices had to be made, however, not only in terms of money and time, but also in terms of human lives.

Until a generation or two ago plants were the primary source of health care for entire populations in most African countries and such plants still remain important sources of drugs for nearly 80 per cent of the population in contemporary Ethiopia. In spite of this we still know very little, whilst many Africans who receive their education abroad believe that imported drugs are always superior, even if these are unaffordable to the large majority of the population. The investigation of the efficacy and safety of traditional remedies is no doubt a big challenge; but to turn our backs on this challenge is not only abandoning our scientific responsibilities but also an indirect perpetration of those harmful practices which we always bitterly criticize and which we try to eliminate. A substantial body of knowledge on the practice of traditional medicine and particularly on medicinal plants has already been lost because most of it was transmitted by word of mouth from generation to generation. This situation has been further aggravated by the expansion of modern education, which is effectively putting an end to family and traditional ties and hence to the passing on of traditional knowledge. Thus, we now have the last generation of practitioners within the age brackets of 40 and 80 years. The ethno-medical information that has been built around numerous plants is, therefore, on the verge of collapse. The most unfortunate consequence of this is not only that we shall lose a crucial guide for potential sources of new drugs against many diseases that either do not succumb to or show resistance to modern drugs, but also the actual medical care that still serves a substantial sector of the population in the developing African countries.

In the industrial countries, with the expansion of conventional medical services, the herbal health care system has become insignificant. However, it would be unrealistic to assume that the same trend of events is taking place in the developing nations. With their ailing economies and ever dwindling external aid, it is extremely difficult for these nations to achieve reasonable modern health care to satisfy even the most acute demands of their people. The endless civil wars and international conflicts, and the recurrent drought, often seriously affect the distribution of national wealth that is intended for the improvement of health service facilities. Furthermore, the population in several of these countries doubles every 25 years, consequently increasing the competition for the already meagre resources of modern health services. These and other similar circumstances, therefore, strongly remind us that we have little or no choice except to develop scientifically our traditional herbal medical practice, to serve as a partner to the conventional health care delivery system.

The traditional health care system

Unlike its Western counterpart, which essentially concentrates on symptoms of illnesses, traditional health care often considers a healthy existence or its absence as a product of the social and physical environments and, therefore, the effort is directed not only towards curing the malfunctioning organ but also to bring the patient into harmony with these environments.

It is important to note that the choice or even the preference of traditional herbal remedies is not dictated by socio-cultural reasons alone. Modern drugs are not only frequently unavailable to most of the rural people, who often live where there is little or no means of communication, but they are also beyond their economic reach. Even the fortunate few who can get hold of the drugs from clinics or rural drug vending shops have very little or no idea how to use them and what dosage to use. The obvious outcome of this is serious side-effects and resistance to many of the drugs (especially to antibiotics) by the disease organisms or their vectors. As a result of these we see a growing mistrust and apprehension towards most of the manufactured drugs and an increasing tendency to resort to traditional herbal preparations as a sole or a complementary source of treatment.

The herbalists are often general practitioners who can handle a series of health problems, particularly those that commonly occur within their own areas, thereby effectively minimizing the wastage of time and money. Of course, the herbalists have their own limits; they are honest enough to admit this and refer the patient to a more experienced practitioner or to a modern medical care centre.

In the herbalogical medical care system prescription charges are often very small and are not necessarily paid instantly in cash but also in kind or even in a form of manual work over a long period of time. In addition to house calls and routine care for the outpatient, the herbalists sometimes manage with up to 10 inpatients who are provided not only with free prescriptions but also with food and shelter for several days or even years.

Traditional drugs of plant origin are easily available and the patient seldom needs to wait for days or even indefinitely before the commencement of treatment, as is now commonly the case with some modern pharmaceutical preparations. The herbalist informs the patient about his or her sickness, how the drug should be taken and the precautions that are needed – all in the local language, which is easily understood by the patient. Since the practitioners are permanent residents within the community, they can perform follow-up programmes on a daily basis to evaluate the effect of the drugs. If the medicine administered is slow in its action or non-effective, the practitioner increases the dosage or may prescribe a more potent drug with faster action. Therefore, the patient is not simply left at the mercy of modern prescriptions that have to be obtained from clinics or drug stores located far away, as a result of which further consultations are difficult or even impossible.

In the traditional medical care system, the patient can have a choice of one or both of the following sources. The first depends on previous information or experience about the medical virtues of a particular species of plant, which the patient is able to obtain from the local markets or from his backyard, to treat less severe cases. Thus anyone who suffers from influenza, fever, stomach discomfort or minor injuries often relies on home treatments by using *Artemisia afra, A. abyssinica, Lippia abyssinica, Plantago lanceolata,* etc. The second choice involves either 'general practitioners', who handle more complicated cases, or 'specialists', who deal with certain diseases such as rabies, cataract, jaundice, malaria, haemorrhoids, etc. The experts in this category, comprising the herbalists, bone setters, traditional birth attendants and faith healers, acquire their theoretical and practical knowledge during in-service training of up to 30 years with one of their parents or teachers who established their own reputations by undergoing similar apprenticeships. The theoretical aspect of traditional medicine comprises a belief about the etiology and syndromes of diseases, diagnoses and the drugs employed to cure them. Mythological concepts are sometimes also included and naturalistic approaches alone are regarded as incomplete, especially in disorders such as epilepsy, insanity, hysteria, etc. The practical part of the system consists of all the necessary steps that must be taken to cure or prevent diseases.

Source of remedies and methods of treatment

The traditional pharmaceutical preparations are of mineral, animal and vegetable origin. Widely used minerals for external applications include sulphur, copper sulphate, sodium chloride and thermal spring waters. Drugs of animal origin are mostly obtained from Coleoptera, chameleon, python, rock rabbit, hyena, greater kudu and elephant. Ninety-five per cent or more of the traditional medical recipes, however, are of plant origin (Table 1). Depending on the availability of materials and their therapeutic actions, fresh or dried plants may be used (Abebe, 1986). High-altitude specimens are thought to be milder and slower in their action than comparable plants from the lowlands. The season and the time of day of plant collecting are sometimes considered as important factors in determining potency. Thus fruiting plants collected during the dry season are believed to be more effective than those obtained during rainy periods. Similarly, dawn collections are preferred to those carried out later in the day (Abebe, 1987). In general, no specific kinds of collecting tools are required, even though some herbalists believe that

Table 1. *Plants used in Ethiopian traditional medicine*

Scientific name	Vernacular name	Uses[a]	Parts of plant used	Form of application	Active substances
Achyranthes aspera	telenji	ei, me, tn, wo	Leaf		Saponin, alkaloid, sterol, fat, KCl
Ajuga remota	anamero	ar, di, ja, ma, rw	Leaf	Leaf juice	
Anethum graveolens	insilal	ax, bh, du, go, ka	Leaf, seed	Leaf juice, seed powder	Limonene, D-carvone
Argemone mexicana	madafe	cn, co	Seed	Oil	Argemone oil, toxic alkaloids, sanguinarine, berberine and protopine
Asparagus africanus	sereti	ap, di, go, lc, vo	Leaf, root	Powder mixed with honey	
Bersama abyssinica	azamir	ar, ha, pw	Root, bark, leaf, seed	Powder	Leaf and root contain insecticide; mortal to cows in Uganda
Brassica nigra	senafichi	fl, id	Root	Fresh root eaten with salt	Fixed oil, glucoside sinigrin
Brucea antidysenterica	abalo	cq, ec, ga, ha, kd, ku, mi, rp	Seed, leaf	Oil of seed, powder of leaf	Bruceantine, fixed oil, volatile acids, bitter principle, resin, phytosterol
Calotropis procera	tobiawu	am, ha	Seed	Powder	
Calpurnia aurea	digita	ad, di, ec, go, ha, hy, is, ja, ve, wo	Seed, leaf	Powder	Calpurnin, tannin

Table 1 (*cont.*)

Scientific name	Vernacular name	Uses[a]	Parts of plant used	Form of application	Active substances
Carissa edulis	agam	by, ep, ia	Root	Powder	Root contains cardiac glycoside carissin
Capparis tomentosa	arangama guracha	he, ld, lk, px, so, ta	Root	Powder	Sulphur oil, stachydrine, alkaloid
Cassia occidentalis	arda bofa	dt, pb	Leaf	Powder	Oxymethyl anthraquinone, tannin, fatty substance, gum, glucose, mucilage, albumin, emodin, fixed oil
Clematis sinensis	azo-hareg, fiti	cc, ec, el, ha, he, le, ls, nk, sr, ta	Root, leaf	Powder	
Clerodendrum alatum	misirich, marasisa	am, as, go, kd, sm	Root, leaf, bark	Powder	
C. myricoides	misirich, marasisa	ap, ep, md	Root, leaf	Smoke inhaled	
Croton macrostachys	bisana	ar, cq, el, ep, fu, hc, he, ja, kd, ku, le, sf, wo	Leaf, seed	Powder	Crotin, crotepoxide resin
Cucumis aculeatus	yemdirimbway	di, go, le, mg, sn, sv, wo	Root	Powder	Cucumine-like principle
C. dipsaceus	anchote	go, nk	Root	Juice	Saponin
Cyphostemma niveum	aserkush tebetebkush	it, nk, sl, sn, sw, wo	Root, bark, leaf	Powder	—

Datura stramonium	astenagir	am, da, ec, ep, es, he, hy, ko, lh, mi, nb, pk, so, sq, ta	Root, leaf, seed	Powder	Daturine, hyoscyamine, atropine, scopolamine, solanine, saponin, carotene, Vitamin C, malic acid, tomatidine
Dracaena steudneri	etse patos	as, ra	Bark	Powder	Saponins, resin
Erythrina brucei	korch	ab, hf, it, ja	Leaf, seed	Leaf juice, powder of seed	
Heteromorpha trifoliata	jib merkuz	ja, ra	Root, leaf	Powder	Volatile oil
Hypericum quartinianum	amija	ja	Leaf	Juice	
Jatropha curcas	habatalumuluk	ar, db, go, he, hy, me, sa, sr	Seed	Powder	Fixed oil, phytosterolin, curcin, tannin, steroidal sapogenin, hydrocyanic acid
Kalanchoe lanceolata	endahula	ar, bs, di, sw, vo	Leaf	Pulverized fresh leaves	
K. marmorata	titu	ar, ad, ht, ko, sk	All parts	Powder	
Lepidium sativum	feto	et	Seed, leaf	Decoction	Cress oil, iodine, uric acid, fixed oil, sanapin, benzyl-isothiocyanate
Lagenaria siceraria	kil	ar, er, ha, md	Root, leaf, fruit, seed	Powder	Fruit contains niacin, riboflavin, thiamine, amygdaline; seeds contain saponin and fixed oil
Lawsonia inermis	henna	rp	Leaf	Decoction	Acid, resin, fats, a yellow pigment, lawson volatile oil, fixed oil, brown oil
Momordica foetida	buke seytana	ap, co, el, ga, kr, rp, sa, sf	Root, leaf	Sap and/or powder	Momordicin

Table 1 (*cont.*)

Scientific name	Vernacular name	Uses[a]	Parts of plant used	Form of application	Active substances
Myrsine africana	kechemo	ah, ar	Fruit	Powder	Embelin, cuercitol, anmyrsine-saponin
Ocimum suave	anchabi	ar, as, cc, ed, he, ra, sf, sx	Leaf	Sap or decoction	Eugenol
Osyris compressa	keret	ga, ja	Leaf	Powder	Tannin, sandalwood oil, sesquiterpene, glucoside osyritin
Phytolacca dodecandra	endod	ad, ar, da, ec, go, ja, ra	Root, leaf, fruit	Powder	Tannin, saponin, alkaloid, volatile oil
Plumbago zeylanicum	amera	ay, cg, ec, go, ha, ku, lb, nk, rp, sr, sw, sy, wo	Bark, root, leaf	Powder	A toxic principle, plumbagin, fixed oil, volatile oil
Polygonum barbatum	gumamila	ah, sr	Leaf	Sap or powder	Hydrocyanic acid
Portulaca oleracea	kentela	ao, du, fe, ga, he, kd, ve	Whole plant	Juice	Cyanic acid, alkaloid, saponins, oxalic acid, Vitamin C, potassium salts, fixed oil, volatile oil
Rumex bequaertii	tult	ae, an, ar, as, bd, co, di, fe, go, ha, hm, ld, me, ox, px, ra, sw, vo, wo	Root, leaf	Powder	
R. nervosus	embuacho	ae, ar, it	Root, stem, leaf	Infusion	

Securidaca longipedunculata	etse menhae	bu, ep, go, ne, le, ma, rp	Root	Powder	Methyl salicylate, saponin, tannin, steroid, glucoside
Spilanthes mauritiana	yemeder berbere	am, co, lt, sa, sh, sz, ta, tn, vo	Whole plant	Juice	Volatile oil which is toxic to fish, acid amide, spilanthol, sterol
Stephania abyssinica	kalala	ar, ch, di, fi	Root, leaf	Powder or fresh pulverized	Morphine
Syzygium guineense	dokma	as, bb, di, sr, wo	Bark	Powder	Leaf yields 27.5% cellulose and 8.4% albumin
Tamarindus indica	roka	ad, ap, ar, bp, cn, hy, sw, ve	Fruit, seed	Infusion	Tannin; tartaric, acetic, citric, malic and succinic acids, sugars, pectin, Vitamins A, B, B_2, C, active carotenoid
Thalictrum rhynchocarpum	sir bizu	ap, as, co, le, sg, sx	Root	Powder	Thalicarpine, which shows anti-tumour activity, has been isolated from *T. dasycarpum*
Verbascum sinaiticum	ketetina	ad, ae, ap, as, cl, di, du, ec, ep, hf, it, ja, kd, kw, ld, me, nk, sm, st, sx, sy, vo	Root, leaf	Juice or powder	Ganglion-blocking alkaloids were isolated from *V. nobile*
Verbena officinalis	atuch	ae, as, di, pa, sa, ve, vo	Root, leaf	Juice or powder	Verbenaline, mucilage, tannin, essence, bitter substance
Vernonia amygdalina	girawa, ebicha	pa, tn	Stem, leaf	Juice	Vernodalin, vernomygdin

Table 1 (*cont.*)

Scientific name	Vernacular name	Uses[a]	Parts of plant used	Form of application	Active substances
V. hymnolepis	murukruk	ae, am, hf, ka, sn, ve	Root	Powder	
Warburgia ugandensis	bifti	co, el, ep, ra, rp, sx	Bark	Powder	Tannin, mannitol, resin
Withania somenifera	gizawa	by, ep, he, md, me, pd, sa, sq, su, sx	Bark	Powder	Scopolamine, withaferin, somniferin, Vitamin C, tannin, fatty acids, pungent volatile oil
Zehneria scabra	hareg resa	di, fe, ma, mn, mt, sy	Root, leaf	Juice or decoction	

[a] ab, abortifacient; ad, amoebic dysentery; ae, anti-emetic; ah, anthelmintic; am, anti-asthmatic; ao, abdominal complaints; ap, aphrodisiac; ar, against *Ascaris*; as, anti-spasmodic; au, against anus prolapse; ax, against anorexia; ay, vitilago; bb, broken bone; bc, breast cancer; bd, boil dressing; bh, bilharzia; bp, bile problem; bu, burns; by, against evil eye; cc, common cold; cg, anti-chapp; cl, chill; cn, against constipation; co, cough; cq, *Tinea corporis*; ct, cataract; di, diarrhoea; dt, acute febrile illness; du, diuretic; ec, eczema; ed, eye disease; ei, epistaxis; el, elephantiasis; ep, epilepsy; es, ear pus; et, against emaciation; fe, fever; fi, fungal infection of face; fl, flatulence; fu, fire burn; ga, gastritis; go, gonorrhoea; ha, haemorrhoids; hc, habitual miscarriage; he, headache; hf, heart failure; hm, haemostatic; hp, heart pain; hy, hypotensive; ia, insanity; id, indigestion; is, insecticide; it, insect repellent; ja, jaundice; ka, heartburn; kd, kidney disease; ko, favus; kr, gastroenteritis; ku, against *Tinea versicolor*; kw, kwashiorkor; lb, lung tuberculosis; lc, lactogogum; ld, liver disease; le, leprosy; lh, alopecia; lk, hysteria; ls, leishmaniasis; lt, loose teeth; ma, malaria; md, madness; me, menorrhagia; mg, migraine; mh, pneumonia; mi, mental illness; mt, chloasma; nb, numbness; nk, tuberculosis and/or cancer; ox, oxytoxic; pa, poison antidote; pb, poisonous reptile bite; pd, prevention of epidemics; pk, pain-killer; pw, pin worm expellent; px, placenta expeller; ra, rabies; rp, rheumatic pain; rw, roundworm expeller; sa, stomach-ache; sc, scabies; sf, stomach distention; sg, spinning head; sh, sore throat; sk, swollen breast; sl, swollen scrotum and penis; sm, stomach trouble; sn, snake bite; so, aphasia; sq, stomach burn; sw, swelling; sx, spasm; sy, syphilis; sz, scorpion bite; ta, tooth-ache; tm, tumour; tn, tonsillitis; up, uterine prolapse; ve, vermifuge; vo, vomiting; wa, warts; wo, wound.

using a horn-handled knife or olive sticks is more likely to increase the strength of the drug. Prayers and even sacrifices may be considered as important conditions, not only to enhance the therapeutic effect but also to remove the poisonous effect of the plant.

The name of a medicinal plant in Ethiopia sometimes takes the name of the disease or of its operative agent. Two or more allied species with overlapping distributions are sometimes given the same name and may be used interchangeably. Among herbalists with a church education plant species often are given vernacular names composed of two words. The first is applicable to all species regardless of their affinities or differences. The second is a specific epithet which usually depicts the characteristic of the taxon when employed as a drug. Thus, 'etse sioul', which literally means 'plant of hell', indicating the burning effect or sensation that is produced by applying *Ranunculus multifidus*. Similarly, 'etse yihayu' means 'restorative plant', describing the effectiveness of *Habenaria* spp. in overcoming impotence.

Depending on its size and its therapeutic action, the whole plant, or different parts of it, is prepared in powder forms, infusions, decoctions, etc., to be employed in the treatment of a variety of diseases. The dosages of the preparations are often measured in a glass for liquids, a pinch or teaspoon for powders or a handful for seeds, roots, leaves, etc. The patient's age, sex, physical condition and stage of the illness are the major factors which determine the type of remedy and the dosages to be prescribed. The formulations prepared from different parts of the same plant may be employed for diseases with different symptoms or could even be used to produce opposite effects (Abebe, 1987). The fruit of *Ficus vasta*, for example, is claimed to have a laxative property, while its root is believed to stop diarrhoea. Different or similar parts of up to 12 species may sometimes be mixed in a given proportion to treat diseases with either clear-cut symptoms or those which manifest mixtures of syndromes. Synergistic and/or antagonistic effects of the various constituents are, therefore, well recognized by the herbalists.

Treatment with herbs does not always have to follow the normal procedures or administrative routes; it may also be applied by, for instance, just tapping the forehead a few times with a fresh stem of *Malva verticillata* to stop epistaxis (nose bleeding), or simply cutting the stem of *Rumex bequaertii* and simultaneously calling the name of the patient who suffers from excessive menstruation. In the traditional health care system preventive and prophylactic treatments are also very well known. Preventive remedies against epidemic diseases, snake and mosquito bites may be carried out by an individual person, or the plant may be grown around the house to protect the whole family. Although not common, prophylactic treatments are employed against rabies, malaria and even tapeworms. Rejuvenants and restorative drugs of plant origin are also known to counter the effect of ageing and to overcome signs of malnourishment, infertility, amenorrhoea, etc. (Abebe, 1986). Certain plants are also claimed to have the ability to boost the memory or intellectual power of teenagers. In fact, there seems to be very little that cannot be influenced by the application of plant materials, be it to attract the opposite sex, stop the rain, prevent attack by an enemy or beast, etc., although many of these claims have yet to be scientifically tested.

The fields for which plants are employed by the herbalists of Ethiopia are as varied as the species themselves. Among these the most important and with high potential for future application are the traditionally claimed drugs for human and veterinary health, insecticides, herbicides and water clarifiers. An insight into traditional medicine will also bring to light more of the abortifacients, teratogens, allergenics, hallucinogenics and other toxic plants which are of enormous significance to the health workers.

Concluding remarks

In the traditional medical system, the knowledge of the plants employed in the cure and prevention of disease is based on repeated observations and is passed on from one generation to the next. As a result all those plants or plant parts that have adverse and serious side-effects are well recognized by the herbalists and are eliminated from the list of therapeutic agents. Even if their use is justified, they are given to the patient under strict supervision and with the antidotes ready to counter their potential harmful effects.

Behind the façade of methods based on superstition, the traditional healing procedures more often embody rational principles and effective drugs against the major diseases afflicting the large sector of society. Exaggerating its weak points out of all proportion could never change the objective situation or the attitudes of over 80 per cent of the population, that considers the traditional medical practice to be its vital health care system.

Given their rapid rate of population growth and their weak economic position, many Third World countries seem to have little or no choice except to develop their traditional medical systems scientifically, in order to achieve maximum health coverage. With the right research approaches, effective, safe and cheap drugs of plant origin will no doubt be relatively easily established as substitutes for imported and often expensive modern medicines. The possibility of discovering superior and even completely new therapeutic agents against the diseases that are less or not at all amenable to existing pharmaceutical preparations is also very high. Therefore, if we are to fulfil our immediate objective of maximizing health care coverage and contributing to the ceaseless worldwide scientific effort directed to the discovery of new drugs, an open mind towards traditional medicine must certainly be maintained.

References

Abebe, D. (1986). Traditional medicine in Ethiopia: The attempts being made to promote it for effective and better utilization. Unpublished report, Coordinating Office for Traditional Medicine, Addis Ababa.

Abebe, D. (1987). Plants in the health care delivery system of Africa. *Proceedings of the 14th International Botanical Congress, 24 July–5 August 1987*, Berlin.

Bannerman, R. H., Burton, J. & Wen-Chen, C. (eds) (1983). *Traditional Medicine and Health Care Coverage.* World Health Organization, Geneva.

7

Traditional aromatic and perfume plants in central Ethiopia (a botanical and ethno-historical survey)

E. GOETTSCH

Introduction

The amazing variety of incense, perfumes and other aromatic materials gained our interest and attention when we collected spices in the marketplaces of Addis Ababa and its surroundings.

Aside from incense and myrrh very little is generally found in research literature about the use of plants as perfumes and aromatics in Ethiopia. In this paper those plants and plant products will be treated, which were found in markets in central Shewa, the administrative region around Addis Ababa and in the capital itself. Some plant products have also been reported from the Bale administrative region.

The subject will be treated in three sections: the first will deal with incense and myrrh and will consider their importance in international trade since ancient times. The other two sections will cover aromatic plant materials of different uses and perfume plants, respectively.

Incense and myrrh

From time immemorial the fragrant smoke of burning resins and the aromatic odours of ointments and balms have been used by Man in religious rituals.

In the ancient Mediterranean civilizations incense (or frankincense, as it is also called) and myrrh were considered at times to be more precious than gold (Gauckler, 1970). The importance of incense in those days is documented by the fact that the first great trade route in history is called the 'incense road', covering a distance of about

5000 km from the kingdoms of southern Arabia ('Arabia Felix') to the cultural centres to the east of the Mediterranean Sea. This trade was already flourishing by about 2000 BC. Both products are mentioned in the Old Testament and were introduced into church ceremonies at the beginning of Christianity (Abercrombie, 1985).

In Ethiopia the use of incense and myrrh for ritual purposes goes back at least to the Aksumite Empire, *ca.* 500 BC (Goldschmidt, 1970) and has ever since been continued by the Orthodox Church where it is still very popular.

In recent years, however, the use of incense in church ceremonies has considerably decreased worldwide; but a growing amount is needed in the industrial sector, e.g. in pharmaceutics and cosmetics. The world markets for incense and myrrh are dominated today by South Yemen, Ethiopia and Somalia. There is a specially strong demand for southern Arabian frankincense because of its superior quality. The best material since ancient times is produced in the Dhofar province of Oman, where soil and climatic conditions are ideal (Zohary, 1983; Abercrombie, 1985).

In Ethiopia, trade in these goods is handled by the Ethiopian Forest and Wildlife Products Processing and Marketing Corporation. The annual production is estimated to be well over 30 000 tonnes (Ahmed Taib, 1982), most of which is consumed locally. In 1983–4 3300 tonnes were exported mainly to Western Europe, the Middle East and China.

In commerce myrrh is sold under the trade names gum oppoponax and gum myrrha, and incense as gum olibanum. From the viewpoint of strict scientific definition these so-called gums are in fact resins. The term resin is not easy to define in a precise manner, for natural resins differ greatly among themselves. They have certain properties in common, however, which make them easily recognizable: resins are insoluble in water but dissolve readily in alcohol, ether, carbon bisulphide and certain other solvents. When heated they first soften and then melt to a more or less clear, sticky fluid. They burn with a smoky flame, are resistant to most natural reagents and they do not decay (Howes, 1949).

Myrrh

Myrrh is a natural exudate of trees of the genus *Commiphora*. In Ethiopia about 48 species of *Commiphora* can be found. The species used for the production of myrrh are *C. myrrha* (Nees) Engl., *C. africana* (A. Rich.) Engl., *C. erythraea* (Ehrenb.) Engl., *C. gileadensis* (L) C. Chr., *C. abyssinica* (Berg) Engl., *C. hodai* Sprague, *C. kua* (R. Br. ex

Royle) Vollesen, *C. quadricincta* Schweinf., *C. schimperi* (Berg.) Engl. and *C. truncata* Engl. (Vollesen, 1989). Except for *C. gileadensis*, which occurs only in Eritrea and the Harerge region below an altitude of 750 m above sea level, the rest of the *Commiphora* species mentioned are relatively widespread in Ethiopia, occurring in the Eritrea, Tigray, Gojam, Gondar, Welo, Shewa, Arsi, Sidamo, Harerge, Bale and Gamo Gofa regions up to an altitude of 2000 m above sea level, although they are also quite common in the lowlands below 100 m. The resins yielded by these species differ from one another in taste and odour. True myrrh is produced by *C. gileadensis* and *C. abyssinica*. Other types of myrrh of different composition are known by their traditional trade names 'bissabol' (*C. erythraea*), 'harabol' or 'perfumed bdellium' (*C. myrrha*) and 'bdellium' (*C. africana*) (Uphof, 1968).

Myrrh ('kerbe' in Amharic), has a powerful scent when it is burnt. The traditional use of 'kerbe' as an incense is not very popular because of this strong scent, and also because of the fact that myrrh is associated with witchcraft. On the other hand, however, prostitutes in Addis Ababa are said to attract visitors by burning 'kerbe' in front of their houses.

Some myrrhs are used by traditional healers as universal remedies. The oil of myrrh has a rich odour and is used as a balm for ritual ceremonies, as a disinfecting ointment and as a perfume (Gauckler, 1970; Zohary, 1983).

Incense

Gum olibanum or true frankincense is an oleo-gum-resin obtained from trees of the genus *Boswellia* by tapping them. So far six *Boswellia* species have been reported to occur in Ethiopia. The two most common species are *B. papyrifera* (Del.) Hochst. and *B. rivae* Engl. (Cufodontis, 1953–72; von Breitenbach, 1963; Atkins, 1964; Werner, 1974; Ahmed Taib, 1982; Vollesen, 1989). *B. papyrifera* is the most common species, known in Amharic as 'itan zaf' (incense tree). It is found in the lowland areas of Gojam, Shewa, Gondar, Tigray and Eritrea (up to 1800 m), whereas *B. rivae* is found between 250 and 800 m in Sidamo and Harerge regions (Maslekar, 1975; Vollesen, 1989). In Konso (Gamo Gofa region) *B. rivae* has been found by the author up to an altitude of 1050 m.

There is a great demand for frankincense in the local markets since large quantities are used in church ceremonies and it is also burnt in private houses during the coffee ceremony. In certain areas magicians use it in their rituals.

The traditional grading of incense is done by referring to (a) the colour, (b) the origin or (c) the use of the resin. The following grading system was and still is used by various merchants in the marketplaces of Addis Ababa, although it was not possible to determine the species from which the particular incense derives:

1.	nech itan	white incense (best quality)
	tikur itan	black incense (said to be produced by old trees)
	kai itan	red incense (inferior in quality, contains pieces of bark)
2.	Tigray itan	used as incense
	Ogaden itan	used as incense and for perfumery
	Asmara itan	black incense
	Bahar itan	imported from outside Ethiopia (e.g. from Aden)
3.	set itan	'Ladies' incense'; dresses are dried in its smoke
	mitan itan	incense mixture, containing different aromatic materials and serving as a cheap substitute for frankincense

Incense also provides the raw material for some manufactured aromatics, and Ethiopian Muslims, for example, who are influenced by Arabian culture, have a preference for these. There are two popular substances of this kind:

1. 'Libanja', the most important incense in this group, is imported from South Arabia, Djibouti, Somalia and perhaps Kenya. This substance has a mineral-like appearance. It contains mainly a refined incense, but it was not possible to obtain any further information on the other ingredients.
2. 'Misketi', a strong aromatic mixture, is produced in Dire Dawa (Harerge region) and preferred especially by Muslims. It is made of 'Libanja', sugar, powder of sandalwood (from India) and 'Miski', a cheap perfume. When burnt 'Misketi' produces a strong sweet scent; it is mainly used during coffee or chat (*Catha edulis* Forssk.) ceremonies.

Other aromatic plant materials

(a) *'Birgud'*

Another important group of aromatic plant materials to be found in Ethiopian marketplaces is called 'birgud' (Amharic).

It was not possible to obtain any clear information on the nature of 'birgud'; the name applies to both a dark resin and a woody bark. Both materials give a pleasant smell when burnt.

According to Wolde-Michael (1980) 'birgud' is the Amharic name for *Cinnamomum cassia* Blume, which grows in south-east China. Its bark was already used as an incense in ancient rituals, which are, for example, repeatedly described in the Bible (Zohary, 1983). In spite of the fact that 'birgud' bark is quite popular in Ethiopia, there is no evidence that this material is produced in Ethiopia. It is probably imported, but the question of its origin remains open. Types of 'birgud' found in the marketplaces are:

1. Small pieces of a crumbling woody bark: not yet identified.
2. A dark resin: the cheaper type of this resin is called 'Chigga birgud', whereas the other type is one of the most expensive natural aromatics available.
3. 'Arussi birgud', pieces of a woody bark from Arussi. Regarding appearance and scent it might be the bark of *Juniperus procera* Hochst. ex Endl.

(b) *Other resins*

Other kinds of resin used as an incense include:

1. 'Wunsi' (Amh.): a black resin, not yet identified but because of its scent it might be produced by *Juniperus procera* Hochst. ex Endl.
2. 'Hunsi' (Amh.) or 'Ancha' (Orom.): bought at Goro-market (Bale); not yet identified. The resin resembles very much that of *Boswellia*. According to a local healer the smoke has medicinal properties.

(c) *Plants producing scent when burnt*

In Ethiopia a number of different plants are known which give a pleasant odour when put into the fire. Some of these plants are listed in Table 1.

Perfume plants

Perfume plants contain different types of essential oils. Either the whole plant is used as a perfume or the essential oil is extracted from it. Many species of perfume plants have been introduced to Ethiopia where they are now grown successfully (e.g. lavender, *Geranium* spp., mimosa, etc.). In this section only the traditional perfume plants will be listed.

In many parts of Ethiopia it is a traditional fashion to butter the hair. This habit is mainly restricted to women, but men may also be accustomed to do so (e.g. Karayu and Afar). In order to overcome the often rancid odour of the butter, perfume plants are mixed with it.

Table 1. *List of aromatic plants which are burnt to produce a pleasant smell*

Amharic name	Scientific name	Use and description
1. Karbaricho	*Echinops* spp.	The root is burnt; the smoke is said to drive out evil spirits and vermin. Today mainly used because of its pleasant smell.
2. Afer kocher syn. Nech krinfud	*Hedychium spicatum*	Tree originating from eastern India growing in Ethiopia. Sliced roots burnt during coffee ceremony; clothes are dried in the smoke; expensive.
3. Bukbuka		Mixture of sandalwood powder and 'afer kocher'. The sandalwood could be imported but could also be produced by the East African sandalwood (*Santalum album* L.), a tree that grows in some areas of eastern Ethiopia. The use is the same as that of pure 'afer kocher', but this mixture is much cheaper.
4. Ye-Jima inchet	Unidentified; 'wood from Jima'	Wood, containing an essential oil, pleasant smell when burnt.
5. Ye-Aden chiraro	Unidentified; 'dry twigs from Aden'	Gives pleasant smell when burnt.
6. Weyra	*Olea europaea* subsp. *africana*	Scented when burnt. The pleasant-smelling smoke is led into containers for milk, home-made beer and yoghurt.
7. Semat	Unidentified	Large tree; bark contains a milky sap. Found in Abbai and Takazze Gorge; thin strips of bark are plaited into a strand which is burnt like an incense stick.
8. Chiz inchet	'smoke-producing wood'	Collective term for mixtures of aromatic woods and perfume plants. The smoke gives a pleasant smell and – depending on the ingredients – evil spirits can also be driven out. Normally these mixtures are very cheap. Ingredients can be: Karbaricho, Kuni, Gizawa, or Weyra, Ades, Birgud and Itan.
9. Tinjut	*Otostegia integrifolia* or *O. steudneri*	Small herbaceous plant. Dried leaves are burnt in containers for local beer and milk; gnats are expelled by the smoke.
10. Mitin chito	'Perfume mixture'	Simply a mixture of wood powder, a little oil and cheap perfumes. Gives strong odour when burnt. Preferred by Muslims and Gurage.

Table 2. *List of traditional Ethiopian perfume plants*

Amharic name	Scientific name	Use and description
1. Koseret or Azkuti	*Ocimum* spp.	At least five species of *Ocimum* are found in Ethiopia, *O. basilicum* L. being the best known ('basobila'). Some *Ocimum* spp. contain an important essential oil which allows them to be used as perfumes.
Kasse	*Ocimum ladiense*	Fresh plants are spread on the floor of the house. Also used to scent butter.
Kasse	*Ocimum sacrum*	Probably indigenous to Ethiopia; used as an incense, also expels gnats.
2. Ariti (Tikur, nech)	*Artemisia rehan*	The plant is mostly sold fresh; the crushed leaves of *Artemisia afra* Jacq. ex Wild. are used as a perfume. Dried leaves are put between cloth. Ariti is burnt for its aromatic smoke; fresh plants are spread on the floor of houses.
3. Tej sar	*Cymbopogon citratus* (Lemon grass)	The plant contains an essential oil which is of some economic importance (e.g. used in cheap perfumes, insecticides, etc.). Traditionally the plant is burnt for its pleasant scent. The fresh plant is spread on the floor. 'Tej sar' is also used as a medicine and as a flavouring agent.
4. Kuni	*Cyperus bulbosus*	The roots of this plant are ground and mixed with butter to improve its smell. 'Kuni' is a cheap perfume in the highlands.
5. Ades	*Myrtus communis*	The plant contains myrtle oil. Leaves are ground and mixed with butter which is put into the hair by traditional women.
6. Tungug	Unidentified	'Tungug' is probably made of a grass; it has the strong and pleasant smell of fresh hay. The plant is mixed with butter, which is rubbed into the hair.

Most of the plants mentioned here are exclusively, or at least to a certain extent, used to scent butter (e.g. Koseret, Kasse, Ariti, Tungug, Ades, Kuni). The perfume plants found in the marketplaces in Addis Ababa are listed in Table 2.

Acknowledgements

I wish to express my thanks to Ato Wolde Michael Kelecha, formerly associated with the Forestry and Wildlife Development Authority. Without his extensive materials the sections on incense and myrrh could not have been written.

References

Abercrombie, Th. J. (1985). Arabia's frankincense trail. *National Geographic Magazine*, **168**, 474–512.

Ahmed Taib (1982). *The Swiss, Italian and Finnish Markets for Ethiopian Gum Olibanum*. Programme for Development Cooperation, Market Research Report No. 5, The Helsinki School of Economics.

Atkins, W. S. (1964). *The Future of the Natural Resin Industry in Ethiopia*. A report for the Technical Agency of the Imperial Ethiopian Government, Addis Ababa.

Cufodontis, G. (1953–72). *Enumeratio Plantarum Aethiopiae, Spermatophyta*. Bulletin du Jardin Botanique National de Belgique, Bruxelles.

Gauckler, K. (1970). Die kostbarsten Drogen der Alten Welt: Weihrauch, Myrrhe, Balsam. *In*: M. Lindner (ed.), *Petra und das Koenigreich der Nabataeer*. Abhandlungen der Naturhistorischen Gesellschaft, Nuernberg.

Goettsch, E. (1985). Aromatic and perfume plants in Central Ethiopia. *PGRC/E–ILCA Germplasm Newsletter*, **8**, 11–16.

Goldschmidt, C. (1970). Die Weihrauchstrasse: Zur Geschichte des aeltesten Welthandelsweges. *In*: M. Lindner (ed.), *Petra und das Koenigreich der Nabataeer*. Abhandlungen der Naturhistorischen Gesellschaft, Nuernberg.

Howes, F. N. (1949). *Vegetable Gums and Resins*. Chronica Botanica Company, Waltham, Massachusetts.

Jansen, P. C. (1981). *Spices, Condiments and Medicinal Plants in Ethiopia, their Taxonomy and Agricultural Significance*. PUDOC, Wageningen.

Maslekar, A. R. (1975). A report on rapid aerial survey for *Boswellia papyrifera* (incense tree), Tigrai province. Addis Ababa (mimeographed).

Uphof, J. C. T. (1968). *Dictionary of Economic Plants, 2nd edn*. Verlag von J. Cramer, Lehre.

von Maydell, H. J. (1981). Baum- und Straucharten der Sahelzone unter besonderer Berücksichtigung ihrer Nutzungsmöglichkeiten. GTZ, Eschborn (mimeographed).

Vollesen, K. (1989). 123. Burseraceae. *In*: I. Hedberg and S. Edwards (eds), *Flora of Ethiopia, vol. 3*. The National Herbarium, Addis Ababa University, Ethiopia, and the Department of Systematic Botany, Uppsala University, Sweden, pp. 442–78.

von Breitenbach, F. (1963). *The Indigenous Trees of Ethiopia, 2nd edn*. Ethiopian Forestry Association, Addis Ababa.

Werner, F. (1974). Memorandum on the collection of incense within Ethiopia, Addis Ababa (mimeographed).

Westphal, E. (1975). *Agricultural Systems in Ethiopia*. PUDOC, Wageningen.
Wolde-Michael Kelecha (1980). *A Glossary of Ethiopian Plant Names, 3rd edn.* Addis Ababa (mimeographed).
Zohary, M. (1983). *Pflanzen der Bibel*. Calwer Verlag, Stuttgart.

8

Spice germplasm in Ethiopia

E. GOETTSCH

Introduction

Although spices are considered as minor crops their significance for Ethiopia can hardly be overestimated. Spices are needed every day in considerable amounts for the preparation of the main dish of the day.

Most of the spices needed in Ethiopia are grown as field or garden crops, although some grow in the wild. Classical spices are also used but have to be imported, mainly from India. The following 12 spices, which originated in Ethiopia or were introduced very long ago and are considered to be of importance, are dealt with in this chapter:

1. *Capsicum annuum* (red pepper); Amh.: berbere
2. *Trigonella foenum-graecum* (fenugreek); Amh.: abish
3. *Nigella sativa* (black cumin); Amh.: tikur azmud
4. *Trachyspermum ammi* (Ethiopian caraway); Amh.: nech azmud
5. *Coriandrum sativum* (coriander); Amh.: dimbilal
6. *Aframomum korarima* (false cardamom); Amh.: korarima
7. *Cuminum cyminum* (cumin); Amh.: kamun
8. *Foeniculum vulgare* (fennel)
   Pimpinella anisum (anise); Amh. for both: insilal
9. *Ruta chalepensis* (rue); Amh.: tena-addam
10. *Ocimum basilicum* (basil); Amh.: basobila
11. *Piper longum* (Indian long pepper); Amh.: timiz
12. *Rhamnus prinoides* (buckthorn); Amh.: gesho

Although 'gesho' is not a typical spice, it is included in this list, since it is of extreme importance in the flavouring of beverages during their preparation (Jansen, 1981).

In a broader sense, shallots (*Allium cepa*) and garlic (*A. sativum*) can be considered as spices. They were introduced very long ago and recently genetic erosion has started in areas where improved varieties are coming into use. Nevertheless, since the two species can be regarded as both spice and vegetable, the latter use being the more important, they are not treated here.

Also not included in this list are *Lepidium sativum* (garden cress, feto) and *Tamarindus indica* (tamarind); both are considered by Westphal (1975) as spices but they are primarily medicinal plants.

Zingiber officinale (ginger) was probably introduced into Ethiopia in the 13th century but its use was and still is very limited (Jansen, 1981). Apparently there is only very little diversity (ginger is propagated vegetatively), so that collecting by the Plant Genetic Resources Centre/Ethiopia (PGRC/E) is not worth while and the plant is not included here.

Also the spices *Myrtus communis* (Amh.: ades), *Lippia javanica* (Amh.: kasse), *Mentha* spp. (Amh.: nana), *Rosmarinus officinalis* (Amh.: siga metbesha) and *Thymus schimperi* (Amh.: tosign) are not listed because they grow abundantly in the wild, they are not endangered and their use is very limited.

Another spice, *Brassica nigra* (black mustard), Amh.: senafich, is not treated here, because it is mainly considered to be an oil crop and collection through PGRC/E has already taken place.

In the past almost all the important classical spices had to be imported into the country. Trials are now under way by the Institute of Agricultural Research (IAR) to introduce at least some of them into Ethiopia as crops.

Turmeric (*Curcuma longa*) and cardamom (*Elettaria cardamomum*) have been cultivated successfully quite recently. Trials with black pepper (*Piper nigrum*) carried out at the IAR Station in Jima are promising. Cinnamon (*Cinnamomum zeylanicum*) and nutmeg (*Myristica fragrans*) could be introduced in the future. Only clove (*Syzygium aromaticum*) does not find a suitable habitat in this country.

General remarks concerning the collection of spice germplasm in Ethiopia

As mentioned earlier spices play a very significant rôle in the daily food preparation of Ethiopia. So far, small-scale production or harvesting of wild plants has been sufficient to satisfy the demand of the people. It is only very recently that the social, economic and technical situation of Ethiopian agriculture has changed drastically. With the introduction of improved farming methods, the destruction of natural habitats and the introduction of advanced cultivars into the country, genetic erosion is very likely to occur even in this group of minor crops.

Species of spices which still also exist as wild types often show remarkable degrees of disease resistance. For example, a severe attack of an unidentified fungal disease was observed in cultivated fennel but did not attack wild plants (Jansen, 1981). Thus, collection of wild germplasm and its careful screening afterwards should go hand in hand.

The ecology, use and need for conservation of the main Ethiopian spices

Regarding production and cultivated area, only *Capsicum*, *Rhamnus* and *Trigonella* are of significance. There has been little export of spices so far. Thus in 1981 only about 900 tonnes were exported, predominantly red peppers (National Bank, 1982). All other spices are mainly grown as garden crops, although in certain areas there may be field production.

Five of the 12 spices dealt with also grow in the wild. *Aframomum* and *Rhamnus* are indigenous spices. It should be mentioned that *Rhamnus* is widespread in Africa but so far only Ethiopians are known to use it as a spice. *Aframomum* may also occur in south Sudan, but so far no use has been reported from there (Jansen, 1981). Use and cultivation of *Nigella* and *Trachyspermum* are also typical of Ethiopia, but they are used elsewhere too.

Out of the 12 spices only *Capsicum* is of New World origin. Besides *Aframomum* and *Rhamnus*, *Trachyspermum* may also be of Ethiopian origin (Wolff, 1927). *Ocimum* and *Piper longum* originated in southern tropical Asia. The remaining seven species have at least one suggested centre of origin in the Mediterranean region (including Egypt and the Near East). Considering this fact, it is becoming clear that they have probably been used in Ethiopia since ancient times (Uphof, 1968; Zeven & de Wet, 1982). For *Coriandrum*, *Nigella* and *Trachyspermum* an especially wide variation can be observed in the country (Jansen, 1981).

In the following paragraphs the ecology and use of selected species will be described. Some remarks about the extent to which their genetic diversity is threatened at the moment are also included. If not mentioned separately, reference has been made to Siegenthaler (1963); Jansen (1981) and Goettsch (1984). The total number of accessions held by PGRC/E and given in the following pages refers to 30 June 1986.

Capsicum annuum. Capsicum is the most important spice in the country. According to Alkaemper (1972) *ca.* 2.5 per cent of the total arable land (*ca.* 230 000 ha) is cropped with *Capsicum*. Fruits of red pepper can be found in almost every market in the country.

Capsicum grows chiefly between 1500 and 2000 m above sea level but is also found from 1000 to 3000 m. The main centres of production are Ghion, Bako (Shewa) and the state farms in the Middle Awash Valley. Red pepper is the main constituent of most kinds of 'wot', a sauce essential in the daily meal. In addition, it is used to flavour meat, and medicinal uses are also known. Since 1964 the Ethiopian Spice Extraction Company has been buying an increasing amount of red pepper to extract the pigment, which is used as a natural colouring agent.

Even in a small indigenous random population of *Capsicum* bought in Addis Ababa (Mercato-market) genetic diversity was very high (Engels & Goettsch, 1984; see also Jansen, 1981). So far, 126 accessions are held by PGRC/E. Considering the importance of this spice for local consumption and its increasing significance for export, this might be regarded as insufficient. Further collecting and screening of red pepper should therefore be regarded as very important.

Trigonella foenum-graecum. Regarding production, *Trigonella* is the second most important spice in Ethiopia. It is grown in all provinces at altitudes between 1800 and 2200 m and is found for sale in almost every market.

Fenugreek is an important spice for the preparation of 'wot'. It is prepared as an appetizer, serves as a milk substitute for babies and is used as a treatment against rheumatism (Westphal, 1974). PGRC/E at present holds 427 accessions (213 being their own collections and 214 donated). In spite of the large number of collected accessions, important crop areas for fenugreek are under-represented (e.g. Sidamo, Bale, Eritrea, Gamo Gofa, Harerge, Welega). Further collection should concentrate on filling these gaps, but since breeding activities have been stopped for the moment and the plant is not endangered at all, fenugreek does not require a high priority.

Nigella sativa. Small-scale production of black cumin is widespread all over the country between 1500 and 2500 m. *Nigella* is cultivated as a crop in the provinces of Gondar (Dembia, Gondar), Shewa (Alem-Gena), Bale (Dinsho), Harerge (Chercher highlands) and Kefa (Jima region). The seeds are used in Ethiopia in the preparation of bread, berbere-sauce ('wot') and local beverages. *Nigella* seed powder is added to berbere-sauces to reduce the pungency of the pepper. In addition there are medicinal uses.

Twenty-eight accessions have been collected so far, which is not sufficient. According to Jansen (1981) the genetic diversity is high. Economically there is an increasing demand for *Nigella* (including export possibilities to neighbouring countries). Thus intensified collecting activities are very advisable.

Trachyspermum ammi. Ethiopian caraway is found in almost every market. It is grown at altitudes between 1500 and about 2200 m as a small-scale crop. Cultivation as a field crop is known from Bale, Gondar, Eritrea, Gojam and Shewa. *Trachyspermum* seeds are mainly used in the preparation of berbere-sauce and bread. Some medicinal uses are also reported.

Vavilov (1951) considered Ethiopia to be a centre of diversity for *Trachyspermum*, where the plant was introduced very long ago. *Trachyspermum* is definitely one of the more important and typical Ethiopian spices and is even grown as a field crop. The plant has some economic future, justifying an increased collecting activity.

Coriandrum sativum. Cultivation of coriander as a garden crop is widespread all over the country (altitude range 1500–2500 m). The plant is grown as a crop in Eritrea, Harerge, Shewa, Kefa, Welega and Gondar. Coriander plays an important role in the Ethiopian domestic spice trade and its seeds are used for the flavouring of berbere-sauce injera, cakes and bread (Kostlan, 1913). In Kefa, seeds are added to cheese and to a porridge made of *Colocasia esculenta* (taro).

Coriander again shows a high diversity. PGRC/E holds 38 accessions, mainly from Gondar and Welega. Coriander has good export potential if the quality can be improved (fungal resistance, yield, etc.), a task which can only be fulfilled with a large variety of local germplasm at hand.

Aframomum korarima. The use of korarima is known only from Ethiopia where it grows in the forests of Kefa, Sidamo, Ilubabor and Welega. The plant grows naturally at (1350–)1700–2000 m altitude, with high humidity and annual rainfall ranging from 1300 mm to more than 2000 mm with no real dry season. Korarima grows in almost the same habitats as natural coffee.

Cultivation of the plant has been reported not only from places where it grows wild, but also from the Lake Tana area, Eritrea and Gelemso (Harerge).

Korarima is very important for flavouring foods. It is used in the preparation of all kinds of 'wot', coffee and sometimes bread. Compared with other *Aframomum* species the seeds of korarima have a less pungent, milder and sweeter flavour. This spice could be developed into an important article of commerce but further experiments with cultivation need to be initiated. There is a demand for korarima in the neighbouring countries and in Arabia where it has long been highly prized as a spice (Russ, 1945). There is little doubt that markets could be found in Europe and America as well.

Hardly anything is known about *Aframomum*. PGRC/E holds only 16 accessions and nothing can be said about the diversity of the species. Korarima is one of the spices in which genetic erosion could be a real danger since its natural habitat, the humid mountain forests of south-western Ethiopia, will be decimated at an increasing rate in the future. In order to meet these problems the range of diversity must be known as a precondition for concentrated cultivation. Thus, comprehensive collection of germplasm from the wild is urgently needed.

Cuminum cyminum. In Ethiopia cumin seeds are found in almost every market. Small-scale cultivation is widespread at altitudes ranging from 1500 to 2200 m. The ground seeds are mainly used to flavour different kinds of 'wot', and only small quantities are required.

Cuminum was introduced into Ethiopia a long time ago but so far no reliable information is available on its range of diversity. PGRC/E holds six accessions. Since *Cuminum* is produced almost everywhere in the country and there is no demand for improvement, its collection is of minor importance.

Foeniculum vulgare and *Pimpinella anisum*. Both plants are common in the highland flora of all regions where they are widespread perennial weeds. They are occasionally cultivated (altitude range 1500–2500 m). The ground seeds are a constituent of 'wots'. More important is their use in the preparation of alcoholic beverages such as 'katikala', 'arake', and 'tedj' (a honey wine).

Both plants are common perennial weeds in the highlands, growing abundantly in the wild so that concentrated collection is not important at the moment. At present PGRC/E does not have any accessions.

Ruta chalepensis. Rue is a widespread herb cultivated in gardens in almost every province of the country (altitude range 1500–2000 m). The plant is used as a culinary herb. The seeds are needed to flavour 'wots', the leaves are also used as a condiment in coffee and tea. *Ruta* is important for the local market only. The plant is extremely widespread and at the moment there is no need for improvement, so that collection is of minor importance.

Ocimum basilicum. Basil is found in Ethiopia in cultivation as well as in the wild. The plant is cultivated on a small scale near houses in all provinces. It has a wide altitudinal range from sea level to 2500 m and it even withstands mild frosts (Jansen, 1981).

Basil is an important and frequently used spice for the preparation of all kinds of 'wot' and for the flavouring of butter. At present no genetic erosion has to be feared in basil. A future prospect could be the extraction of perfume oil from the plant but at the moment this is of lesser importance. PGRC/E at present holds 12 accessions.

Piper longum. This plant is said to be indigenous to Ethiopia, but the assumption is more than doubtful. Probably in this case *P. longum* was confused with *P. guineense*, which is at least of (West) African origin. *P. longum* certainly originates in India (Uphof, 1968; Zeven & de Wet, 1982), but there is no doubt that it was introduced in ancient times into Ethiopia, where it is now found to be growing in the wild. Like korarima the plant grows in almost the same habitats as coffee. So far there is only small-scale production. No reliable information is available on the growing season and other aspects of its husbandry.

P. longum is found in Kefa, Ilubabor and Welega, probably also in parts of Gamo Gofa (altitudinal range *ca.* 1500–2000 m). The inflorescence of the plant (a spike) is used for the preparation of 'wot'. The taste of *P. longum* is quite equal to that of black pepper, for which it serves as a substitute. It is preferred by the local consumer because of its lower price and greater availability. Thus in future *P. longum* could play a more important role in the local spice trade. Little is known about *P. longum* in Ethiopia. Its range of diversity – if there is any – has still to be described. Like korarima its natural habitat will be endangered at an increasing rate in the future. Collecting and screening of this interesting plant seems to be fully justifiable. PGRC/E holds three accessions.

Rhamnus prinoides. Buckthorn or 'gesho' is found growing in the wild all over Ethiopia between 1500 and 2500 m, but it is cultivated as well, sometimes even on a larger scale as a field crop. *Rhamnus* covers about 5000 ha of the land under permanent production (Jansen, 1981). It is a woody bush, whose leaves are used like hops for the preparation of alcoholic beverages such as 'talla' and 'tedj', which are common household drinks in the country. 'Gesho' is widespread all over the country. It serves the needs of the people so well that at least at the moment no improvement is needed.

Conclusions

(a) Collecting of *Capsicum* is required because the introduction of improved varieties will reduce the local diversity.

(b) *Aframomum korarima* and *Piper longum* should be collected for three reasons:

1. Conservation aspect: their habitat will become endangered at an increasing rate.
2. Scientific aspect: very little is known about the diversity and other properties of these interesting spices.
3. Economic aspect: korarima especially deserves to be concentrated on since it has a promising potential for export. *P. longum* is an important substitute for black pepper.

(c) *Nigella, Trachyspermum, Coriander* and to a lesser extent *Ocimum* are important spices which deserve attention even if they are not endangered. These spices are highly diversified and promising economically.

(d) *Trigonella* is well represented in the collection, which represents the genetic diversity found in the country, so that further activities should be limited to filling existing gaps.

(e) For the remaining spices *Cuminum cyminum, Pimpinella anisum, Foeniculum vulgare, Ruta chalepensis* and *Rhamnus prinoides* there is no need for immediate or intensified action.

References

Alkaemper, J. (1972). Capsicum – Anbau in Aethiopien für Gewürz- und Färbezwecke. *Bodenkultur*, **23**, 97–107.

Engels, J. & Goettsch, E. (1984). Capsicum in Ethiopia: some notes on its diversity. *PGRC/E–ILCA Germplasm Newsletter*, **6**, 12–15.

Goettsch, E. (1984). Proposal for further spice collecting at PGRC/E. Plant Genetic Resources Centre, Addis Ababa, 26 pp. (mimeographed).

Jansen, P. C. M. (1981). *Spices, Condiments and Medicinal Plants in Ethiopia, their Taxonomy and Agricultural Significance*. PUDOC, Wageningen.

Kostlan, A. (1913). Die Landwirtschaft in Abessinien. I. Teil: Acker- und Pflanzenbau. *Beiheft Tropenpflanzer*, **14**, 182–250.

National Bank (1982). *Annual Report, 1981*. Addis Ababa.

Russ, G. W. (1945). Reports on Ethiopian forests. Reprinted by Wolde-Michael Kelecha (1979), Forestry and Wild-life Development Authority, Addis Ababa.

Siegenthaler, I. E. (1963). *Useful Plants of Ethiopia*. J.E.C.A.M.A. Experimental Station Bulletin, no. 14. Alemaya Agricultural College, Ethiopia.

Uphof, U. C. T. (1968). *Dictionary of Economic Plants, 2nd edn*. Verlag von J. Cramer, Lehre.

Vavilov, N. I. (1951). The origin, variation, immunity and breeding of cultivated plants. *Chronica Botanica*, **13**, 1–366.

Westphal, E. (1974). *Pulses in Ethiopia, their Taxonomy and Agricultural Significance*. PUDOC, Wageningen.

Westphal, E. (1975). *Agricultural Systems in Ethiopia*. PUDOC, Wageningen.

Wolff, H. (1927), *Cuminum* and *Trachyspermum*. *In*: Engler (ed)., *Das Pflanzenreich, vol. 4*, Paper 90:228.

Zeven, A. C. & de Wet, J. M. J. (1982). *Dictionary of Cultivated Plants and their Regions of Diversity*. PUDOC, Wageningen.

9

A diversity study in Ethiopian barley

J. M. M. ENGELS

Introduction

Barley (*Hordeum vulgare* L.), one of the oldest of cultivated plants, has been grown in Ethiopia for at least 5000 years (Harlan, 1969; Doggett, 1970). Generally, Ethiopia is considered as a secondary gene centre, or a centre of diversity, for barley and not as a centre of origin (Tolbert *et al.*, 1979). However, in recent studies some evidence has been presented to suggest that Ethiopia might be a centre of origin (Bekele, 1983b; Negassa, 1985) as was originally suggested by Vavilov. The diversity in Ethiopian barley germplasm accessions has been presented in a number of studies (Ward, 1962; Tolbert *et al.*, 1979; Bekele, 1983a,b; Negassa, 1985) which were based mainly on discrete (non-continuous) characters. In studies on disease resistance in Ethiopian barley it was found that Ethiopian barley germplasm possesses resistance genes for almost all major diseases (Moseman, 1971; Lehmann, Nover & Scholz, 1976). In addition, high protein and lysine contents have been found in some Ethiopian genotypes (Munck, Karlsson & Hagberg, 1971).

In this chapter a detailed analysis is presented of the phenotypic diversity in the barley germplasm collection of the Plant Genetic Resources Centre/Ethiopia (PGRC/E), which possesses considerably more accessions than have been used in earlier diversity analyses. The results of previous studies on Ethiopian barley germplasm will also be summarized, particularly the ones on diversity indices.

Materials and methods

The records of 3765 accessions in the PGRC/E barley collection were surveyed. These records originated from germplasm collecting missions (e.g. passport data) and from the routine

Table 1. *Characters used and their respective classes*

Character	Character states	Codes used
1. Kernel row number	6 rows	6
	2 rows[a]	2
	irregular	irr.
2. Spike density	lax	1
	intermediate	2
	dense	3
3. Number of spikelets per spike	<15	1
	15–20	2
	20–25	3
	25–30	4
	≥30	5
4. Caryopsis	covered	covered
	naked	naked
5. Kernel colour	white–brown	1
	purple–black	2
6. Thousand grain weight	<25 g	1
	25–35 g	2
	35–45 g	3
	45–55 g	4
	≥55 g	5
7. Number of days to maturity	<100 days	1
	100–115 days	2
	115–130 days	3
	130–145 days	4
	≥145 days	5
8. Plant height	<60 cm	1
	60–90 cm	2
	90–120 cm	3
	120–150 cm	4
	≥150 cm	5

[a] Both categories, sterile and rudimentary lateral florets, are combined.

characterization activities of PGRC/E. The latter have been carried out since 1982 at Holetta (2400 m above sea level) and in this analysis data from the years 1982–5 were used. The data originated from unreplicated small plots (up to 2 sq m) and were generally based on the means of five plants or ears per accession, or on plot means. The majority of the accessions are morphologically rather uniform since they were selected from landraces based on their agro-morphological characteristics.

Data were analysed for the characters, presented in Table 1. Some characters with continuous variation were included in order to examine their value in diversity studies. The choice of the characters used was based on the following criteria: their use in previous diversity analyses; their consistency over the years in the characterization work; and their reliability in scoring. The phenotypic frequencies of the characters were analysed by the Shannon-Weaver information index (H') in order to estimate the diversity of each character within each administrative region and within each geographic region. The country was therefore arbitrarily divided into four ecogeographic zones, the northern administrative regions (Eritrea, Tigray, Gondar and Welo); the western administrative regions (Gojam, Welega, Ilubabor and Kefa); the southern administrative regions (Gamo Gofa, Sidamo and Bale) and the central and eastern administrative regions (Shewa, Arsi and Harerge).

The index was calculated as presented by Negassa (1985):

$$H' = -\sum_{i=1}^{n} p_i \log_e p_i$$

where p_i is the proportion of accessions in the ith class of an n-class character. In order to keep the values of H' in the range of 0–1 each value of H' was divided by its maximum value, $\log_e n$. The standard error was calculated as follows: SE $= S^2/r - 1$ where S^2 is the variance of the means and r the number of means. In order to determine whether the variance of the diversity was due to differences between or within administrative regions a hierarchical ANOVA was conducted with the normalized data for each character.

Results and discussion

The percentages of the phenotypic classes of accessions for each administrative region and the weighted mean percentages for the ecogeographic regions are presented in Table 2. In general, the mean percentages per ecogeographic area do not show marked variation. The same is true for the frequencies per administrative region for the majority of the characters. However, some exceptions are Eritrea, Tigray, Welega, Ilubabor and Kefa for kernel row number which have high frequencies (>72 per cent) for two-rowed barley. The spike density frequencies increase from north to south and from west to east, and show a clinal variation. The covered barleys are more concentrated in northern and western Ethiopia. The purple to black coloured kernels are more frequent in the south-west (Welega, Ilubabor, Kefa, Sidamo and Arsi). Gojam and Welega show the highest grain weights, although the differences are not significant. The average number of days to maturity is lower than the Ethiopian

Table 2. *Percentage of phenotypic classes of entries for each administrative region and a weighted mean percentage of each ecogeographic region for eight characters*

Administrative region	Number[a] of entries	Kernel row number			Spike density			Spikelets per spike				
		6	2	irr.	1	2	3	1	2	3	4	5
Eritrea	55	18	82	0	38	58	4	0	18	58	22	2
Tigray	287	17	81	2	48	44	8	0	17	52	28	3
Gonder	635	47	49	4	52	41	6	2	19	54	23	2
Welo	143	46	37	17	43	48	9	4	31	43	21	1
Region	1120	38	57	5	49	44	7	2	20	52	24	2
Gojam	135	38	52	10	54	40	6	7	21	51	19	2
Welega	72	13	87	0	53	44	3	3	17	51	26	3
Ilubabor	14	21	79	0	64	36	0	0	0	57	43	0
Kefa	102	26	72	2	41	56	3	6	15	42	35	2
Region	323	28	67	5	50	46	4	5	17	48	28	2
Gamo Gofa	88	47	50	3	41	46	13	2	14	45	35	4
Sidamo	58	41	55	4	59	36	5	0	12	50	36	2
Bale	104	44	44	12	39	49	12	2	24	64	9	1
Region	250	44	49	7	44	45	11	2	18	54	24	2
Shewa	952	39	57	4	42	46	12	7	26	45	20	2
Arsi	389	46	48	6	29	63	8	4	18	44	29	5
Harerge	159	58	34	8	39	36	25	8	28	49	13	2
Region	1500	43	52	5	38	49	13	6	24	45	22	3
Unknown	572	43	52	5	34	53	13	2	23	45	25	5
ETHIOPIA	3765	40	55	5	42	48	10	4	22	48	23	3

[a] These numbers vary insignificantly from character to character. Only TGW had a total number of 482 entries.

Table 3. *Mean squares and percentage of total variance from the hierarchical analysis of variance for H' of the individual characters*

Variance source	DF	Kernel row number		Spike density		Number of spikelets per spike		Caryopsis	
		MS	%	MS	%	MS	%	MS	%
Ecogeographic regions	3	0.0424	29.2	0.0199	48.5	0.0097	6.9	0.0131	24.7
Administrative regions within ecogeographic regions	10	0.0309	70.8	0.0063	51.5	0.0111	93.1	0.0119	75.3

Caryopsis		Kernel colour		Thousand grain weight (TGW)					Days to maturity					Plant height				
covered	naked	1	2	1	2	3	4	5	1	2	3	4	5	1	2	3	4	5
100	0	87	13	0	50	50	0	0	0	71	18	11	0	0	24	76	0	0
97	3	82	18	2	15	56	27	0	3	51	34	11	1	2	45	51	2	0
95	5	70	30	6	29	59	6	0	2	22	60	15	1	0	37	59	3	1
98	2	76	24	–	–	–	–	–	1	71	19	8	1	1	43	52	3	1
96	4	75	25	3	20	56	21	0	2	38	46	13	1	1	39	56	3	1
95	5	77	23	0	12	23	62	4	4	31	47	18	0	4	18	65	13	0
99	1	65	35	0	14	50	36	0	15	51	25	9	0	0	26	65	9	0
100	0	64	36	–	–	–	–	–	20	67	0	13	0	0	29	71	0	0
95	5	65	35	–	–	–	–	–	9	51	21	18	0	0	28	68	4	0
96	4	70	30	0	13	32	53	2	9	43	32	16	0	2	23	66	9	0
99	1	71	29	0	21	41	26	12	5	56	33	6	0	1	18	77	3	0
100	0	67	33	0	8	38	38	16	16	21	40	24	0	0	22	76	2	0
100	0	82	18	6	41	41	6	6	12	36	31	20	1	1	31	63	5	0
100	0	75	25	2	24	40	23	11	10	40	34	16	0	1	34	71	4	0
96	4	72	28	1	30	47	21	1	4	32	43	20	1	1	39	54	6	0
99	1	67	33	0	16	58	24	2	3	17	36	43	1	2	24	65	8	0
98	2	80	20	0	20	80	0	0	1	48	33	18	0	1	35	43	21	0
97	3	72	28	1	26	51	21	1	3	30	40	26	1	1	35	56	8	0
96	4	69	31	4	22	54	18	2	1	23	43	31	2	2	31	60	6	1
97	3	72	28	2	23	50	23	2	3	33	41	22	1	1	34	58	6	1

Kernel colour		1000[a] gram weight		Number of days to maturity		Plant height	
MS	%	MS	%	MS	%	MS	%
0.0187	29.9	0.0329	46.3	0.0197	28.1	0.0150	36.2
0.0132	70.1	0.0164	53.7	0.0096	61.9	0.0080	63.8

[a] DF of TGW are 3 and 7, respectively.

average in Eritrea, Tigray, Welo, Welega, Ilubabor, Kefa, Gamo Gofa and Harerge. Apart from a possible drought escape mechanism through early maturity, which might have evolved in the northern administrative regions, there seems to be another natural selection pressure in the southern and eastern administrative regions. Finally, the barleys from southern Ethiopia, as well as from Eritrea, have a higher average straw length than the barleys of the rest of Ethiopia. These differences between administrative regions, and to a certain extent between ecogeographic regions, are supported by the percentages of the total variance for each of the two areas (Table 3).

The diversity indices showed relatively wide variations between characters (Table 4). The caryopsis (covered or naked barley) and plant height were the least diverse characters studied. Spike density caused the highest diversity, followed by kernel colour. This agrees with the earlier findings of Negassa (1985) except for spike density, which showed little variation in his study. This may have been due to his smaller sample size or to some selection of the germplasm during collecting in the field.

The pooled diversity indices over characters within administrative regions and within ecogeographic regions are relatively uniform. The least diverse administrative regions are Eritrea and Ilubabor (Table 4). One of the reasons for this could have been the small number of samples in the study from these regions. On the other hand, there might be greater natural selection in barley in both regions due to generally low rainfall in Eritrea and high rainfall in Ilubabor. The highest diversity index was found for the Shewa administrative region, followed by Gojam and Arsi. Only the indices for Shewa and Eritrea are significantly different ($t = 2.35$, $P < 0.05$).

Of the four ecogeographic regions, the central and eastern regions showed the highest diversity index. All the other ecogeographic regions were almost equally diverse and were not significantly different from each other ($t = 1.84$, $P > 0.05$). This finding is also confirmed by the low percentage of the total variance caused by the 'among ecogeographic regions' source (Table 5). Although the percentage of the total variance for 'among administrative regions within ecogeographic areas' is slightly higher (5.7 per cent) it can be concluded from this table that by far the highest variance is due to 'among characters within administrative regions'.

The overall diversity index for Ethiopian barley is relatively high for almost all the characters as well as for the pooled index over characters. These results are similar to other studies (Qualset & Mose-

Table 4. *Estimates of the diversity indices (H') for the various administrative regions, for the four ecogeographic regions and of the mean diversity ($\bar{H}'$) and its standard error over all characters as well as results of some other authors*

Administrative region	Kernel row number	Spike density	Spikelets per spike	Caryopsis	Kernel colour	TGW	Days to maturity	Plant height	$\bar{H}' \pm$ SE	$\bar{H}' \pm$ SE (Negassa, 1985)
Eritrea	0.43	0.76	0.65	0.00	0.55	0.44	0.49	0.34	0.46±0.08	
Tigray	0.49	0.83	0.69	0.19	0.69	0.65	0.70	0.53	0.60±0.07	0.65±0.04
Gonder	0.77	0.80	0.70	0.28	0.89	0.61	0.66	0.51	0.65±0.07	0.54±0.11
Welo	0.93	0.85	0.77	0.14	0.80	—	0.53	0.56	0.65±0.10	
Region	0.76	0.82	0.72	0.25	0.81	0.67	0.69	0.56	0.66±0.07	
Gojam	0.85	0.78	0.77	0.28	0.78	0.63	0.72	0.60	0.68±0.06	0.60±0.07
Welega	0.36	0.74	0.76	0.08	0.94	0.62	0.74	0.52	0.60±0.10	0.66±0.05
Ilubabor	0.46	0.59	0.43	0.00	0.94	—	0.53	0.38	0.48±0.11	
Kefa	0.60	0.74	0.80	0.28	0.94	—	0.74	0.47	0.65±0.08	
Region	0.70	0.77	0.77	0.25	0.88	0.65	0.77	0.56	0.67±0.07	
Gamo Gofa	0.75	0.91	0.75	0.08	0.87	0.81	0.63	0.40	0.65±0.10	0.71±0.08
Sidamo	0.75	0.76	0.66	0.00	0.91	0.77	0.83	0.38	0.63±0.11	0.54±0.06
Bale	0.89	0.89	0.60	0.00	0.69	0.76	0.85	0.53	0.65±0.10	0.60±0.04
Region	0.82	0.88	0.71	0.00	0.81	0.85	0.78	0.47	0.67±0.11	
Shewa	0.75	0.89	0.81	0.24	0.84	0.70	0.76	0.57	0.70±0.07	0.64±0.03
Arsi	0.80	0.77	0.81	0.08	0.91	0.64	0.75	0.55	0.66±0.09	0.63±0.08
Harerge	0.81	0.99	0.77	0.14	0.71	0.31	0.67	0.69	0.64±0.10	
Region	0.78	0.89	0.81	0.19	0.85	0.79	0.77	0.59	0.71±0.08	
ETHIOPIA	0.77	0.86	0.79	0.19	0.85	0.74	0.77	0.59	0.70±0.08	0.68±0.02
Ethiopia (Tolbert *et al.*, 1979)	0.91			0.53	0.82				0.51±0.01	
Ethiopia (Qualset, 1975)[a]	0.71	0.37		0.53	0.93		0.81[b]		0.67±0.10	
Ethiopia (Qualset & Moseman, 1966)[a]	0.75			0.59	0.92				0.75±0.10	

[a] Calculated by author.

[b] This character is in fact heading time.

Table 5. *Hierarchical analysis of variance of the diversity index (H') and the percentage of the total variance*

Source	DF	SS	MS	%	F	
Among ecogeographic regions	3	0.1166	0.0389	1.9	1.12	NS
Among administrative regions within ecogeographic regions	10	0.3470	0.0347	5.7	0.52	NS
Among characters within administrative regions	84	5.6229	0.0669	92.4		

man, 1966; Qualset, 1975; Negassa, 1985) despite their use of different types and numbers of characters and different sample sizes (varying from 485 to 3765 samples) and the use of classified quantitative characters in this study. Tolbert *et al.* (1979) reported lower values, probably due to the use of growth habit (winter or spring) and awn type (rough or smooth), as these characters do not vary much or at all in Ethiopian barley. Thus the conclusion by Tolbert *et al.* that Ethiopia is a secondary centre of diversity may be revised if other more relevant characters are used in diversity studies. In the present study awn type was omitted as all accessions were rough.

The use of quantitative characters which were scaled in an arbitrary way seems to be justifiable and useful as the barley accessions show considerable variation within and between administrative regions for these characters (Table 3) and they are, in general, of more interest to the plant breeder than the discrete or qualitative characters. However, because of the arbitrary decision on the number of classes per character and their influence on the magnitude of the diversity index, a comparison of the indices from different studies is meaningless.

The results of this analysis have shown that Ethiopia is a centre of diversity for barley and that this diversity is rather evenly distributed over the barley growing areas of the country, although there are some concentrations for individual characters. Furthermore, the initial results of a study of the diversity index by altitudinal strata of Ethiopia as a whole have shown an obvious relationship between the diversity index and the altitude. The index is highest around 2500–2600 m above sea level and decreases with increasing or decreasing elevation (Engels, 1990).

Acknowledgements

The author would like to thank all colleagues involved in the data collection during the collection and characterization work at PGRC/E as well as Mrs Karin Ralsgård for the data on thousand grain weight and Dr John Lazier for his critical comments.

References

Bekele, E. (1983a). Some measures of gene diversity analysis on landrace populations of Ethiopian barley. *Hereditas*, **98**, 127–43.

Bekele, E. (1983b). A differential rate of regional distribution of barley flavonoid patterns in Ethiopia, and review on the centre of origin of barley. *Hereditas*, **98**, 269–80.

Doggett, H. (1970). *Sorghum*. Longman, London.

Engels, J. M. M. (1990). The genetic diversity in Ethiopian barley in relation to altitude. *In*: S. Iyama and G. Takeda (eds), *Proceedings, 6th International Congress of the Society for the Advancement of Breeding Research in Asia and Oceania, 21–25 August 1989*. SABRAO, Tsukuba, Japan, 107–10.

Harlan, J. R. (1969). Ethiopia: a centre of diversity. *Economic Botany*, **23**, 309–14.

Lehmann, C. O., Nover, I. & Scholz, F. (1976). The Gatersleben barley collection and its evaluation. *In*: H. Gaul (ed.), *Barley Genetics, vol. III. Proceedings 3rd International Barley Genetics Symposium, Garching, 1975*. Karl Thiemig, Munich, pp. 64–79.

Moseman, J. G. (1971). Co-evolution of host resistance and pathogen virulence. *In*: R. A. Nilan (ed.), *Barley Genetics, vol. II. Proceedings 2nd International Barley Genetics Symposium, Pullman, 1969*. Washington State University Press, Pullman, Washington, pp. 450–6.

Munck, L., Karlsson, K. E. & Hagberg, A. (1971). Selection and characterization of a high protein, high-lysine variety from the world barley collection. *In*: R. A. Nilan (ed.), *Barley Genetics, vol. II. Proceedings 2nd International Barley Genetics Symposium, Pullman, 1969*. Washington State University Press, Pullman, Washington, pp. 544–58.

Negassa, M. (1985). Patterns of phenotypic diversity in an Ethiopian barley collection, and the Arussi–Bale Highland as a centre of origin of barley. *Hereditas*, **102**, 139–50.

Qualset, C. O. (1975). Sampling germplasm in a centre of diversity: an example of disease resistance in Ethiopian barley. *In*: O. H. Frankel and J. G. Hawkes (eds), *Crop Genetic Resources for Today and Tomorrow*. Cambridge University Press, Cambridge, pp. 81–96.

Qualset, C. O. & Moseman, J. G. (1966). Disease reaction of 654 barley introductions from Ethiopia. USDA/ARS Progress Report (unpublished).

Tolbert, D. M., Qualset, C. O., Jain, S. K. & Craddock, J. C. (1979). A diversity analysis of a world collection of barley. *Crop Science*, **19**, 789–94.

Ward, D. J. (1962). Some evolutionary aspects of certain morphological characters in a world collection of barley. *USDA Technical Bulletin 1276*.

10

Sorghum history in relation to Ethiopia

H. DOGGETT

Introduction

Hypotheses on crop development in Africa are long on theory and short on fact. Let me present my own ideas. Harlan & Stemler (1976) have presented a theory that sorghum developed in the southern Sudan–Chad region. My problem with that is the answer to the question 'how'? It is true that as soon as Man began to sow the seeds of wild grasses, selection and sowing over the years would result in cultivated types being developed. That assumes that Man somehow learnt the idea of agriculture. Perhaps he did, but it is not clear how this happened in the rainfed savannahs. There were lots of grasses there anyway, and lots of grass seed. Why sow more? How would man have learnt to sow seed? How would he have distinguished between the masses of grass seedlings coming up with the rains and those which he had put in? By clearing a separate plot of land for sowing? That presumes the idea of agriculture. Having worked in the tropics for many years, I find this altogether too difficult to imagine.

It is more likely, to my mind, that agriculture was discovered along rivers and that the discovery was a rather rare event. One should always look at the possibility of the spread of the idea of agriculture from elsewhere before concluding that it had been discovered all over again. My scenario for the discovery of agriculture in a situation of this kind is presented below.

Discovery of agriculture

The three oldest civilizations of the Old World all arose along rivers: in due course, each spread out along its respective river valley for hundreds of miles (Fairservis, 1971). Rivers and seasonal streams provide sites from where the idea of cultivation may have emerged. Many patches of silt, exposed as the rains ended and the rivers fell, would have been weed-free at first. People gathering seeds of wild grasses for food, who also fished, could well have noticed that seeds dropped on these patches sometimes grew into mature plants on residual moisture. From this, the use of sickles for harvesting would have favoured the variants with persistent spikelets. Gradually, the idea of deliberately sowing these riverine flats with a seed so harvested, and replanting the following season, would have led to the accumulation of non-shedding types. People would gradually have become accustomed to the regular discipline of seed-time and harvest on silt flats needing no land preparation and no weeding. This would have provided an additional resource; fishing, food-gathering and hunting would have continued as before. Once seeding became an established practice, it is not difficult to imagine a gradual awakening of interest in crop improvement as more desirable types were noticed.

This reconstruction of the possible origins of agriculture also provides an explanation for the way in which people became locked into the hard labour and drudgery involved. So long as people were using the natural resources of hunting, fishing and food-gathering, the population could not increase beyond the number those natural resources would carry. Improved harvesting and grass-seed processing technology made better use of the resource base but did not enlarge it. Learning to seed the silt flats deliberately was a different matter. This enlarged the resource base and provided a way to feed an expanding population. As the population grew, more silt flats could be seeded. In due course the population expanded beyond the point of no return. No longer were hunting, fishing and food-gathering sufficient. From then on, the pressures demanded the extension of irrigation, preparation and weeding of land to imitate the conditions on the silt flats, leading on eventually to the development of rainfed agriculture. For that, there was basic crop husbandry to be learnt: clearing the land, tillage, the time and method of sowing, and weed control. All this had to be done initially with stone axes and sticks as the only tools.

Movement of crops

Crop movement presents a problem. Near the beginnings of agriculture, the idea of agriculture needed to move together with the new crops. Movement of the agriculturalists themselves is one obvious method; they may then have acted as focal points for the teaching of the new technology. Their new neighbours would come to learn this remarkable new technology and both crops and methodology would spread from such focal points. Conquest must also have been a vehicle for the transfer of crops and technology and the success of the agriculturalists may well have made them liable to attack from jealous neighbours. The long-distance transfer of individual crops other than through actual carriage by agriculturalists is difficult to imagine, unless the recipients were themselves agriculturalists (Gramly, 1979).

It seems most unlikely that settled arable agriculturalists ever moved until forced by circumstances to do so. They then took their technology and high-yielding varieties as a package with them, having first located a site which they considered suited to their crops and methods. Harlan & Stemler (1976), referring to the spread of agriculture to the west and to the east, noted that 'what moved out of the nuclear area (West Asia) was a complete system including barley, emmer wheat, einkorn wheat, lentil, vetch, pea, chickpea, faba bean, rape, flax, vegetables, spices, tree and vine fruits, sheep, goats, cattle and an array of agricultural techniques'. Doubtless they moved very much as the people under pressure in the Sahel zone are moving today. The man of the family goes south, living as best he may. He prospects, and if he finds a suitable area he returns to help his family pack up and they move, taking with them their tools, seed and livestock, together with their accumulated agricultural knowledge and wisdom. Groups of several households may emigrate together for mutual protection and support.

Agricultural development in north-east Africa

The people

Language and archaeological studies (Hiernaux, 1974; Ehret, 1979) show that a long-headed, long-faced people had been present in north-east Africa since the latter Pleistocene, in much of the area labelled today as Sudan, Ethiopia, Somalia, Tanzania, Kenya, Rwanda and Burundi. They were Africans, not 'Mediterranean types'. (In this chapter, 'Ethiopia' will be used for the area south of the confluence of the Blue Nile and the Atbara, and east of the White Nile along the rivers of the plain). The Afroasiatic language group arose in the Ethiopian area, extending roughly from the Amba Farit mountains on the west, past Lake Tana and reaching almost as far as the Lake Nasser of today. On the eastern side it followed the Red Sea hills and the shores of the Red Sea. This language group later developed into Semitic, Berber, Ancient Egyptian, Cushitic, Omotic and Chadic. The proto-afroasiatic people lived at least 15 000 years ago. Proto-cushitic was being spoken at least 9000 years ago. Descendants of these people spread widely across Africa and into the Mediterranean region. The Semitic people moved out from Africa, some of them returning later. These people were strongly associated with grass-seed collecting and pottery. It is tempting to suggest that the value of fermentation in utilizing grass seed for food was known to them. Injera may have a long history.

The Nile

The Nile has always been a major route into Africa and the agriculture of south-west Asia developed within travelling distance of the Nile. It was certainly of greater antiquity than agricultural development in Africa. There have been big climatic changes along the Nile. During the last Ice Age, the Mediterranean climate was forced south into Africa and the Mediterranean plants survived there, including some of the grasses from which crop plants were later derived. Wendorf & Schild (1984) have excavated key sites along the Nile and demonstrated human activities along the river over thousands of years, including a strong indication of arable agriculture. They thought that a very early cultivated barley had been discovered but subsequent tests showed that the find was an intrusion. We do know that the climate was growing warmer. We also know that there was an ancient trade route up the Nile and the Atbara through to the frankincense and myrrh products of Saba. The presence of barley has been demonstrated in Egypt in the Fayum, dating probably to the fifth millennium BC (Arkell & Ucko, 1965). The Nile has silted some of the areas that archaeologists would like to explore, and the use of barley in Egypt could be yet earlier.

Agricultural development in Ethiopia

The author suggests that barley was taken into Ethiopia by people from Egypt, or from cultivators along the banks of the Nile if barley was indeed grown there. The cool conditions favouring the cultivation of barley were moving northwards towards the Mediterranean and temperatures along the Nile were rising. Some people had already become dependent upon barley, using the technology that was common along the river prior to the climatic change. Small groups of these people would have moved into the hills, following the barley climate as it receded. This probably took place along the Blue Nile, the Atbara and other tributaries of the Nile system. Alternatively, barley may have been introduced up the Nile from West Asia later. According to Ethiopian tradition, barley is a very ancient crop in that country. There is a great diversity of barleys in Ethiopia and Helbaek (1960, 1966) drew attention to the whole series of forms grown at the beginning of agriculture in Egypt. He also recorded *Hordeum irregulare* from the Fayum and this group all seem to have originated in Ethiopia. Helbaek also noted resemblances between the ancient emmer wheats of Egypt and some modern Ethiopian types. Harlan (1969) recorded the great variability in the barleys and tetraploid wheats in Ethiopia. The barley-with-emmer combination of the Ethiopian highlands was important in ancient Egypt, dating back to *ca*. 4500 BC.

Situation in the hills

Settled cultivators are sitting targets. After harvest, they have a stock of food which others would gladly seize. The arable agriculturalists, therefore, have, in the past, occupied defensive positions on the hills. They may well have cultivated in the valleys as well, returning to their defended communities daily before nightfall.

The early cultivators in the hills were caught between increasing population size, on the one hand, and the climatic and ecological limitations of barley culture on the other.

The climate became warmer and drier and population numbers increased. The early agriculturalists responded to these challenges in two ways: (a) by domesticating new crops adapted to warmer or more difficult conditions than those suited to barley; (b) by developing a more intensive agricultural system.

Domestication of new crops

Crops that can survive well on difficult soils in the barley zone include niger seed (noog, *Guizotia abyssinica*), teff (*Eragrostis tef*) and linseed (*Linum usitatissimum*).

Niger seed was almost certainly ennobled in Ethiopia. Teff could well have originated from one of the preferred grasses of the grass-seed collection days, taken into cultivation as a result of learning the principles of agriculture and subjected to selection for persistent spikelets. To the casual observer, teff is a wild grass. Linseed was probably introduced as an edible oilseed crop. The fibre (flax) was used by the Egyptians, especially for fabric with which to bind the dead.

Two cereal crops extending from below the 'barley line' in the highlands to the lowlands are finger millet (*Eleusine coracana*) and sorghum. There is little doubt that finger millet was developed from *E. africana*. One archaelogical find probably dates to the third millennium BC (Mehra, 1962; Phillipson, 1977a; Hilu, de Wet & Harlan, 1979). Thus a range of crops was developed, or introduced from West Asia, adapted to a wide range of soils and climatic conditions. The development of sorghum will be considered below.

Development of cultural methods, with soil and water management

Important developments in soil and water management, essential to reduce the effects of population pressure on land in the Ethiopian hills, may be illustrated by looking at the current agriculture of the Konso. These people have lived in south-west Ethiopia for a long time, although they claim to have inherited at least some of their practices from the Mādo people who preceded them. We may speculate about the order in which the various practices were developed, but the whole 'package of practices' is impressive. The Konso lived in relative isolation (apart from contacts through the market systems) prior to 1896 and may be regarded as inheritors of an ancient agricultural tradition developed over many centuries.

In order to minimize erosion, the soil is retained by the construction of many hundreds of miles of stone terraces, which follow the contours. They are built as dry stone walls, the soil being cut away vertically and the wall built against the vertical face. Only undressed stones from the ground near the place of construction are used, but with great skill and neatness. The terrain is steep; terraces are often about 2.5 m wide and 1.5 m high. The wall projects above the level of the field it is retaining. After heavy rain, a length of wall may collapse, but is immediately rebuilt by the owner who will rush out naked in a rainstorm to see what is happening to the water on his land. The land itself has a ridge on the outside and other ridges are made at right angles to it, forming a series of boxes to hold the water, as with tie-ridging or basin listing.

Any streams are used to irrigate the fields and are walled to protect the fields from flood water. Elaborate stone leats are constructed to allow the water to pass through a series of walled gardens. Such irrigable streams are rare: most of the stone-lined drains carry storm water and are used as paths, especially for cattle. The runoff is carefully channelled through leats on the land and the owner will be there during heavy rain to see that water is being well distributed over his land. Water for domestic use is obtained from wells or from the few permanent streams. Huge reservoirs have been constructed to conserve rainwater for cattle; dams may be as much as 12 m high and more than 60 m in length, containing many hundreds of thousands of litres. Towns are usually situated on high ground and the stream beds are in the valleys. Water may be collected from points half-an-hour's walk from the town and 60 m below it.

Soil fertility is maintained by the liberal use of manure, which is applied once before sowing and frequently during the growing season. Human manure is used. In each town, there are a number of places, generally along the outer walls, for defecation. The faeces dry quickly in the sun; they are collected and mixed with animal manure and then periodically taken to the fields. This may well be a further indication of the age of agriculture in the area. It is hard to believe that the organized use of human manure would have been adopted and retained as an ancient custom if animal manure had been readily available. The manure is collected outside the homesteads and left to rot; in some areas, pits are dug in which the dung can mature.

The people live in walled towns with gates built in defensive positions. Only in recent years have the gates been neglected and security relaxed. The cattle (including sheep and goats) are penned within the homesteads and are partly stall-fed with fodder cut and carried from the valleys and lowlands. They are taken out under careful supervision along certain walled paths to the grazing area. Only a few pastures are found near the town; the greater part of the available land is situated some distance away and the cattle are grazed there. Many of the distant fields are terraced, but not manured, and rotation with fallow is practised; the grazing of the cattle doubtless contributes to fertility maintenance (Hallpike, 1970, 1972).

Ploughing was introduced by the Amhara. Traditional cultivation used a three-pronged hoe of a type found formerly in ancient Egypt.

Konso cropping pattern

The plateau of the Takadi area to the west is only a few hundred feet above the Garati area to the east, yet the cropping is different. Wheat and barley are the main crops on the Takadi plateau; linseed, sorghum and finger millet are also grown there. Sorghum is grown mainly on the lower ground – the Garati area – and ripens several weeks before the same crop on the plateau. Most sorghum is interplanted with finger millet. Maize has now also become an important crop. Some ensete is grown, but it is not very popular. The long-established ensete cultivation of south-west Ethiopia should be noted. This vegeculture could be very ancient and perhaps people from this area first made contact with the barley crop and seed-crop agriculture further up the Nile. Barley is a common crop along with ensete at 3000 m (Hallpike, 1970; Westphal, 1975).

Origin of the sorghum crop

Area of origin

Mann, Kimber & Miller (1983) outlined the current hypotheses on the time and place of the origin of sorghum cultivation. There can be no doubt that the cultivated sorghums of today arose from the wild *Sorghum bicolor* subsp. *arundinaceum*. There is no evidence of cultivated types ever having arisen from the rhizomatous diploid or tetraploid *Halepensia*. The wild forms of *S. bicolor* were confined to Africa until recent historical times and it is certain that the crop was domesticated on the African continent.

The Saharan and north-eastern regions of Africa (at least) enjoyed a pluvial period (with interruptions) prior to 3000 BC. The sorghums of those days were doubtless adapted to wetter conditions; many of the wild types still are. De Wet, Harlan & Price (1970) listed the distribution and habitat of 16 of Snowden's wild 'species'. Seven of them belonged to wet or humid areas; five were characteristic of hot, dry regions; all belonged to damp places, swamp and stream margins, or irrigation ditches.

Wild sorghum occurs in Ethiopia up to about 2300 m above sea level. It is fairly common at 1500–1700 m and shatter canes (derivatives of wild × sorghum crosses) are the most serious weed around 1700 m in the central plateau, where they are known as 'keelo' (the fool). In October 1982, the author sampled wheat fields along the road from Debre Zeit (1800 m) to Nazret (1600 m) and for some 27 km down the road towards Awasa. In a distance of 77 km, stops every few kilometers showed that 97 per cent of these wheat fields contained at least a few plants of wild sorghum. My suggestion that wild sorghum, occurring as a weed of cultivation, attracted the attention of early agriculturalists remains a possibility (Doggett, 1965).

South-western Ethiopia provides the type of sites where sorghum could have been ennobled. The agriculturalists of south-western Ethiopia would have lived near their main crops and needed the high altitude, also for defence. Lowland crops would have been cultivated along the rivers, as is the present-day practice of the people of Gamo Gofa, Kefa and Ilubabor. The Konso also continue this ancient agricultural way of life. They grow cold-tolerant crops in the highlands. The Konso grow at least 24 varieties of sorghum, including race guinea. Race caudatum and type bicolor are the main forms grown. Sorghums are the staple crop and are grown on the higher ground, but much is also grown on the lower ground. Sorghum is often ratooned (Hallpike, 1970, 1972).

Gebrekidan (1970), on a trip to the Konso area, noted wild species along the roads and the abundance of shatter canes. He recorded the importance of bicolor types and the fact that the cultivated forms of sorghum all had very thin, grass-like stems with both compact and loose panicled forms represented. These were all growing in the highlands and were cold-tolerant when tested at the University in Harare.

Time of origin

The sorghum races guinea and durra reached India; the more recent races caudatum and kafir did not. Links through Arabia to India may have been severed in the third century BC when the 'Abyssinians' invaded south-west Arabia.

Finds of sorghum in India have been reported from Jorwe, *ca.* 1000 BC, together with finger millet (Kajale, 1977); from Pirak, on the edge of the Indus plain, *ca.* 1350 BC; at Ahar near Udaipur, *ca.* 1500 BC; and at Imagon, near Ahmadnagar, between 1800 and 1500 BC (calibrated) (Allchin & Allchin, 1982). These dates give estimates of the latest date by which sorghum had reached India. The crop may well have arrived there earlier.

There is one early date for sorghum in the Sudan. Excavations at Kadero, 18 km north-east of Khartoum and some 6 km to the east of the Nile, were dated to the second half of the fourth millennium BC. There, numerous grindstones and abundant sherds carrying impressions of grass seeds were found. Klichowska (1984) regarded one group of 15 impressions as *S. vulgare*, and measured the mean dimensions as 3.4 × 3.6 mm. A second group, her *Sorghum* cf. *vulgare*, had 11 impressions averaging 3.7 × 3.4 mm. These dimensions lie well outside the grain sizes of wild sorghum and within the ranges of several cultivated types. (As a comparison, the bicolor sorghum c.245, recorded by Clark & Stemler (1975), measured 3.0–3.4 × 2.3–2.9 mm.) Similarly, the dimensions for 20 impressions of *Eleusine* grains averaged 2.1 × 2.0 mm, which lies within the range of dimensions of cultivated finger millet grains. Unless the impressions much exaggerated the size of the original grains that made them, there were cultivated sorghums and finger millets in the Kadero material (Harlan, 1969).

Another discovery of impressions of cultivated sorghum dates back some 4500 years and was found in Abu Dhabi, near the mouth of the Persian Gulf (Cleuziou & Constantini, 1982). Cultivated finger millet has been found in Ethiopia probably dating to the third millennium BC (de Wet *et al.*, 1984).

Phillipson (1977b) accepted the third millennium BC as a general estimate of the period during which many African crops were brought under cultivation. At present, there are no grounds for modifiying the opinion that sorghum was first developed in the north-east quadrant of Africa some 5000 years ago, probably in the Ethiopia–Sudan region (Doggett, 1965).

Development of the sorghum crop

The use of the names of the different sorghum races in the following paragraphs is based on the classification of Harlan & de Wet (1972).

Bicolor sorghums

The earliest stages of ennoblement would have produced bicolor types. Bicolor sorghums with small grains and often loose panicles are frequently found in wet conditions. With these loose panicles and very dark, small grains often covered by the glumes, they are adapted to high-altitude, humid conditions. The grains dry quickly and are little troubled by birds or grain moulds. These forms were developed in ecological situations similar to those which they now occupy. They are used for beer, for their sweet stems and for special kinds of food preparation. We possess these samples from the past history of the crop because they are still worth growing in appropriate habitats for particular uses. It is not known when the bicolors were carried to India, but they were presumably among the earliest to arrive there.

Guinea sorghums

The wild sorghum race *arundinaceum* occurs all along the northern forest margin of western Africa, in northern Uganda and on through southern Sudan into Ethiopia. Sorghums of Snowden's *Guineensia* are found along the same belt. The association between the wild *arundinaceum* and many of the cultivated *guineense*, and the derivation of one from the other, need not be questioned. Race guinea contains cultivars adapted to high rainfall areas and others adapted to the decrue agriculture of Mali. In so far as the Shallu derivatives are representative of the guinea sorghums, they are adapted to humid rather than to dry conditions, with a low level of drought tolerance.

Where were the guineas developed? There is no evidence of annual seed crop cultivation in West Africa before 1500 BC, so the development of guinea sorghum in western or central Africa, followed by its movement to India, seems unlikely. The most probable area of origin lies in western and south-western 'Ethiopia', stretching as far as the White Nile. An island of guinea sorghum survives in the highlands of south-west Ethiopia, grown by the Konso people practising an ancient agriculture, and isolated today from the nearest guinea sorghum on the plain to the south by a distance of some 300 km.

The guinea sorghums spread to India and south to South Africa. They also spread across Africa along a rather narrow belt of land. *S. roxburghii* is the principal member of the *Guineensia* occurring in eastern Africa and also in India. Coastal sea traffic sailing on the monsoons has connected the two continents for over 2000 years. Guinea sorghums occur very widely in coastal areas of the Old World, having been spread by ship along the coasts of South-East Asia.

Durra sorghums

The country possessing the best relics of the development of durra sorghums is Ethiopia. Bicolor types occur in western Ethiopia, typically in high rainfall, highland areas. Forms with larger, more 'cultivated type', grains are found in the less humid, less rainy parts of the country. They combine some of the features of bicolor and durra, for example, the local cultivars 'Fundishu' and 'Zangada'. Other forms would be classified as Harlan and de Wet's race durra-caudatum, especially in Gondar, Simen and Tigray. In the dry areas, durra types with their large grains and compact panicles predominate, notably in the Harerge administrative region on the Chercher highlands.

Race durra probably developed through introgression with the wild type *aethiopicum*, which occurs in the drier areas of Ethiopia. Above average levels of drought resistance and seed size are characteristics of this race. As the Ethiopian climate dried and early bicolors were moved eastwards into drier areas, still under selection by able agriculturalists, introgression with wild forms adapted to drier conditions occurred. Intermediate types between the races are also found.

K. E. Prasada Rao (personal communication) reviewed the collection in Ethiopia (4000 entries) and noted 25 local races of durra, eight of durra-bicolor, three of durra-guinea and one of caudatum-bicolor. There were also three races each of bicolor, caudatum and guinea-caudatum. All the durras and durra relatives had been collected from high and medium altitudes. Panicle type is governed by the humidity of the environment at flowering and ripening time and very dense panicles are found in types which flower and ripen grain under really dry conditions. Compact durras are outstanding in such situations, but open-panicled durras occur in the higher rainfall areas of Ethiopia. The whole sequence, wild type–bicolor–durra-bicolor–durra, can be seen clearly in Ethiopia. From Ethiopia, durra spread westwards through the Sudan and across Africa, occupying the dry belt below the southern margin of the Sahara.

The semi-nomadic peoples of Somalia still grow durra sorghums in a well developed agricultural system, which is certainly old, and could be very old. The durras were probably carried from the Horn of Africa through Yemen and Saudi Arabia. From there, they may have moved into Iran and Afghanistan or West Asia, or through Oman and across the Baluchistan. They were then carried either through the Punjab to northern India, or through Sind to peninsular and southern India.

K. E. Prasada Rao (personal communication), collecting in the hills of Madhaya Pradesh occupied by 'Tribals' practising a simple agriculture, noted that the 'pig-mouth' durras grown by the Tribals in the 800 mm rainfall zone are similar to the 'Zurru' durras of Tigray region in Ethiopia.

Caudatum sorghums

Caudatums are grown by the Konso, as noted above. Both caudatums and half-caudatums are grown in Ethiopia, especially in the lowlands. The fact that they do not occur in India shows that this race is younger than guinea or durra and also suggests that caudatums could not arise from all the combinations of durra and guinea crosses that must have occurred there. It is probable that continued interaction with the wild sorghum gene pool was necessary for caudatums to appear. Caudatums are often associated with pastoralists.

The caudatums encountered wild and bicolor sorghums in the Sudan plains, as well as durras and probably guineas, and introgression occurred. Many must have been subjected to bulk mass selection every time the pastoralists harvested them. Often they grew in the presence of *Striga*, shoot-fly, other pests, diseases and increasingly tough environmental conditons as the climate continued to dry. The casual agricultural standard of the pastoralists encouraged intercrossing. Generations of crude mass selection in these populations, harvesting the larger grained survivors, resulted in the development of really tough caudatums, able to yield in spite of the ills listed above and also possessing a degree of bird resistance.

Stemler, Harlan & de Wet (1975) gave an excellent account of the association between these sorghums and speakers of the Chari–Nile languages, linked to their migrations. They noted the importance of caudatum sorghums to the people of northern Cameroun and around Lake Chad and suggested that cultivars in the savannah belt from the eastern side of Lake Chad through the southern half of the Sudan were probably growing caudatum sorghum by about AD 1000 and in a belt between 9°50 and 12° N. That must surely be correct. The arrival of caudatums much further west in Africa could be relatively recent and they are still spreading southwards into the guinea zone. Essentially, this race occupies the belt across West Africa between the durra and the guinea races.

On the eastern side, the Chari–Nile speakers carried these sorghums down to Lake Victoria. The Karamajong still use them in the traditional way, but some of the Luo in Kenya are now settled agriculturalists, growing these sorghums in addition to other types. Further south, some of these caudatums were moved through into Tanzania, where the Wasukuma use them on heavy land in the quelea bird areas of the eastern Shinyanga district. The Wasukuma agriculture contains an important component of Ethiopian agricultural methodology. This is particularly true of the peoples on Ukara Island in Lake Victoria. Caudatums were well represented in Snowden's material from Tanzania. More recently, on two collecting trips in the Dar-es-Salaam–Dodoma–Mwanza–Musoma areas, Prasada Rao & Mengesha (1979) collected 154 cultivated sorghums which contained 27 caudatums, 30 durra-caudatums and 5 guineas. High-altitude durra-caudatums are found in western Uganda, Rwanda and Burundi and have probably been there for a considerable time.

Kafir sorghum

Harlan & de Wet (1972) took Snowden's *S. coriaceum* and *S. caffrorum* as their race kafir, which gives a rather distinct group occupying the region from Tanzania to southern Africa. Snowden reported kafir collections from Tanzania, Zambia, Zimbabwe, Angola and South Africa. In the Transvaal, there are some excellent kafir cultivars growing under high-altitude conditons which set seed equally successfully when planted in the highlands of Ethiopia. The sorghum crop is deeply involved in many of the traditional ceremonies of the Wasukuma and other Bantu tribes.

Race kafir spread neither to West Africa nor to India and it does not seem to have reached either Ethiopia or the Sudan. This race is associated with the Bantu and their spread into southern Africa.

Spread of the Ethiopian crops and technology

Mid-African drainage basin

Sutton (1974) presented evidence that, between the ninth and third millennia BC, the wetter conditions prevailing resulted in the water levels in the lakes and rivers being high with some of the internal basins temporarily linked, especially in the 'Middle African' belt. From western Ethiopia, the all-important idea of agriculture, together with the early cultivated sorghums, moved along the margin of the aquatic culture towards the west, no doubt often used in the decrue agriculture which is still practised along the rivers of south-west Ethiopia as they enter the plains.

Eastern and southern Africa

Crop movement through eastern and southern Africa was the result of human migrations. The Konso are the inheritors of the crops and technologies of early agriculturalists. These latter were under relentless pressure and met the challenges through skill and sheer hard work. Gebrekidan (1970) wrote: 'The Konso are probably one of the hardest, if not the hardest, working people in Ethiopia.' This must also have been true of the early agriculturalists. They developed new crops and intensified their production methods, yet population pressures on the land grew. Eventually, the point was reached when some of the people had to move. These people were highly professional agriculturalists whose survival depended on hard work and 'getting it right'. They had their selected seeds and their agricultural methodology – high-yielding varieties and a package of practices. When forced to move, they took their seeds, their animals and their knowledge with them. They looked for sites where familiar types of agriculture could be practised.

Such movements were probably made by small groups of people who colonized suitable hills and established scattered nuclei of developed agriculture. As these settlements in turn became overcrowded, some were forced to move on, so that those further afield were the most recently settled. Agriculture was new and populations were sparse.

East Africa

The Cushitic and Omotic speakers of south-western and southern Ethiopia expanded into East Africa, moving initially down the Rift Valley from Lake Turkana, and could well have come from the Kefa, Gamo Gofa and Sidamo areas. They were followed by other waves of immigrants from southern Ethiopia, moving through the mountains of western Kenya and adjacent Tanzania, bounded by Mounts Kenya and Kilimanjaro on the east, Mount Elgon and Lake Victoria on the west and Lakes Eyasi and Manyara to the south. This was not an immigration of arable agriculturalists on a broad front. Much of East Africa was livestock country and many of the immigrants were pastoralists. The arable agriculturalists, with their mixed farming, would have consisted of small groups who colonized scattered defensive positions on hills, where they established their terraces and settled agriculture. Doubtless they influenced the indigenous inhabitants in due course. Dates on pottery at several sites from West Kilimanjaro go back to 3000 BC. Phillipson (1977b) wrote: 'It is tempting to suggest that a gradual spread of Cushitic speakers . . . began at some time.'

Movement southwards

It is likely that the movement south was continued by small groups of skilled agriculturalists, carrying the Cushitic/Omotic traditions. Some of the terraced sites appear to have been established initially in the Stone Age. There are undressed stone terraces at Inyanga, Zimbabwe, but no dates. Gramly (1979) argued that crop production must have preceded iron-working, to provide the profit motive for the ironsmiths. Craftsmen need a fair return for their products. An initial immigration of small groups of professional agriculturalists, with their crop cultivars and their technology, followed by the spread of agriculture to the surrounding people and the incoming Bantu, represents the probable prelude to the Iron Age in southern Africa.

Gramly (1979) suggested that the iron technology carried by groups of blacksmiths moved independently of any movement of pottery or crops. The smiths probably did contract work at various sites with the local agriculturalist, buying hoes and similar tools from them. Better tools would have increased the area under cultivation. Much more remains to be discovered; but clearly race kafir spread into suitable agricultural areas as the cultivating population of southern Africa increased. Introgression with the local wild germplasm must have been occurring all the time.

One of the marker crops of this movement was the yellow-flowered niger seed or noog (*Guizotia*), which occurs sporadically from Ethiopia to Malawi. Niger seed is rapidly disappearing south of Ethiopia. Finger millet was another crop of the complex carried south.

The Sudan

The Nuba Mountains were probably occupied early on. Settled agriculture there is old, with terracing on the steeper slopes and animal manure with crop residues in use on the fields. In the hills, the Nuba have adapted the natural defences of their rocky outposts by building houses in remote areas and surrounding them with stone walls. From a distance, southern Nuba hill communities resemble medieval European castle fortresses built of stone. Cultivated sorghums show much diversity and the crop is ancient there, as is sesame (Bedigian & Harlan, 1983). The same is likely to be true for other suitable hill sites in the Sudan.

The passage to India

The movement of agriculture to Arabia must have occurred at an early date. The old terraced agriculture, using the same crops, is still practised there. The overland routes from Saba into Asia are very ancient and are based on camel caravans. Imports from East Africa and Somalia usually came to India through Aden. Sea traffic was also important. There is evidence of sea trade between the Kulli culture of South Baluchistan and early dynastic Sumer soon after 2800 BC. The port of Dilmum (probably Bahrain Island) traded extensively along the coast, possibly as far as Lothal in the Gulf of Cambay. Before the turn of the era, there was trade between the port of Dhufur in Saba, and India. Frankincense and myrrh from Saba were highly valued and spices from India were prized in Arabia and west Asian countries.

Agriculture in the Indian sub-continent

The Great Indian Desert forms a barrier to movement into India from the west. The route to the north of this desert goes from the North-West Frontier across northern Punjab to the Ganges and Jamuna rivers. That to the south skirts between the desert and the Rann of Kutch and links Sind to Gujarat, Malwa, southern Rajasthan, Maharashtra and peninsular India. There were important ports for the coastal traffic at Lothal and Rangpur in the Gulf of Cambay.

Agriculture moved into the north from Iran and was based on the west Asian crops, with the addition of cotton. The first settled agricultural communities date to the period 8000–5000 BC.

Peninsular India

In the southern Deccan there was an independent indigenous culture prior to 3000 BC and so contemporary with the urban phase of the Harappan civilization. A neolithic culture developed in Karnataka, the first phase of which dated to the period *ca.* 3000–2000 BC. From then on, more permanent settlements were discovered, often located on the crowns and slopes of granitic hills, dated between *ca.* 2100 and *ca.* 1700 BC. There was a terraced agriculture, with dry stone retaining walls for the terraces. Kodekal and Utnur are representative of the first period. The people there had domesticated cattle with sheep and goats. Many rubbing stones and querns were found, indicating either seed collecting or grain cultivation. The second period occurs at Utnur, but more information has come from Piklihal and Hallur, dating from *ca.* 2500 BC and *ca.* 2200 BC, respectively, and continuing until the early Iron Age. In the second period, circular hutments of daub and wattle on a wooden frame were found, with mud floors. Tool types in the third period are reminiscent of those found in the Banas culture at the Malwa and Maharashtra sites. Grinding stones were found, with mullers, as well as large pots, buried up to their necks, which probably served as storage jars. Cattle raising continued and crops grown probably included finger millet. This was identified in Karnataka at Tekkalaksta I. The dates lie between 2100 BC and 1500 BC (Allchin & Allchin, 1982).

The ancestry of the earliest settlements of the southern Deccan was independent of those further north. They date to the end of the third millennium BC and a range of food grains was cultivated. Allchin & Allchin (1982) noted that local variations in grain utilization at the present day were already reflected during the Neolithic–Chalcolithic period. They wrote: 'It is difficult to believe that the Dravidian languages do not owe their origin to the same people who produced the neolithic culture there.' Certainly today finger millet is strongly associated with the Dravidian speakers. Sankalia, Deo & Ansari

(1971) and Dhavalikar (1979) postulated four claimants as the originators of these cultures: (a) immigrants from west Asia or Iran; (b) aboriginal tribes who were chased up into the hills by the Aryans *ca.* 1000 BC; (c) unknown indigenous people who merged completely with the Sanskrit peoples; (d) a primitive indigenous people from western Asia who developed these early farming communities. To this list might be added the possibility of immigrant agriculturalists from the Horn of Africa, carrying the seeds of their crops with them. Southern and central Indian agriculture has a strong component of the Ethiopian crops – finger millet, sorghum, niger and cowpea. Harlan (1969) noted the resemblance between the Ethiopian emmer wheats and the southern Indian 'Khapil' wheat. Both have more than two vascular bundles in the coleoptile. Niger is widely grown among the Tribals; there were 427 different tribes in India in 1961, with 255 languages. Not all are primitive. Some use a terraced agriculture with finger millet and niger seed as major crops. Sesame might well have arrived in the south from Ethiopia, rather than from west Asia.

The following comment of Seegeler (1983) is worth mentioning: 'It is curious to note an old tradition of the region of Wolcait reported by Baldrati (1950). According to this legend, an Ethiopian queen had occupied a vast territory in India in the very remote past. She made groups of Ethiopians emigrate to India. How far this story has historical background is unknown, but it is striking that there are so many similarities between the crops of the traditional agriculture in Ethiopia and India, and that there are groups of Jaferbad in Kathiawar who consider themselves to be of Ethiopian origin.'

One of the oldest of the Afroasian language group's crops must surely be sesame. Bedigian & Harlan (1983) recorded the ritual use of this crop for births, marriages and deaths among the peoples of the Nuba Mountains in the Sudan. The same is true for the speakers of the Chadic languages in West Africa, for the peoples of Nepal, and for the Tamils of southern India, where, to this day, sesame seeds are placed in the mouths of the dead.

References

Allchin, B. & Allchin, R. (1982). *The Rise of Civilisation in India and Pakistan.* Cambridge University Press, Cambridge.

Arkell, A. J. & Ucko, P. J. (1965). Review of predynastic development in the Nile Valley. *Current Anthropology*, **6**, 145–56.

Bedigian, D. & Harlan, J. R. (1983). Nuba agriculture and ethnobotany, with particular reference to sesame and sorghum. *Economic Botany*, **37**, 384–95.

Clark, J. D. & Stemler, A. B. L. (1975). Early domesticated sorghum from central Sudan. *Nature (London)*, **254**, 588–91.

Cleuziou, S. & Constantini, L. (1982). A l'origine des oasis. *La Recherche*, **13**, 1181–9.

de Wet, J. M. J., Harlan, J. R. & Price, E. G. (1970). Origin of variability in the *spontanea* complex of *Sorghum bicolor*. *American Journal of Botany*, **57**, 704–7.

de Wet, J. M. J., Prasada Rao, K. E., Brink, D. E. & Mengesha, M. H. (1984). Systematics and evolution of *Eleusine coracana*. *American Journal of Botany*, **73**, 550–62.

Dhavalikar, M. K. (1979). Early farming cultures of Central India and early farming cultures in the Sudan. *In*: D. P. Agrawal and D. K. Chakrabarti (eds), *Essays in Indian Protohistory*. B.R. Publishing Corporation, New Delhi, pp. 229–247.

Doggett, H. (1965). The development of cultivated sorghums. *In*: Sir J. B. Hutchinson (ed.), *Essays on Crop Plant Evolution*. Cambridge University Press, Cambridge, pp. 50–69.

Ehret, C. (1979). On the antiquity of agriculture in Ethiopia. *Journal of African History*, **20**, 161–72.

Fairservis, W. A., Jr (1971). *The Root of Ancient India*. Allen and Unwin, London.

Gebrekidan, B. (1970). A report on crop collecting trip to Konso wereda in Gamo-Gofa, 24 Dec. 1969–2 Jan. 1970. Addis Ababa University, Addis Ababa (mimeographed).

Gramly, R. M. (1979). Expansion of Bantu speakers versus development of Bantu language and African culture *in situ*: an archaeologist's perspective. *South African Archaeological Bulletin*, **33**, 107–16.

Hallpike, C. R. (1970). Konso agriculture. *Journal of Ethiopian Studies*, **8**, 31–43.

Hallpike, C. R. (1972). *The Konso of Ethiopia*. Oxford University Press, Oxford.

Harlan, J. R. (1969). Ethiopia: a centre of diversity. *Economic Botany*, **23**, 309–14.

Harlan, J. R. & de Wet, J. M. J. (1972). A simplified classification of cultivated sorghum. *Crop Science*, **12**, 172–6.

Harlan, J. R. & Stemler, A. B. L. (1976). The races of sorghum in Africa. *In*: J. R. Harlan, J. M. J. de Wet and A. B. L. Stemler (eds), *Origins of African Plant Domestication*. Mouton Publishers, The Hague, pp. 465–78.

Helbaek, H. (1960). The paleoethnobotany of the Near East and Europe. *In*: R. J. Braidwood and B. Howe (eds), *Prehistoric Investigations in Iraqi Kurdestan. Studies in Ancient Oriental Civilization*, **31**, 99–118. University of Chicago Press.

Helbaek, H. (1966). Commentary on the phylogenesis of *Triticum* and *Hordeum*. *Economic Botany*, **20**, 350–60.

Hiernaux, J. (1974). *The People of Africa*. Weidenfeld and Nicolson, London.

Hilu, K. W., de Wet, J. M. J. & Harlan, J. R. (1979). Archaeobotanical Studies of *Eleusine coracana* spp. *coracana*. *American Journal of Botany*, **66**, 330–3.

Kajale, M. D. (1977). On the botanical findings from excavations at Daimabad, a chalcolithic site in western Maharashtra, India. *Current Science*, **46**, 818–27.

Klichowska, M. (1984). Plants of the neolithic Kadero (central Sudan). *In*: L. Krzyzaniak and M. Kobusiewicz (eds), *Origin and Early Development of the Food-producing Cultures in North-eastern Africa*. Polish Academy of Sciences, Poznamm, pp. 321–40.

Krzyzaniak, L. (1978). New light on early food production in central Sudan. *Journal of African History*, **19**, 159–72.

Krzyzaniak, L. (1984). The neolithic habitation at Kadero (central Sudan). *In*: L. Krzyzaniak and M. Kobusiewicz (eds), *Origin and Early Development of the Food-producing Cultures in North-eastern Africa*. Polish Academy of Sciences, Poznamm, pp. 309–20.

Mann, J. A., Kimber, C. T. & Miller, F. R. (1983). The origin and early cultivation of sorghums in Africa. *Bulletin 1454*, Texas A&M Agricultural Experimental Station, Texas.

Mehra, K. L. (1962). Natural hybridization between *Eleusine coracana* and *E. africana* in Uganda. *Journal of the Indian Botanical Society*, **41**, 531–9.

Phillipson, D. W. (1977a). The excavation of Gobedra Rock-shelter, Axum. *Azania* **12**, 53–82.

Phillipson, D. W. (1977b). *The Later Prehistory of Eastern and Southern Africa*. Heinemann, London.

Prasada Rao, K. E. & Mengesha, M. H. (1979). Sorghum and millet germplasm collecting in Tanzania. *Genetic resources progress report 9*, ICRISAT, Hyderabad.

Sankalia, H. D., Deo, S. B. & Ansari, Z. D. (1971). *Chalcolithic Navdatoli*. Deccan College Post-graduate and Research Institute, Poona.

Seegeler, C. J. P. (1983). *Oil Plants in Ethiopia, their Taxonomy and Agricultural Significance*. PUDOC, Wageningen.

Stemler, A. B. L., Harlan, J. R. & de Wet, J. M. J. (1975). Caudatum sorghums and speakers of Chari–Nile languages in Africa. *Journal of African History*, **16**, 161–9.

Sutton, J. E. G. (1974). The aquatic civilisation of Middle Africa. *Journal of African History*, **15**, 527–46.

Wendorf, F. & Schild, R. (1984). Some implications of late palaeolithic cereals; exploitation in Upper Egypt at Wadi Kubbaniya. *In*: L. Krzyzaniak and M. Kobusiewicz (eds), *Origin and Early Development of Food-producing Cultures in North-eastern Africa*. Polish Academy of Sciences, Poznamm.

Westphal, E. (1975). *Agricultural Systems in Ethiopia*. PUDOC, Wageningen.

11

Prehistoric Ethiopia and India: contacts through sorghum and millet genetic resources

K. L. MEHRA

Introduction

Several publications have dealt with India's cultural contacts with western, central and south-east Asian countries, but little information is available on India's contacts with African countries (Asthana, 1976). This is understandable because little archaeological work has been done on Neolithic to Iron Age sites in Africa, compared with Asia. Even in India, where several Neolithic–Chalcolithic sites have been excavated, archaeologists have continued to look for some kind of west Asian similarity/influence in interpreting their findings. Thus, even (a) the finds of human skeletons showing Hamitic-negroid features associated with the Langhanag (Gujarat) microlithic culture (Sankalia, 1962); (b) terracotta head-rests discovered in Neolithic burials at Narsipur (*ca.* 1800 BC), Hammige, Hallur (*ca.* 1800 BC) and Paklihal in the Kaveri and Krishna basins, showing affinity with similar objects found in Africa and Egypt (Nagarajarao, 1975); and (c) archaeological finds of African crop plants (Vishnu-Mittre & Savithri, 1982) have been ignored. Evidences for indigenous origin(s) of few, or even several, Neolithic–Chalcolithic cultures of India have been recently discussed but with bitter controversy (Possehl, 1982). African millets were incorporated into the cropping system of Chalcolithic farming communities of India, and these may provide evidence of contacts between India and Ethiopia where agriculture was practised (*ca.* third millennium BC). This chapter deals with recently published archaeobotanical findings pertaining to Ethiopian cultural contacts with prehistoric India through the contribution of millet genetic resources. Some aspects of origin, domestication and evolution of millets in Africa and the impact of millet introduction on the agricultural history of India are also highlighted.

Origin, domestication and evolution

Finger millet

Two main groups of cultivars are recognized: (a) the African highland type, which originated in Africa from its wild progenitor, *Eleusine africana*, through selection for plump grains and non-shattering habit, is adapted for cultivation in the highlands, and has long spikelets, long glumes and grains enclosed within the florets; (b) the Afro-Asian type, which is adapted for growing in the lowlands, has short spikelets and short glumes in which the mature grains are exposed distally, and which evolved from the highland type through selection of plants with short glumes and exposed grains. This view is supported by the genetics of glume length. Glume length factors G-1, G-2 and G-3 are responsible for glume length, with any one or no factor producing long, any two factors expressing medium-long and all three factors together producing short glumes (Ayyanger & Warrier, 1936). The African highland type is split into the races *coracana* (most primitive), *elongata, plana* and *compacta* (de Wet *et al.*, 1984), the last three races being products of natural and human selection under domestication, to suit the crop to diverse farming systems and ethnic preferences. The Afro-Asiatic type is also designated as race *vulgaris* (de Wet *et al.*, 1984). All these races occur at present in Africa and in India. A distinct East Indian group is also recognized (Hilu & de Wet, 1976) and clinal variation occurs among cultivars in India, with extremes showing a decrease in south Indian influence and an increase in eastern Indian influence (Hussaini, Goodman & Timopthy, 1977). Indian cultivars also show variation in number of days to maturity and plant height in relation to increase in latitude, longitude and altitude, with north-east Indian cultivars being short statured and of late maturity (Kempanna, 1975).

Introgressive hybridization occurs between *E. africana* and all races of cultivated finger millet (African highland and Afro-Asiatic types) and also between races, wherever sympatric, producing rich variation, in the highlands of Ethiopia southwards to Zimbabwe (Mehra, 1962; de Wet *et al.*, 1984; Apparao & Mushonga, 1987). Stabilized weedy derivatives of introgressive hybridization between cultivated and wild types resemble cultivated kinds, except being free-shattering, and occur in disturbed habitats and cultivated fields. Such hybrids involving *E. africana* have not so far been reported from India. In crosses between long glume × long glume types, some hybrids are likely to have one or two glume length factors but such medium-long types (with two factors) are likely to segregate or, when crossed with long-glumed *E. africana*, would again produce the long glume type. Thus, evolution of the short glume Afro-Asiatic type would take a longer time in the African highlands in the presence of *E. africana* as compared with the time it would have taken in India, from where this progenitor has not been reported. It seems more likely that the major part of the evolution of the Afro-Asian type occurred in India from where it was taken to East Africa along with durra sorghums during the Islamic expansions. Thus, India first received the African highland type from Ethiopia, and from this the Afro-Asiatic type was developed in India.

The Ethiopian highlands (Mehra, 1963a,b), the eastern Sudan zone, the highlands stretching from Ethiopia to Uganda (Harlan, 1971) and the highlands of East Africa (Hilu & de Wet, 1976) have been suggested as likely areas of finger millet domestication. Cultivated finger millet (race *plana* of the African highland type) has been identified from archaeological remains found in Axum, in the northern highlands of Ethiopia (*ca.* third millennium BC: Hilu, de Wet & Harlan, 1979) and in Kadero, Sudan (*ca.* second half of the fourth millennium BC: Klichowska, 1984). Since the sample belonged to race *plana*, an advanced race, the domestication of finger millet must have occurred much earlier (de Wet *et al.*, 1984). Finds of finger millet in India have been reported from Hallur, Karnataka (*ca.* 1800 BC); Paiyampalli (*ca.* 1400 BC), Eastern Ghats; Daimabad (*ca.* 1400–1100 BC, topmost Jorwe Ware) and Songaon (wild, 1290 BC) in Maharashtra; and Surkotda, a Harappan site (*ca.* 1660 BC), associated with sherds of the white-painted black and red Ware of the Ahar I culture (Kajale, 1974, 1977; Allchin & Allchin, 1982; Vishnu-Mittre & Savithri, 1982). Thus finger millet, of the African highland type, seems to have been domesticated in the fourth millennium BC in the Sudan–Ethiopian region and it began to be grown in India from about 4000 BP.

Pearl millet

Pearl millet (*Pennisetum americanum* subsp. *americanum*) was domesticated, in a diffuse belt extending from Sudan to Senegal (Harlan, 1971), from its wild progenitor subsp. *monodii*, with which it produces a weedy derivative subsp. *stenostachyum*, in western Sudan, northern Nigeria and western Senegal (Brunken, 1977; Brunken, de Wet & Harlan, 1977; Marchais & Tostain, 1985). Four distinct races, *typhoides*, *nigritarum*, *globosum* and *leonis* are known, but only race *typhoides* was introduced into India. Racial variation in West Africa is due mainly to incorporation of genes from closely related wild taxa with which pearl millet hybridizes (Billiard *et al.*, 1980), but in the absence of wild relatives in India such a system does not operate. Variation among populations in India is mainly due to natural and human selection under domestication. Thus Indian collections are much less variable than African.

Finds (charred lumps) of pearl millet from archaeological sites in India are from Rangpur III (1700–1400 BC), Saurashtra, associated with Lustrous Red Ware (Ghosh & Lal, 1963; Rao, 1963). It seems to have been introduced into India around 4000 BP.

Sorghum

Harlan (1971) proposed that domestication of sorghum took place first along the broad band of savannah between the Sudan and Nigeria. The first domesticated sorghums were bicolor-like and those originated from its progenitor ssp. *arundinaceum* race *verticilliflorum*. From this region sorghum culture spread to tropical West Africa where race guinea developed; to Southern Africa where race kafir developed; to the Sudan–Uganda area where race caudatum developed; and to Asia and Ethiopia where race durra developed, respectively, through hybridization with different close relatives. Durra sorghums probably developed outside Africa from the race bicolor that was introduced into Sind, Punjab and North-West India (Stemler *et al.*, 1977) and the developed durras were reintroduced into Ethiopia and Arab colonies in East Africa during the Islamic expansion.

Of these sorghums, bicolors were the first to be incorporated in Indian agriculture, followed probably by guineas, especially *S. roxburghii* adapted to coastal areas. Next to develop were durras, but caudatums and kafir were not brought into India.

Archaeological finds of sorghum in India are from:

1. Ahar, Rajasthan, along with rice (*ca.* 1500 BC: Vishnu-Mittre, 1969), associated with Northern Black Polished Ware;
2. Daimabad, near Pravara river, Maharashtra, late Harappan phase (*ca.* 1800–1500 BC: Vishnu-Mittre & Savithri, 1982), and another sample, topmost Jorwe phase (*ca.* 1400–1100 BC: Kajale, 1977);
3. Inamgaon, near Ghod river, Maharashtra, in the early phase

of habitation (*ca.* 2000–1600 BC, associated with Malwa Ware: Allchin & Allchin, 1982), and another sample associated with Jorwe Ware (*ca.* 1100–800 BC, calibrated by the radio-carbon method: Vishnu-Mittre, 1974), and recent finds of sorghum in Sangrur district, Punjab (2nd millennium BC: Saraswat, 1986). These dates are the latest so far known, but sorghum may have been incorporated into Indian agriculture around 4000 BP.

Mehra (1963b) proposed that finger millet was taken along the Sabaean Lane route from Ethiopia to India during pre-Aryan times. Archaeological finds of sorghum (sherd impressions) at Hili, Abu Dhabi (*ca.* 2700 BC: Cleuziou & Constantini, 1982) and at Pirak, Baluchistan (*ca.* 1900 BC: Constantini, 1979) now confirm that African millets came to India from Africa through the Sabaean Lane–Baluchistan route. Movement through the sea route might also have been possible because finger millet was identified from Hallur and Paiyampalli, pearl millet was cultivated at Rangpur III (1700–1400 BC) and maritime commercial contacts between Gujarat and Mesopotamia may have existed from at least the beginning of the second millennium BC (Oppenheim, 1954; During Caspers, 1971).

Discussion

Before the incorporation of African sorghum and millets into the farming systems of India, the crop-based agricultural economies of different civilizations/cultures of different regions of prehistoric India were as follows:

1. The Harappan civilization (*ca.* 2350–1750 BC) in the plains of the Indus River system was based on wheat, barley, rape, lentil, pea, chickpea, sesame and cotton; while in Gujarat, Saurashtra and the plains of the Yamuna–Ganges rivers (Neolithic–Chalcolithic cultures), rice was cultivated in addition to these crops (Allchin & Allchin, 1982; Vishnu-Mittre & Savithri, 1982).

2. Contemporary with the Harappan civilization but independent of its influence were the Neolithic cultures of Karnataka (3000–1000 BC), whose subsistence economies were based on *Paspalum scorbiculatum* and *Dolichos biflorus* (Vishnu-Mittre, 1969; Kajale, 1977). The discovery of terracotta head-rests from several sites and finds of finger millet are suggestive of African contacts and incorporation of finger millet in the farming system of the region from *ca.* 1800 BC (Nagarajarao, 1975).

3. Neolithic–Chalcolithic cultures (2100–1000 BC), late contemporaries of the Harappan civilization in Rajasthan, Madhya

Pradesh and Maharashtra, cultivated wheat, barley, *Linum*, grass pea, field pea and lentil in winter and rice, green gram, black gram and species of *Dolichos* and *Lathyrus* during the summer rainy season (one or more crops at one site: Vishnu-Mittre & Savithri, 1982). The legumes of Indian origin and African sorghum and millets were added to the cultivated plants grown in India from the earlier period.

The subsistence economies of Rajasthan, Madhya Pradesh and Maharashtra were based mainly on minor millets and agro-pastoral systems. Finds of species of *Panicum*, *Setaria* and *Echinochloa* are from Surkotda, Maharashtra (*ca.* 1660 BC: Vishnu-Mittre & Savithri, 1982), associated with the Ahar culture. When more productive African millets and sorghum became available for cultivation under different rainfall regimes and soil types, they rapidly started replacing the minor millet cultivation, and the subsistence economy gave way to a food surplus producing economy. African millets can be stored for a long period without damage in traditional storage systems. They also produce fodder for cattle besides seed for human consumption. In areas with higher annual rainfall, sorghum and finger millet were sown in the summer rainfall season along with Asiatic Vignas, while in low rainfall areas pearl millet was cultivated. Sorghum and millet culture moved from Rajasthan to Madhya Pradesh, Maharashtra and southwards to Gujarat, associated with Ahar, Malwa and Jorwe cultures and to Gujarat during the late Harappan period. Pearl millet cultivation was taken up in the drylands of Gujarat and Saurashtra during phase III of Rangpur, associated with Lustrous Red Ware (Ghosh & Lal, 1963; Rao, 1963).

African sorghum and millet culture thus played an important role in the agricultural history of India following the opening up of opportunities for rainfed agriculture and mixed (agriculture and animal husbandry) farming systems. This led to a change in the settlement pattern. Instead of urban centres with neighbouring food-producing villages, several small villages began to emerge over a large stretch of the land. The new system progressed rapidly because the centrally controlled production and distribution system, so characteristic of the Indus valley civilization, did not operate. Similarly, the incorporation of pearl millet in the dryland agriculture of Gujarat seems responsible for the sudden increase in the number of settlements during Rangpur phases B and C.

African sorghum and millets further evolved under cultivation and natural selection to suit different ethnic preferences and farming systems, including the cultivation of sorghum and finger millet during the winter months in south India and of finger millet in the hilly areas. But since the wild relatives (progenitors) of sorghum, pearl millet and finger millet, with which these crops hybridize in nature even today occur in Africa and not in India, rich variation in several characters continues to be generated in Africa. Therefore, African genetic resources of all these crops are more variable than those from India. Since the wild relatives of these crops are not found in India, it seems that when these crops were brought into India the seeds of closely related wild/'companion weedy' species did not accompany the initial samples of introduction. These crops provide examples of crop plant evolution in which the evolutionary process has occurred in Africa in the presence of their wild relatives and in India in their absence.

We do not, at present, have any non-biological evidence except the terracotta head-rests for establishing contacts between Africa and India during the period under reference. The archaeobotanical evidence is strong, though it needs to be substantiated by future archaeological findings in East Africa and India. African sorghum and millets increased the agricultural prosperity of prehistoric India and continue to do so even today, for which India is grateful to Africa.

References

Allchin, B. & Allchin, R. (1982). *The Rise of Civilization in India and Pakistan.* Cambridge University Press, Cambridge.

Apparao, S. & Mushonga, J. N. (1987). *A catalogue of passport and characterization data of sorghum, pearl millet and finger millet germplasm from Zimbabwe.* IBPGR, Rome.

Asthana, S. (1976). *History and Archaeology of India's Contacts with Other Countries from Earliest Times to 300* BC. B.R. Publishing Corporation, Delhi.

Ayyanger, G. N. R. & Warrier, U. A. (1936). Inheritance of glume length in Ragi, *E. coracana,* the finger millet. *Madras Agricultural Journal,* **24**, 132–4.

Billiard, J., Guyen, V. E. & Parnes, J. (1980). Analyse des relations génétiques entre formes spontanées et cultivées chez le mil à chandelle. *Annales de l'Amélioration des Plantes,* **30**, 229–51.

Brunken, J. N. (1977). A systematic study of *Pennisetum* Sect. *Pennisetum* (Gramineae). *American Journal of Botany,* **64**, 161–76.

Brunken, J., de Wet, J. M. J. & Harlan, J. R. (1977). The morphology and domestication of pearl millet. *Economic Botany,* **31**, 163–74.

Cleuziou, S. & Constantini, L. (1982). A l'origine des oasis. *La Recherche,* **13**, 1181–9.

Constantini, L. (1979). Plant remains from Pirak. *In*: J.-F. Jarrige and M. Samtoni, *Texte Fouilles de Pirak, vol. 1.* Publications de la Commission des Fouilles Archaeologiques.

de Wet, J. M. J., Prasada Rao, K. E., Brink, D. E. & Mengesha, M. H. (1984). Systematics and evolution of *Eleusine coracana. American Journal of Botany,* **71**, 550–62.

During Caspers, E. C. L. (1971). New archaeological evidence for maritime trade in the Persian Gulf during the late protoliterate period. *East and West*, **1–2**, 21–44.

Ghosh, S. S. & Lal, K. (1963). Plant remains from Rangpur. *In*: S. R. Rao (ed.), *Excavations at Rangpur and Other Explorations in Gujarat*. Ancient India, 18–19, New Delhi, pp. 161–75.

Harlan, J. R. (1971). Agricultural origins: centers and noncenters. *Science*, **174**, 468–74.

Hilu, K. W. & de Wet, J. M. J. (1976). Racial evolution in *Eleusine coracana* ssp. *coracana* (finger millet). *American Journal of Botany*, **63**, 1311–18.

Hilu, K. W., de Wet, J. M. J. & Harlan, J. R. (1979). Archaeobotanical studies of *Eleusine coracana* ssp. *coracana* (finger millet). *American Journal of Botany*, **66**, 330–3.

Hussaini, S. H., Goodman, M. M. & Timopthy, D. H. (1977). Multivariate analysis and the geographic distribution of the world collection of finger millet. *Crop Sciences*, **17**, 257–63.

Jarriage, J. (1982). Excavations at Mehrgarh: Their significance for understanding the background of the Harappan civilization. *In*: G. L. Possehl (ed.), *Harappan Civilization: A Contemporary Perspective*. Oxford & IBH Publishing Co., New Delhi, pp. 79–84.

Kajale, M. D. (1974). Ancient grains from India. *Bulletin of the Deccan College Research Institute*, **34**, 55–74.

Kajale, M. D. (1977). On the botanical findings from excavations at Daimabad, a chalcolithic site in Western Maharashtra, India. *Current Science*, **46**, 330–3.

Kempanna, C. (1975). *Variability Pattern and its Impact on the Structure of Yield in* Ragi. University of Agricultural Science, Bangalore, India.

Klichowska, M. (1984). Plants of the neolithic Kadero (central Sudan). *In*: L. Krzyzaniak and M. Kobusiewicz (eds), *Origin and Early Development of the Food-producing Cultures in North-eastern Africa*. Polish Academy of Sciences, Poznamm, pp. 321–40.

Marchais, L. & Tostain, S. (1985). Genetic divergence between wild and cultivated pearl millet (*Pennisetum typhoides*) II. Characters of domestication. *Zeitschrift für Pflanzenzüchtung*, **95**, 245–61.

Mehra, K. L. (1962). Natural hybridization between *Eleusine coracana* and *E. africana* in Uganda. *Journal of the Indian Botanical Society*, **41**, 531–9.

Mehra, K. L. (1963a). Differentiation of the cultivated and wild *Eleusine* species. *Phyton*, **20**, 189–98.

Mehra, K. L. (1963b). Considerations on the African origin of *Eleusine coracana* (L.) Gaertn. *Current Science*, **32**, 300–1.

Nagarajarao (1975). Ancient links between India and Africa. *Sunday Standard* (Magazine section), 5 January 1975. New Delhi, p. 1.

Oppenheim, A. L. (1954). Sea faring merchants of Ur. *Journal of the American Oriental Society*, **74**, 6–17.

Possehl, G. L. (ed.) (1982). *Harappan Civilization: a Contemporary Perspective*. Oxford & IBH Publishing Co., New Delhi.

Rao, S. R. (ed.) (1963). *Excavations at Rangpur and Other Explorations in Gujarat*. Ancient India 18–19, New Delhi, pp. 5–207.

Sankalia, H. D. (1962). From food production to urbanization in India: *In*: T. N. Madan and Gopala Sarang (eds), *Indian Anthropology, Essays in Memory of D. N. Majumdar*. Asia Publishing House, Bombay, pp. 66–104.

Saraswat, K. S. (1986). Ancient plant economy of Harappans from Rohira, Punjab (*c*. 200–1700 BC). *The Palaeobotanist*, **35**, 32–8.

Stemler, A. B. L., Harlan, J. R. & de Wet, J. M. J. (1977). The sorghums of Ethiopia. *Economic Botany*, **31**, 446–60.

Vishnu-Mittre (1969). Remains of rice and millet. *In*: H. D. Sankalia, S. B. Deo and Z. D. Ansari (eds), *Excavations at Ahar (Tambavati), Poona*. Deccan College Post-graduate and Research Institute, Poona.

Vishnu-Mittre (1974). The beginning of agriculture, palaeobotanical evidence in India. *In*: J. B. Hutchinson (ed.), *Evolutionary Studies in World Crops: Diversity and Change in the Indian Sub-continent*. Cambridge University Press, Cambridge, pp. 3–30.

Vishnu-Mittre & Savithri, R. (1982). Food economy of the Harappans. *In*: G. L. Possehl (ed.), *Harappan Civilization: A Contemporary Perspective*. Oxford & IBH Publishing Co., New Delhi, pp. 205–27.

12

Konso agriculture and its plant genetic resources

J. M. M. ENGELS AND E. GOETTSCH

Introduction

Konso is the name of a relatively small area (approximately 500 sq km) situated in south-west Ethiopia at a latitude of 5°15′ N and a longitude of 37°30′ E, which is populated mainly by the Konso people. The topography is characterized by rugged and stony highlands, cut by deep valleys that enter into the heart of the country. The main agricultural area ranges in altitude from 1400 to 2000 m above sea level and the climate is of the dry montane type with temperatures ranging from below 15 °C at night to 32 °C during the day at the hottest time of the year. The Konso Highlands run across the Rift Valley in an east–west direction and are situated in the dry belt of Ethiopia with an unreliable rainfall not exceeding 800 mm per year. There are two rainy seasons: the big rains are concentrated in March and April and the small rains fall around October and November. In general, the rains come in the form of violent thunderstorms which seldom last more than two hours (Hallpike, 1972). The Sagan River forms the eastern and southern borders of Konso, while to the north the great plains of Gomida and Lake Shamo and, more to the west, the Gidole mountains and the Woito Valley form natural boundaries.

The Konso are a small tribe of about 60 000 people (Minker, 1986). Their language belongs to the East Cushitic group (Hallpike, 1970). The Konso have evolved their remarkable (agri)culture in a high degree of isolation during the many centuries they have occupied the area. Their neighbours are mainly pastoralists (e.g. the Borana in the south) or agriculturalists (among others the Gauwada tribe in the west) and most of them belong to the Oromos. 'The Konso are markedly shorter and more negroid than the neighbouring Borana. They are clearly an amalgam, both physically and culturally, in which other stocks than Galla are represented' (Hallpike, 1972).

The Konso live in densely populated towns, each inhabited by an average of some 1500 people (Minker, 1986). These towns are surrounded by striking stone walls, with narrow corridors connecting the walled or fenced homesteads. Within a homestead one can normally find the main round hut, the kitchen, the grinding house, sleeping huts, stores and one or more stables, all located close together in order to leave some space for the gardens. Almost all towns and villages are situated on mountain ridges or on steep slopes and therefore do not have water. The 6–8 m deep wells are found in the deep gorges next to the dry riverside and are carefully protected against floods or dirty river water. During the dry season the women and girls are frequently occupied for several hours per day in order to obtain sufficient water (Kuls, 1958).

The agricultural system

The soils in Konso are, in general, of volcanic origin and in certain parts of the country basalt and tuff layers 100 m or more thick can be found on top of crystalline formations. The terrain is extremely mountainous and stony and the paucity of rainfall makes water and soil conservation of prime importance. The visitor quickly realizes that the concern for sufficient water for people, cattle and crops has a dominant place in the daily life of the Konso, and this can also be frequently observed in the landscaping. The soil is preserved by the construction of stone terraces, hundreds of kilometres long and often several metres high. Because of the steep slopes, the terraces are generally narrow, only a few metres wide. The dry-stone walls are built along the contour lines of the hillsides and their principal function is to prevent the rainstorms from washing away the soil and crops. At the same time they assure an adequate supply of water for the crops by retaining the water within the terraces. In the flatter areas and in the bigger terraces the land is subdivided into big plots, up to 9 sq m, each surrounded by 10–20 cm high earth walls, frequently strengthened with sorghum straw in areas where stone is short (Kuls, 1958). In some parts another practice can be seen where, in the middle of a terrace, a reinforced hollow some 150 cm wide is used to concentrate the meagre water supply. The water is distributed through carefully constructed channels and irrigates fields which are sometimes 100 sq m or more in size. Such terraces are separated from each other by carefully constructed stone walls up to 6 m high. The water inlet from the river-bed can be regulated according to the needs of the crop plants. Because of this long lasting and careful management the soil of the irrigated terraces is generally very fertile.

One of the striking features of Konso agriculture is the use of manure, of both animal and, less frequently, human origin (Hallpike, 1972). The connection between animal husbandry (cattle, goats and sheep) and intensive agriculture is typical of the Konso. Dung is never used as fuel – as can be observed in many other parts of Ethiopia – and the practice of stall-feeding (unknown elsewhere in Ethiopia) all year around is a valuable source of manure. Normally, cattle are kept on pastures near the periphery of the villages from where the dung is collected then left to rot in heaps or pits, together with other organic wastes.

Nearly all the land surrounding the villages is permanently cultivated, the terraces are richly manured and only a few pastures are found. Manure is applied not only before sowing, but also during the growing period of the crops. Since many crop species are interplanted in the same field, crop rotation is not necessary. Only in the more remote fields where manure is not regularly applied can fallow land be observed. Because of the terraces and, sometimes, the very steep slopes, the double-bladed hoe and the less important digging stick are the most commonly used implements. Ploughing has only recently been introduced and is not a common practice.

The agricultural activities are determined by the rainfall pattern. Field preparation starts at the onset of the first rains in January/February and, in general, the fields are worked only once before sowing (Westphal, 1975). Sowing starts as soon as the big rains begin. The seeds of the cereals and pulses are mixed, broadcast and lightly covered with soil. Root and tuber crops as well as cotton seeds are planted earlier and are carefully protected by soil to prevent them being eaten by birds. After planting or sowing there is a laborious period of weeding, and bird and animal scaring as well as protection against insects. From May onwards the various crops are harvested: first the roots and tubers, followed by the cereals and pulses and finally, in mid-September, the sorghum. If sufficient small rains fall, mainly during October and November, a second sorghum crop can be harvested in December/January from the ratooned sprouts of the first season's plants (Kuls, 1958; Hallpike, 1970). From time to time, crop failures are caused by drought; famine is the unavoidable consequence, as was observed in 1984 and 1985. However, because of the extremely high number of food crop species – domesticated, semi-domesticated or wild – even in times of severe drought at least some food will be available (e.g. Araceae-tubers).

The harvested crops are carefully stored in granaries or kept hanging in the huts. Hallpike (1970) reported that the ashes of burnt cow dung were used to protect whole sorghum heads against insects in the granary. The tubers are frequently left in the ground until they are used.

One important factor for each farmer is that he possesses a piece of land situated in each of the areas having a different soil type, e.g. in an irrigable area near to the village, and in a more remote area for the production of cotton. Kuls (1958) reported an average of at least 10 different fields per farmer, each of them seldom bigger than a quarter of a hectare.

Plant genetic resources and their uses

The most striking feature of Konso agriculture is the high number of plant species used and the way they are intercropped (see Table 1). By integrating multi-purpose trees into this system the Konso have created an indigenous type of agroforestry, well adapted to the prevailing dry conditions. Hardly any unused piece of land will be found around the villages and, throughout almost the whole year, the soil is covered with crops, crop residues, stones, trees such as *Moringa*, *Terminalia* and *Balanites* spp., or shrubs (e.g. pigeon pea, cotton, coffee and yams). Both trees and shrubs are important in preventing soil erosion.

Some of these plant species demand a careful management (e.g. *Sorghum* and the cabbage tree, *Moringa stenopetala*), whereas others are wild plants which are tolerated and harvested only in abnormal years (e.g. some Araceae species). Others are used for medical and/or ritual purposes. Some of the crop plants in Konso have been grown since ancient times, e.g. sorghum, Araceae, cabbage tree, cotton and coffee. The Konso believe that God has given these traditional plants to their ancestors in the mythological past. In the meantime the Konso have adapted many crop species from outside. During the Amharic occupation by the end of the last century important species such as maize, potato, sweet potato, teff, linseed, pigeon pea, onion, garlic, chilli pepper, lemon and papaya were introduced. During the Second World War the Italians introduced, for example, sunflower and orange. Missionaries and emigrated members of the Konso tribe

Table 1. *Plant genetic resources found or reported to be used in Konso*[a]

Scientific name	Local name	Common name or family name	Cultivation[b]	Use and remarks
Cereals				
Amaranthus caudatus	Pasa	Amaranth	(1), 2	Seeds used for the preparation of 'dama,[c] beer. Use of leaves as a vegetable
Eleusine coracana	Pareja	Finger millet	1, (2)	Beer, soup, unleavened bread
Eragrostis tef	Kajeta	Teff	1	Cash crop, rare
Hordeum vulgare	Boita, poorta	Barley	1	'Dama, beer, soup, unleavened bread, seeds sometimes roasted
Sorghum bicolor	General name: unta (= cereal). There are many different names according to variety	Sorghum, millet	1, (2)	Staple food ('dama), beer, soup, unleavened bread. Stalks used as fodder; an important fuel
Triticum durum	Kaba, kapa	Wheat	1	'Dama, beer, soup, unleavened bread, seeds sometimes roasted
Zea mays	Paza, pogoloda	Maize	1, (2)	Staple food, ('dama), beer, soup, unleavened bread. Stalks used as fodder; an important fuel
Pulses				
Cajanus cajan	Ashakilta, ohota faranjeta	Pigeon pea	1	Seeds boiled or eaten raw

Table 1 (*cont.*)

Scientific name	Local name	Common name or family name	Cultivation[b]	Use and remarks
Cicer arietinum	Sumpura	Chickpea	1, 2	Boiled or eaten raw. Ground for flour in times of famine
Lablab purpureus	Okala	Hyacinth bean	1, 2	Boiled or eaten raw
Lens culinaris	Sirota	Lentil	1	Boiled
Phaseolus lunatus	Bapello	Lima bean	1	Seeds boiled or eaten raw when young, ground for flour in times of famine
P. vulgaris	Alkoka	Kidney bean	1, 2	Seeds boiled or eaten green. Ground for flour in times of famine
Pisum sativum	Atara	Pea	1, 2	Boiled or eaten raw
Vicia faba	Bakala	Horse bean	1	Seeds boiled or eaten green. Ground for flour in times of famine
Vigna unguiculata	Ohota, okala	Cow pea	1	Boiled or eaten raw
NN	Neeqayta	—	1, 2	Small green bean
Tubers				
Amorphophallus abyssinicus	Saganeida		1, (2)	Used to make 'dama in times of famine since these tubers can survive considerable drought. Also used to make beer
Arisaema sp.	Burie, lameeta pakana	Indian turnip	1, (2)	Contain a poisonous substance which has to be eliminated during preparation

Sauromatum nubicum	Pansala		1, (2)	
Colocasia esculenta	Longa	Taro	1	Tubers used to prepare 'dama. Leaves used to flavour beer
Dioscorea abyssinica	Hidana	Yam	2, 3, 5	Tubers boiled and eaten or pounded and made into 'dama
Ipomoea batatas	Dinitscha faranjeta	Sweet potato	2	Tubers boiled and eaten
Manihot esculenta	–	Cassava	1, 2	Tubers boiled and eaten
Solanum tuberosum	Tinassa	Irish potato	1, 2	Tubers boiled and eaten
Vegetables				
Adenia ellenbeckii	Kaguta	Passifloraceae	4(?)	Leaves used as a vegetable
Allium cepa	Tuma tima	Onion	2	Bulbs used as a vegetable
A. sativum	Tuma ata	Garlic	2	Bulbs used as a condiment/ vegetable
Brassica carinata	Gomano	Ethiopian mustard	(1), 2	Leaves are an important vegetable, seeds are also used
Cucurbita pepo	Potota	Pumpkin	(1), 2, 5	Fruits and leaves are consumed young
Digera alternifolia	Kogata	Amaranthaceae	1, 2	Leaf vegetable
Launaea taraxacifolia	Hangoleita	Compositae	4	Wild growing vegetable
Lycopersicon esculentum	Njannja	Tomato	2	Occurs naturally, fruits are eaten
Moringa stenopetala	Shelagda, tellakata Halako Shiferaw	Cabbage tree	(1), 2	Leaves are a very important vegetable; eaten boiled with 'dama. Leaves especially important during the dry season, medicine

Table 1 (*cont.*)

Scientific name	Local name	Common name or family name	Cultivation[b]	Use and remarks
Pergularia daemia	Korroda	Asclepiadaceae	4	Leaf vegetable, elsewhere as edible fruit
Portulaca quadrifida	Mereita	Portulacaceae	4	Leaf vegetable
NN[d]	Kulbabita		1	Herbaceous 30 cm high plant. Succulent, leaves are eaten as a vegetable
NN	Xagalaa	—	5	Climbing, herbaceous plant. Leaves are eaten
NN	Kutata	—	2, 5	Leaves are eaten
NN	Rasota	Konso gomen	2	Leaves are eaten
Spices				
Capsicum annuum	Parpara (red) Qaara (green), mitmita	Red pepper	1, 2	Fruits eaten fresh and used to flavour food
Coriandrum sativum	Tibichota	Coriander	1	Seeds used to flavour food
Foeniculum vulgare	—	Fennel	2	Seeds used as a condiment
Linum usitatissimum	Talpa	Linseed	1	Seeds used to flavour food, also to prepare a kind of dough
Ocimum spp.	Iffaya	Basil	2	Leaves are used to flavour food, very common
Rhamnus prinoides	—	Buckthorn	2	Leaves and wood used as a condiment to flavour alcoholic beverages

Ruta chalepensis	—	Rue	2	Seeds and leaves used to flavour food
Oil crops				
Carthamus tinctorius	—	Safflower	3	Seeds are consumed and used for oil
Helianthus annuus	Sufeta	Sunflower	1, (2)	Seeds roasted, also infused and liquid drunk
Ricinus communis	—	Castor bean	2	Oil used for lighting and for softening leather
Fruits				
Azanza garckeana	Aureta	Malvaceae	4	Edible fruit
Carica papaya	Papayata	Papaya	2	Edible fruit
Citrus aurantifolia	Loomet	Lime	2	Edible fruit
C. sinensis	—	Orange	2	Edible fruit
Grewia tenax	Kotjata	Tiliaceae	4	Edible fruit
Morus mesozygia	Inch'orre	Mulberry	2	Edible fruit
Musa paradisiaca	Museta	Banana	2	Edible fruit
Rhus natalensis	Kabudeida	Anacardiaceae	4	Edible fruit
Opuntia ficus-indica	—	Prickly pear	2, 5	Fruits are greatly enjoyed by children; frequently used in fences and as fodder
Saccharum officinarum	Sonkara	Sugar cane	1, 2	Shoots are eaten raw
Vangueria madagascariensis	Murganta (?)	Rubiaceae	2, 4	Wild tree, edible fruits; planted in villages
Ximonia coffra	Inginkada	Olacaceae	4	Edible fruit
Ziziphus spina-christi	—	Christ thorn	2	Edible fruit

Table 1 (*cont.*)

Scientific name	Local name	Common name or family name	Cultivation[b]	Use and remarks
NN	Kenenta		2, 4	Wild tree, edible fruits; planted in villages
NN	Maderta	–	2	Edible fruit
Beverages				
Coffea arabica	Punitta	Coffee	1, 2	Beans are roasted with butter and cereals; leaves used to prepare a tea. Important cash crop. Also used for ritual purposes
Fibres				
Gossypium herbaceum	Garatita	Cotton	1, (2)	*G. herbaceum* is the older introduction, *G. hirsutum* gives higher yields. Basis of the important weaving industry. Cash crop
G. hirsutum	Futota			
Narcotics				
Catha edulis	Teemahada	Chat	2	Leaves chewed for ritual purposes
Nicotiana tabacum	Tampota	Tobacco	2	Leaves fermented and smoked
Miscellaneous				
Balanites aegyptica	Hangalta		4, 5	Leaves used as browse for cattle and sometimes as vegetable. Fruits are eaten

Boswellia rivae	Dangarda itana	Incense tree	4	Resin used in ceremonies; exported
Commiphora sp.	—	—	2, 4, 5	Tree (often living) is important to build strong fences, especially in villages
Ensete ventricosum	Dupana	False banana	2	Starch of pseudostem used as staple
Hyparrhenia spp.	—	Thatching grass	4	Thatching of roofs
Lagenaria siceraria	Dahanta	Gourd	1, 2	Fruits produce important containers
Solanum incanum	Kimbilota	Sodom apple	4	Used for ritual purposes
Terminalia brownii	Weybata Lia		1	Widespread cultivated tree, important timber; leaves are harvested as browse for cattle

[a] The information is mainly compiled from Goettsch *et al.* (1984) and Westphal (1975). In addition, Hallpike (1970) has reported the use of 80 wild plant species and trees for food, animal fodder, medicine, building material, magico-rituals and miscellaneous.
[b] Key for 'Cultivation': 1, field; 2, backyard in village; 3, border of terraces in fields; 4, 'wild' plant; 5, fences.
[c] 'dama: sorghum or other starchy products are ground, kneaded into balls and boiled in water.
[d] NN = Not known.

now living in Kenya introduced new bean varieties. Of the more recently introduced species, maize is gaining more and more importance at the expense of sorghum (Hallpike, 1970; Minker, 1986).

In the following section, relevant agronomic and botanical details are given of the important crop plants as well as of some of the striking species which are typical and traditional in Konso.

Cereals

Sorghum (*Sorghum bicolor*)

The major staple crop of the Konso is sorghum (*Sorghum bicolor*). It is believed to be an ancient crop of the region (Doggett, 1988) and a wide spectrum of varieties exists. Harlan & Stemler (1976) report an 'unusual assemblage of sorghums' in a few small areas of Africa. One such region is Konso. The guinea sorghum, one of the major races, is dominant in West Africa but is also found in Konso (Stemler, Harlan & de Wet, 1977). Hallpike (1970) mentions that at least 24 varieties are grown and he listed the names of 17 which are distinguished according to their uses: ground, then boiled in water in the form of kneaded balls – 'dama – to be served with boiled cabbage tree leaves, or made into beer, sour or unleavened bread for travelling (Hallpike, 1970). Two types of beer are produced, normal beer and an unmalted variety, named 'erorda'. The Konso are able to describe each sorghum variety very clearly and some examples are given here, taken from Hallpike (1970).

1. yedoda — not much good for beer or 'dama. Bitter, husks difficult to remove. Quick ripening. Birds do not eat it much.
2. sulida — 'dama and beer are good. Beer lasts one month. Soup and erorda also good. Its husks are rather difficult to remove. Birds like it. A slow grower – five months.
3. ha dida — 'dama good, but beer is not, goes bad in two days. A bitter grain. Birds eat it, and it is rather a slow grower.
4. kulsida — 'dama good, erorda good, and the grains are good roasted. Beer not good, goes off in two days. If grubs eat it, becomes bitter. Birds eat it. Quite a quick grower.
5. tisgara — 'dama good, erorda good, and good roasted. Beer is good, lasts two weeks. Birds cannot get at it since the heads hang down. Slow grower.

6. harboreda	not good for erorda or beer since it has nasty taste. Not much better for 'dama. Birds eat it. Grows very quickly.
7. hargiti	good for all purposes. Beer only lasts a week. Birds eat it. Very quick grower.
8. 'gonada	very nice for all purposes. Beer lasts two weeks. Birds eat it. A slow grower.
9. ken dera	very nice for all purposes. Beer lasts two weeks. Birds like it. A quick grower.
10. obiyada	quite nice for all purposes. Beer lasts two weeks. Birds eat it. A slow grower.
11. rereda	very nice for all purposes. Beer lasts two weeks. Birds eat it. A slow grower.
12. hoiriada	nice for 'dama, erorda not very nice. Beer is nasty. It is only eaten if unaffected by grubs. Birds cannot reach grains. A slow grower.
13. magaloda	only 'dama is made from it (not used even for this if attacked by grubs). Never used for beer as far too bitter. Immune to korba. Birds eat it. A slow grower.
14. 'gamadeda sira	nice for all purposes. Beer lasts one month. Birds eat it. Rather a slow grower.
15. ongo/uwada	good for 'dama if grubs do not spoil it. Not nice for beer, and erorda is not much good either. Birds eat it. A quick grower.
16. o jara	nice for all purposes. Beer lasts a week. Birds eat it. Slightly slow grower.
17. pi jita	very nice for 'dama and beer. Beer lasts two weeks. Birds cannot get at heads which hang down. Rather slow grower.

The following characteristics of the sorghum 'varieties' are considered to be relevant in deciding which of these varieties to grow:

- taste;
- suitability for 'dama/erorda beer;
- growing time (e.g. earliness);
- effect of grubs on taste;
- effect of korba (a fungal disease);
- difficulty of removing husks;
- bird resistance.

From several recent collecting missions organized by the Plant Genetic Resources Centre/Ethiopia to the Konso region it was concluded that some of the varieties described by earlier workers do not exist any more and systematic collecting of indigenous germplasm is therefore an urgent matter.

Finger millet (*Eleusine coracana*)

Finger millet is another traditional crop in Konso and it can almost always be found interplanted with sorghum, especially at lower altitude. The grain of finger millet is mainly used for the preparation of beer, but also for soup and bread. No details are known about the diversity of this ancient crop in Konso.

Barley (*Hordeum vulgare*)

Barley is, next to sorghum, the most important cereal in Konso and is a traditional component of the cropping pattern. Above 1900 m it is the dominant crop and it is used for the preparation of beer and food (e.g. dough balls, soup and bread). Although no striking differences from other Ethiopian primitive varieties have been observed, more systematic collecting and study of the landraces in Konso is required.

Wheat (*Triticum* spp.)

The use of wheat is very similar to that of barley and it is grown under similar conditions. Durum wheat has been collected by the Ethiopian genebank and during the missions it was frequently observed that wheat, barley, finger millet and sorghum were grown in the same field. Minker (1986) has observed that wheat is not as much favoured by the Konso as barley.

Pulses

The production of pulses plays an important role in the Konso agriculture. Of the many species reported or collected, horse bean and the common bean are definitely of pre-Amharic age and several varieties of both crops are known. Other species commonly found in the fields or even sometimes in the backyards are chickpea, garden pea, cowpea, pigeon pea, lentil, mung bean, hyacinth bean and, more rarely, lima bean. Pulses are eaten raw or boiled and in times of famine the seeds are ground and used as flour.

Tuber crops

The cultivation and/or use of root and tuber crops in Konso has a long tradition. They are crushed, ground and formed into flat bread or dough balls. Taro and yams are known from other areas in Ethiopia as well, but the use of wild Araceae is almost entirely restricted to Konso and these species play an important role in the mythology of the region (Hallpike, 1972). In normal years the tubers are left in the ground and only in times of food shortage are they harvested. As a result, famines in Konso are less severe compared with other areas in Ethiopia. The number of tubers left in the ground plays an important role when a field is sold (Minker, 1986).

'Pakana' is the common name for the three different Araceae species. The specific local and botanical names for each of the species are:

Pansala: *Sauromatum nubicum*
Saganeida: *Amorphophallus abyssinicus*
Lameeta: *Arisaema* (*? schimperianum*) (Indian turnip)

Amorphophallus is widespread in the villages as well as in the cultivated fields and their borders. Since its tubers contain a toxic substance they have to be crushed and exposed to the air for oxidation. After this procedure the product can be consumed without any problem. Because of the limited knowledge existing on these species a thorough collection and study would be worth while, also in respect to its potentialities for other dry areas in the country.

Cabbage tree (*Moringa stenopetala*)

The most striking characteristic of the Konso agricultural system is the cultivation of the cabbage tree. The tree is densely planted within the villages and generally more widely spaced in the fields and terraces between 1600 and 1800 m. Its light green leaves and the conspicuous grey bark are characteristic features of the cabbage tree.

Konso can be considered as the area where the tree was first cultivated. From here the cultivation has spread into neighbouring areas where it is being used intensively as well. In the whole region the cabbage tree does not occur in the wild (Minker, 1986). The tree is raised from seed; it requires relatively good soil conditions and prefers wind-protected places. After 5–6 years the first leaves can be harvested. They are boiled and eaten as a vegetable with any warm meal. The leaves are rich in vitamins and are mainly harvested in the dry season when other vegetables are scarce. During the rainy season there are only a few leaves left on the tree and they do not taste good. The leaves are an important trading product in the local markets.

Outside the villages, especially on the terraces, the cabbage tree plays an important role in reducing soil erosion. There are trials under way to use the tree for this purpose in other areas of the country. Furthermore, some very promising medicinal uses have been found. A tea of dried leaves is reported to be very efficient in treating light cases of diabetes and it is said that the extract of fresh leaves can cure indigestion and even the cure of an amoebic dysentery has been reported. Eye inflammations are treated and a root extract helps against unconsciousness (Aschalew Hiude, personal communication). Recently it has been reported that ground seeds can be used to clarify muddy water (Jahn, 1981; Goettsch, 1984). Experiments have shown that this powder has the same effectiveness as the best technical water clarifying agents.

It can be concluded that *Moringa stenopetala* is a greatly underutilized and relatively unknown tree which deserves further investigation. It could play a much more important role in the nourishment of people and in the stabilization of the environment in areas with limited rainfall in the tropical belt between 1400 and 1900 m.

Other traditional crops

Vegetables are highly appreciated by the Konso (Minker, 1986). The traditional group of vegetables is composed of a number of species, cultivated or wild, of which the leaves are used. Some of the *Brassica* species, such as *B. carinata*, are among the most important.

The gourd (*Lagenaria* spp.) is a very old and important plant of which the ripe fruits are used as containers. *Capsicum* was introduced by the Amharas. A number of other spices and herbs are in use. Fruit-bearing trees are mainly introduced (e.g. papaya, orange, lime) whereas traditional fruits are collected from the wild. Coffee growing has a long tradition and trees can be observed quite frequently despite the prevailing unfavourable ecological conditions.

Two types of tobacco are cultivated: one is very strong and is said to be pre-Amharic (Minker, 1986). The Konso are famous for their woven goods. Several varieties of cotton are grown, and at least one of them is pre-Amharic (*Gossypium herbaceum* var. *acerifolium*: Hallpike, 1970).

The remarkable number of different tree species in the man-made vegetation of Konso should be mentioned. All the trees are used somewhere and somehow and a systematic study of their uses would be extremely interesting and promising. The importance of mixed cropping under the ecological conditions of Konso cannot be overemphasized. It plays a significant role in the food and fodder production security of the region because of the different water and temperature requirements. It also is an important factor in soil conservation. Goettsch, Engels & Demissie (1984) reported 40 different species in one village and collected 24 species in a terraced field of about 0.2 ha at an elevation of 1750 m. This is quite exceptional when compared with other montane areas in Ethiopia.

Conclusions

Considering the difficult agro-ecological conditions which prevail in Konso, it is remarkable how many people can be fed from a rather limited area when appropriate farming methods are applied.

The ancient terraces and other constructions, as well as the simple but efficient irrigation methods, are the salient features of Konso agriculture which allow an optimal use of water throughout the year.

The intercropping of various crop and tree species together with the cultivation practices seem to be important factors in food and fodder production security as well as in the soil conservation of the Konso area.

The diversity of crop species and the genetic diversity within many of the crop species make Konso an important area from the germplasm conservation and exploration point of view.

The cultivation of the cabbage tree as well as of certain tuber crops is almost entirely confined to the Konso highlands. These species may have good potential in other similar areas where rainfall is limited and where, so far, only relatively small numbers of crops are grown.

References

Doggett, H. (1988). Sorghum history in relation to Ethiopia. *In*: J. M. M. Engels (ed.), The conservation and utilization of Ethiopian germplasm. Proceedings of an international symposium, Addis Ababa, 13–16 October 1986, pp. 97–115 (mimeographed).

Goettsch, E. (1984). Water-clarifying plants in Ethiopia. *Ethiopian Medical Journal*, **22**, 219–20.

Goettsch, E., Engels, J. M. M. & Demissie, A. (1984). Crop diversity in Konso agriculture. *PGRC/E–ILCA Germplasm Newsletter*, **7**, 18–26.

Hallpike, C. R. (1970). Konso agriculture. *Journal of Ethiopian Studies*, **8**, 31–43.

Hallpike, C. R. (1972). *The Konso of Ethiopia. A Study of the Values of Cushitic People*. Oxford University Press, Oxford.

Harlan, J. R. & Stemler, A. B. L. (1976). The races of sorghum in Africa. *In*: J. R. Harlan, J. M. J. de Wet and A. B. L. Stemler (eds), *Origins of African Plant Domestication*. Mouton Publishers, The Hague, pp. 465–78.

Jahn, S. al A. (1981). *Traditional Water Purification in Tropical Developing Countries. Existing Methods and Potential Application*. GTZ Publication 117, Eschborn, Federal Republic of Germany.

Kuls, W. (1958). *Beiträge zur Kulturgeographie der Suedäthiopischen Seenregion*. Frankfurter geographische Hefte 32, University of Frankfurt, Federal Republic of Germany.

Minker, G. (1986). *Birji – Konso/Gidole – Dullay*. Reihe F, Bremer Afrika Archiv, Band 22, Überseemuseum, Bremen.

Stemler, A. B. L., Harlan, J. R. & de Wet, J. M. J. (1977). The sorghums of Ethiopia. *Economic Botany*, **31**, 446–60.

Westphal, E. (1975). *Agricultural Systems in Ethiopia*. PUDOC, Wageningen.

Part III

Germplasm collection and conservation in Ethiopia

13

Theory and practice of collecting germplasm in a centre of diversity

J. G. HAWKES

Introduction

In the past, collecting activities have concentrated on particular species or on certain genetic characters which plant breeders were seeking. At present, because of the rapid rate of genetic erosion of crops in most parts of the world, exploration trips are now becoming 'rescue operations' in which as much diversity as possible is being collected. The concept of 'now or never' is in the forefront of the collectors' minds.

The methodology of collecting and the scientific basis of sampling have also received considerable attention. Whereas some 30 years ago it was thought sufficient to collect a few seeds from a single plant, write one or two words on a label and put them all into a bag, this method is now thought to be most unsatisfactory.

The genetic resources collector is looking for diversity. Whereas the botanical or horticultural collector was content to collect a few herbarium specimens and a small packet of seeds, to serve as a representative for a species in a particular area, the genetic resources collector needs not uniformity, but diversity. How is this to be accomplished?

Studies of the population genetics of wild species by Allard (1970) and his colleagues showed that more sophisticated methods were needed if a reasonable amount of the genetic diversity of a species was to be captured. Marshall & Brown (1975), Bradshaw (1975) and Jain (1975) developed sampling methods for wild species and stated that such methods should be applicable to cultigens since all these were held to possess some kind of population structure.

Before we proceed to set out the generally agreed sampling strategies for genetic resources aims, it will be as well to discuss the general patterns of diversity within a species. These can be partitioned into three sections.

Firstly, there is the diversity within a species that results from broad environmental differences acting on it over different parts of its distribution range. Where environmental changes over great distances are slight, the changes in genetic diversity will also be slight. Where environmental changes are abrupt, as with changes in altitude or soil type, correspondingly abrupt changes in genotype may be expected.

The second type of diversity lies within a small area where a mosaic of populations results in what is spoken of as interpopulation diversity.

The third type is to be seen within populations, necessitating proper population sampling techniques.

Naturally, genetic studies of natural populations show differences between species in the amount of variation between and within populations. On the whole, inbreeding species should have greater interpopulation variability and less intrapopulation variation than comparable outbreeders. Thus if the breeding system of a species is known, appropriate modifications to sampling methods can be made.

Instruction manuals for collecting genetic resources materials have been on the market for some time. Several chapters in Frankel & Hawkes (1975) and in Hawkes (1983) deal with the practical and theoretical concepts of field sampling, and a comprehensive work by Mehra, Arora & Wadhi (1981) is also available. A field collection manual (Hawkes, 1980), dealing with all practical aspects, was published in 1980.

Before discussing detailed sampling methods, we must remember that not all crops or wild species are propagated by seed. Many others are largely vegetatively propagated and special sampling methods are needed for them. Another special group is formed by the fruit and nut trees, which will be discussed separately. First, however, we shall discuss the exploration and collection of annual seed crops and annual or herbaceous perennial wild species, which also reproduce by seed.

Sampling of seed crops and wild species

It is generally agreed that sampling on a population basis should be random or non-selective, since selective sampling will pick out only those genes with a clear morphological expression in the phenotype and will be likely to omit those controlling disease resistance and other physiological characters.

The standard procedure for a field crop is to take a sample of 50 seeds from each of 50 plants taken at random and to treat the 2500 seeds as a single sample. The plants are sampled by walking up and down through the rows, taking a seed sample every so many paces, the distance between plants being determined by the size of the field and by the reproductive system of the plants themselves. Thus for a wind pollinated allogamous plant such as maize, the distance can be quite great. This method, using 50 seeds from each of 50 plants, was devised by Marshall & Brown (1975); on population genetics theory it should give us all the alleles in the population, at 95 per cent certainty, that are present at 5 per cent frequency or more. This method can also be easily applied to wild species where the collector can walk through a population sampling as non-selectively as possible. If time is pressing or the population is small, a minimum of 25 plants will still give reasonably good results.

We have just dealt with intrapopulation sampling. The selecting of interpopulation sampling sites is also important. In annual crops, where much mixing is probably very common, the sites should be evenly dispersed. On the other hand, where wild or weedy species occur in a series of microhabitats, the sites should be clustered. Thus the clustering or the evenly dispersed sites will cover the geographical range of a species. One must, however, remember that the sampling sites or clusters of sites must be much closer together when soils, climate, altitude and other environmental features vary rapidly. This is especially necessary in highly dissected mountain terrain where a slight difference of altitude or aspect can bring with it a great difference in soil or rainfall, or changes from tribe to tribe (Harlan, 1975).

Having strongly advocated the necessity for random sampling, I may be laying myself open to criticism if I do not suggest that some selective sampling may be useful under certain circumstances. Harlan (1975) argues positively for selective as well as random sampling. Bennett (1970) counsels that after taking a random sample the collector should add extra seeds to the sample from extreme or interesting phenotypes. Marshall & Brown (1975), however, are not completely satisfied about this. They recommend that if extreme or interesting phenotypes are found in a population they should form the basis of a separate sample. Bogyo, Porceddu & Perrino (1980), although recommending random sampling in general, so as to preserve as much genetic variation as possible, add: 'Certainly when collecting for specific characters, such as disease resistance, dwarf growth types, stem strength, etc., collectors will not resort to random sampling. Many of these traits are rare and only large samples will ensure that genes responsible for these traits will be represented in the collection.'

Whilst I am inclined to agree with Bogyo and his co-workers in general terms, it must be pointed out that disease resistance is very difficult to evaluate with certainty in the field. Thus, the absence of the disease on the crop may mean that conditions were not right for the spread of the disease when the sample was made. Again, the presence of a disease infection on a crop might not mean that the crop was susceptible to all pathotypes of the disease but only to the one that happened to be present when the collection was made. An excellent discussion of this subject is given by Dinoor (1975).

Several authors recommend taking a large number of rather small samples (about 25 seeds each from 50 plants per sample) and so capturing the diversity from different localities (see Brown & Munday, 1982).

My conclusion must be that no universal sampling system is perfect but that a generalized random sampling system as proposed by Marshall and Brown is reasonably good for all situations.

The problem of how to capture rare alleles in a population has not yet been answered satisfactorily. Thus Chapman (1984) concludes that alleles which would occur at frequencies of 1 in 10 000 or less are better sought through mutation breeding than in a genebank. However, if we do not find them by our present sampling methods, how are we to know that they exist at all? Nevertheless, he concludes that we would be likely to find such alleles if the genebanks were big enough (say 30 000 samples). This, however, depends very much on the spread of samples in the bank and although Chapman argues that as much diversity might be found by sampling a localized area as by sampling an extended area, it would appear that he bases this argument on rather few data.

After all, the genebank collector is looking not only for a wide range of variation, but also for diverse kinds of variation. Thus barley yellow dwarf virus resistance is found only in Ethiopia and, to a limited extent, in some neighbouring countries. Virus M resistance in potatoes is found only in one small valley system in north-west Argentina. Tungro virus disease resistance in rice is found only in a few accessions of one wild rice species; bacterial wilt resistance in cultivated potatoes is found only in one area of Colombia in the northern Andes. Such examples could be multiplied many times.

Thus, I am convinced that careful, evenly spaced sampling of a crop species throughout its distribution area is an essential part of the work of a germplasm explorer.

Since today we are talking specifically of germplasm collecting in a centre of diversity (Ethiopia), we should ask ourselves whether any special rules apply in this case. Quoting Chapman (1984) again, he says that 'there are no well-defined centres of diversity in any of the crops'. This seems to me to be patently untrue and based on insufficient evidence. One has only to look at the clear centre of diversity for cauliflower in Italy, for *Brassica carinata* in Ethiopia, for potatoes and maize in the central Andes of Peru and Bolivia, for teff in Ethiopia, for chilli peppers in Mexico and many other examples, to realize that there are so many exceptions as to render Chapman's statement almost meaningless.

Sampling vegetatively propagated crops and wild species

Having discussed seed crops at length, let us now turn to a completely different problem, that of vegeculture crops. These are often quite common in tropical areas, particularly in the New World tropics; however, they are by no means absent from Africa.

Most vegetatively propagated (vegeculture) crops still possess the capacity to reproduce by seed, or at least they did so before the advent of plant breeding and selection, and the very primitive forms and old landraces are probably still fertile. Related wild species retain both methods of reproduction as useful alternative strategies, according to whether uniformity (asexual reproduction) or diversity (sexual reproduction) is needed for survival.

In all vegetatively propagated crops that have come to my notice, the farmers seem to have carried out such strong artificial selection that the original population structures have virtually disappeared. For instance, in the ancient centre of diversity of potatoes in the Peruvian and Bolivian Andes, each market area or district contains generally not more than 50–100 distinct morphotypes and sometimes fewer. The number of morphotypes could have been as high as 200 or more before genetic erosion took place, though we cannot be certain of this. Each morphotype probably represents a distinct genotype, but it is possible that some morphotypes may conceal several different genotypes. Adjacent market areas seem to possess much the same spectra of morphotypes but not identical ones. In other words, as we move from market area to market area, we see gradual changes, some morphotypes disappearing and other new ones appearing. Market areas some distance from each other will differ very clearly in their morphotype spectrum (Hawkes, 1975).

This knowledge helps us to build up a sampling strategy which is in fact the reverse of that advocated for sexually reproducing plants. Thus, for yams, potato, cassava, sweet potato, ensete, *Plectranthus edulis* and all other crops of this sort, it is recommended that a selective sampling of every distinguishable morphotype in each area be carried out. This undoubtedly means that there will be a large number of duplicates, though not quite as many as might at first be imagined. In the introduction station the collections are all grown together and all or most duplicates eliminated. It is best always to save two or three duplicates of each morphotype to avoid accidental loss or death through disease or pest infection. At this stage it is generally possible to identify slight differences between what were, at first sight, thought to have been identical clones.

Where sexual reproduction takes place in a vegetatively propagated crop, and if the crop is largely outcrossing, the seed should be sampled on a population basis, even though the plants do not constitute a true population. Most vegetatively propagated crops (perhaps all?) are very heterozygous and release much diversity when they reproduce from seed; in addition, the seed collection will represent a sample of the local gene pool. Such seed samples should be made as a useful parallel adjunct to the clonal collections but should be given distinct collection numbers.

When collecting the related wild species the objectives are slightly different. With these, the population structure is still present and when seed is available the methods used for seed crops can be used. When seeds are not available it is useful to take a randomly collected tuber sample from as many plants as possible (say, up to 25) and treat them all as a single collection. The plants from this should be sib-mated in the experimental field or glasshouse and the seeds mixed in equal proportions from each plant so as to constitute a single sample.

This advice is perhaps often to be regarded as an ideal procedure which can seldom be realized in practice. Often one finds only one or two wild plants in any locality, or what seemed at first sight to be a population proves to be no more than a single clone.

The general advice given to collectors in these circumstances is 'collect what you can, always striving so far as possible to reach the ideal procedures'. After all, one collection, however inadequate, is better than no collection at all.

Fruit and nut tree sampling

The biggest problem with fruit and nut tree genetic resources is that many species possess seeds that cannot be stored in the usual way (recalcitrant seeds). Thus with these materials, whose seeds are short-lived, the collecting strategy must always be closely related to storage. Seeds need to be sown directly or sent back to introduction stations as soon as possible after collecting. If cuttings are taken they must be stored carefully during the collecting trip to avoid drying out. These cuttings must be grafted onto prepared rootstocks or rooted by mist propagation or similar techniques. It must be borne in mind that every genotype will have to be grown into a permanent bush or tree unless multiple grafting techniques are employed.

This means that population sampling is not likely to be possible since it is quite out of the question to grow 2500 plants for every collection made in the field (i.e. 50 seeds from 50 plants). In any case, if collecting is being carried out in tropical forests where the number of species is large but the number of individuals of each species is very limited, it may be difficult to identify populations. In fact, it has been found in South-East Asia that no more than one or two mature individuals are to be found in every 10 hectares.

In Ethiopia, a special collecting and storage procedure has been devised for coffee, in which 15 berries from 20 bushes selected at random within a population are collected, and 10 seeds from each of the bush samples are taken for sowing (Hawkes, Engels & Tadesse, 1986). Of these, three seedlings are planted out in the field. Thus, each collection will in the end contain 30 plants, which represents a fairly reasonable sample of the population (Table 1 and Fig. 1).

In general, for wild fruit and nut trees where the numbers of individuals are scattered and population size is difficult to estimate, the following procedure is recommended:

Collect seeds from 10–15 individuals over an area of about 2–5 ha and combine them into a single sample. If random collections can be made this will be an advantage, but if the individual trees are very scarce, seeds from each tree must be collected.

If no seeds are available, suckers or budwood cuttings, one per tree from 10–15 individuals, should be taken. Repeat this procedure at intervals depending on climatic, altitudinal or soil differences.

For cultivated fruit and nut trees, the first thing to do is to ask the farmers if they propagate their trees by means of grafted or rooted cuttings, or by means of seed. If propagation is by cuttings then treat these as for vegetatively propagated crops. If propagation is by seed,

Table 1. *Sowing and planting scheme for each population*

Code numbers of bushes sampled	1	2	3 ... 18		19	20
Berries	15	15	15 ...	15	15	15
Seeds extracted	30	30	30 ...	30	30	30
Seeds sown	10	10	10 ...	10	10	10
Seedlings transplanted	3	3	3	3	3	3
Numbering (assuming collecting number to be 153)	153-1	153-2	153-3 ...	153-18	153-19	153-20

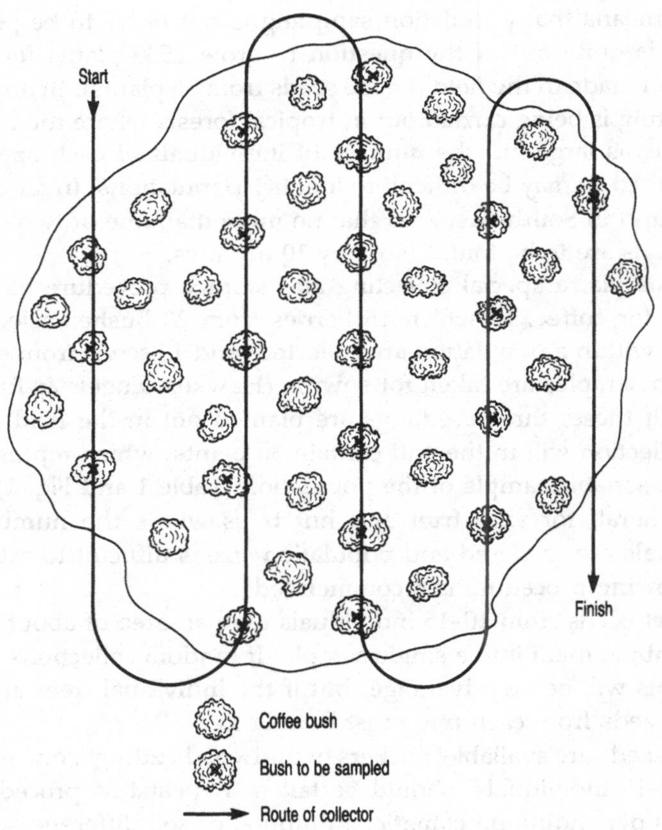

Fig. 1. Area of coffee population to be sampled. 20 plants sampled; 15 berries taken from each plant sampled and kept in separate bag from others; one collector's number given to the whole population sample of 20 plants.

it is likely that each villager will not possess more than one to five trees in his garden. If this is so, it is best to treat the whole village as the collecting site and take a random sample from 10–15 individuals, bulking all these together as one sample. In this case, the trees throughout the village are to be regarded as a single population. If seeds are not available, cuttings should be taken in the same way. It will be advisable to sample as many sites or villages as possible at scattered intervals throughout the region.

Market sampling

When it is impossible to do field sampling it may be necessary to collect market samples. This may occur when lack of time or bad transport facilities prevent the collector getting to the farms. Where the crop has already been harvested and sent to market, taking of market samples may suffice. In any case, a visit to the local markets may give the collector a good idea of what is available in the general area.

Market sampling has its drawbacks, however, and must be approached with caution. In the first place, not all cultivars or populations are sent to market since some will be retained purely for home consumption. Secondly, many of the market samples may be artificial mixtures. Thirdly, they may be standard bred cultivars of no interest to the genetic resources collector. Finally, the information for passport data may be insufficient, lacking, or totally wrong and misleading.

Having said all this, some market sampling may still be valuable, but will be nowhere near the value of field sampling, where the plants in their natural or semi-natural states as population and landraces can be sampled by the collector.

Data recording

Minimum data sheets have been advocated by the International Board for Plant Genetic Resources (IBPGR) and these can be printed and bound up in books of 100 before starting on a collecting expedition. One side should show essential descriptors and have tear-off labels at the base; the other side should be used for optional descriptors (Table 2).

Specialized sheets for particular crops may also be devised but it is inadvisable to ask for too much information on the data sheet. Either the collector will not fill in the information, or he will try to do so and thus spend too much time on this to the detriment of his collecting

Table 2. *Notes on field data recording*

1. Expedition and organization (name, year, etc.).
   (Note: Descriptors 1, 2 and 3 can easily be printed on all collection sheets before beginning the mission.)
2. Collecting team. The names of all team participants.
3. Collector's name. The name of the leader or taxonomist.
4. Collector's number (note: a single sequence is strongly recommended for the whole expedition and any others in which the leader may be participating. The numbers can be stamped in sequence on to the collecting sheets before beginning the mission).
5. Date of collection (day, month, year).
6. Photograph numbers of specimen, habitat, farm field, etc.
7. Latin names of genus, species, subspecies, etc. (written in full).
8. Vernacular or cultivar name.
9. Precise locality (e.g. political division, province, department, etc., including compass direction and approximated distance from nearest village or geographical feature – distance in kilometres along road between two inhabited places).
10. Latitude (degrees and minutes).
11. Longitude (degrees and minutes).
12. Altitude (in metres above sea level).
13. *Types of material (seeds; inflorescences, spikes, panicles, etc.; vegetative storage organs, such as rhizomes, corms, roots, tubers; whole living plants; herbarium specimens).
14. *Sample type (whether populations, pure lines or individuals sampled and whether by random, non-random or both methods).
15. *Status (cultivated, weed, wild).
16. *Sources (e.g. from field, farmer's store, market, shop, garden, wild vegetation, etc.).
17. *Original sources of sample (if from market or store, or if obtained from another genebank, where grown originally. Farmers may know the source of their material, also).
18. Frequency (a rough estimate of frequency of wild species).
19. Habitat (mostly relevant for wild species).
20. Descriptive notes (scoring of morphological features of interest for the particular crop or wild species; thus – plant height, branching, etc., colour of stems, leaves, flowers, fruit. Amount of diversity in population and range shown in voucher specimens).

    Note: tear-off labels at the bottom of the page with collector's name and number are to be placed with the collections (these can be printed and stamped on to the labels before beginning the mission).

In addition, the following specifications may be recorded on the reverse side of the sheet.

21. Uses (i.e. for human or animal food, medicinal purposes, etc.).
22. Cultural practices (irrigated, dry, with or without fertilizer, fungicides, pesticides, etc.).
23. Approximate dates of sowing and harvesting (for instance, 'early April to mid-July').
24. Soil observations (texture, stoniness, depth, drainage, approximate soil colour if thought desirable to record).
25. Soil pH (if thought desirable to record).

Table 2 (*cont.*)

26. Land form observation (aspect, slope).
27. Topography (swamp, flood plain, level, undulating, hilly, hilly dissected, steeply dissected, mountainous, other – specify).
28. Plant community (natural vegetation; for wild species).
29. Other crops grown in surrounding fields or in the rotation.
30. Field observations of pest and pathogen infections.
31. Name and address of farmer (if thought desirable to record).
32. Taxonomic identification (fill in later if identity checked by expert).
33. Expert's name and date.
34. Name of Institution and accession number (where these differ from the collector's name and number).

* Ring appropriate words.

work. When crop genetic diversity is disappearing so quickly, it is best to push on with collecting genetic resources and add only the information required by the minimum data sheet.

Duplication of collections

Many of the early collections made in Ethiopia by Vavilov and his colleagues some 60 years ago have probably been lost. Other collections made by expatriate explorers may still exist and sub-samples of many of these are being returned to the Ethiopian Plant Genetic Resources Centre (PGRC/E).

Although it would seem logical to 'repatriate' materials in this way it may not always be necessary to do so. Thus, if Ethiopian materials are known to exist in other genebanks, are being well looked after, and arrangements are available for obtaining them when required, much duplication of effort would be saved simply by leaving them where they are. The inventories of the other genebanks might indicate that certain samples could be of immediate value to Ethiopian breeders and such samples could be recalled for immediate use. However, the rest could be left where they are, providing that the relevant passport data are available. If no passport data are now attached to them they could just as easily be put into a 'reserve' collection in the other genebank rather than use up much needed space at PGRC/E.

Conclusions

Although we have dealt in this chapter with the theory and practice of germplasm collecting, it must be remembered that before setting out into the field, the collector needs to know what has already been collected by himself and by previous collectors. In view of the fact that the Centre is now celebrating 10 very successful years of its existence, a survey of the present situation should be made for each crop, as follows:

- to assess how many samples are in the genebank;
- to call up information from the data base on where they were collected;
- to produce hand-plotted maps of distribution of the collection sites as well as the altitudinal and ecological data associated with them;
- to identify areas or eco-geographical zones where collections with useful characters of resistance have been found, such as insect, virus and fungus resistance, and adaptation or tolerance to stress conditions. A thorough knowledge of the literature would be an advantage here;
- to identify blank areas on the maps and check from previous experience whether the crop may be found in them, and at what altitudes and ecological zones;
- to sift the international literature for further information on Ethiopian collections, especially if provenance data are available;
- to use the information obtained to formulate a coordinated strategic plan for collecting genetic resources in Ethiopia during the next 5–10 years.

In this way, one can be certain that a major part of the genetic resources of the Ethiopian centre of diversity will be conserved for plant breeding use, now and in the future.

References

Allard, R. W. (1970). Population structure and sampling methods. *In*: O. H. Frankel and E. Bennett (eds), *Genetic Resources in Plants*. Blackwell, Oxford, pp. 97–107.

Bennett, E. (1970). Tactics of plant exploration. *In*: O. H. Frankel and E. Bennett (eds), *Genetic Resources in Plants*. Blackwell, Oxford, pp. 235–7.

Bogyo, T. P., Porceddu, E. & Perrino, P. (1980). Analysis of sampling strategies for collecting genetic material. *Economic Botany*, **34**, 160–74.

Bradshaw, A. D. (1975). Population structure and the effects of isolation and selection. *In*: O. H. Frankel and J. G. Hawkes (eds), *Crop Genetic Resources for Today and Tomorrow*. Cambridge University Press, Cambridge, pp. 37–51.

Brown, A. H. D. & Munday, J. (1982). Population-genetic structure and optimal sampling of landraces of barley from Iran. *Genetica*, **58**, 85–96.

Chapman, C. G. D. (1984). On the size of a genebank and the genetic variation it contains. *In*: J. H. W. Holden and J. T. Williams (eds), *Crop Genetic Resources: Conservation and Evaluation*. Allen and Unwin, London, pp. 102–119.

Dinoor, A. (1975). Evaluation of sources of disease resistance. *In*: O. H. Frankel and J. G. Hawkes (eds), *Crop Genetic Resources for Today and Tomorrow*. Cambridge University Press, Cambridge, pp. 201–10.

Frankel, O. H. & Hawkes, J. G. (eds) (1975). *Crop Genetic Resources for Today and Tomorrow*. Cambridge University Press, Cambridge.

Harlan, J. R. (1975). Practical problems in exploration. Seed crops. *In*: O. H. Frankel and J. G. Hawkes (eds), *Crop Genetic Resources for Today and Tomorrow*. Cambridge University Press, Cambridge, pp. 111–15.

Hawkes, J. G. (1975). Practical problems in exploration. Vegetatively propagated crops. *In*: O. H. Frankel and J. G. Hawkes (eds), *Crop Genetic Resources for Today and Tomorrow*. Cambridge University Press, Cambridge, pp. 117–21.

Hawkes, J. G. (1980). *Crop Genetic Resources Field Collection Manual*. IBPGR and Eucarpia, Wageningen.

Hawkes, J. G. (1983). *The Diversity of Crop Plants*. Harvard University Press, Cambridge, Massachusetts.

Hawkes, J. G., Engels, J. M. M. & Tadesse, D. (1986). Suggested sampling and conservation system for coffee in Ethiopia. *PGRC/E–ILCA Germplasm Newsletter*, **11**, 25–8.

Jain, S. K. (1975). Population structure and the effects of breeding system. *In*: O. H. Frankel and J. G. Hawkes (eds), *Crop Genetic Resources for Today and Tomorrow*. Cambridge University Press, Cambridge, pp. 15–36.

Marshall, D. R. & Brown, A. H. D. (1975). Optimum sampling strategies in genetic conservation. *In*: O. H. Frankel and J. G. Hawkes (eds), *Crop Genetic Resources for Today and Tomorrow*. Cambridge University Press, Cambridge, pp. 53–80.

Mehra, K. L., Arora, R. K. & Wadhi, S. R. (1981). *Plant Exploration and Collection*. National Bureau of Plant Genetic Resources, New Delhi.

14

A decade of germplasm exploration and collecting activities by the Plant Genetic Resources Centre/Ethiopia

ABEBE DEMISSIE

Introduction

The richness of Ethiopia's biological resources is well known. It has been mentioned by several scientists that the country exhibits an extraordinary genetic diversity in cereals such as barley (*Hordeum vulgare*), wheat (*Triticum* spp.), sorghum (*Sorghum bicolor*) and teff (*Eragrostis tef*), oil crops such as castor bean (*Ricinus communis*), sesame (*Sesamum indicum*), and other lesser known but potentially valuable species of plants. Eleven cultivated crop species have been identified as having their centre of diversity in Ethiopia (Zohary, 1970). Vavilov (1951) indicated that some 38 species are connected with Ethiopia as a primary or secondary gene centre.

Owing to the potential and uniqueness of the biological resources of this country, numerous exploration expeditions have been undertaken in the past. The earliest was probably the one made by Schimper in 1840, a year which appears to mark the beginning of botanical collecting in Ethiopia (Gentry, 1971). However, it was after the establishment of the Plant Genetic Resources Centre/Ethiopia (PGRC/E) that systematic collecting was launched on a large scale.

Agents of genetic erosion

The valuable genetic diversity in Ethiopian crop species, as well as in their related wild species, has been built up over the centuries by the natural selective forces of the environment and the farming community. Such diversity exists not only among the different agricultural areas of Ethiopia, but also within each area and even within one farmer's field. This wealth of biological and genetic diversity is seriously threatened by a number of factors, as follows.

Natural calamities

In the last decade there have been several catastrophic droughts which have led to complete crop failures and subsequently severe genetic erosion has taken place in the landraces that have been maintained through many generations by the farmers. During the last few years several regions in Ethiopia have been affected by severe famine and farmers have been forced to consume the seeds normally kept for planting the next season. This erosion is particularly aggravated by the distribution of food grain (mainly cereals) by relief agencies because such grain can replace the native landraces.

Introduction of high-yielding varieties

It has been reported that the percentage of Ethiopian farmers using improved seeds has reached 2 per cent, although in Arsi administrative region (Anonymous, 1984), where research work has been going on for many years, the figure is 21 per cent. Worede (1983) stated that the traditional varieties have been almost completely replaced by modern, uniform, advanced cultivars in areas such as Chilalo in Arsi and Adaa in the central highlands. According to unpublished data from the Ethiopian Seed Corporation (ESC), well over 24 wheat and 9 barley varieties have entered commercial production in Ethiopia since 1968. A recent publication by ESC indicates that 7 wheat, 7 barley, 4 teff, 11 sorghum and 2 faba bean varieties have been released for commercial production (Anonymous, 1981). This is a welcome development approach although, paradoxically, it is a threat to the genetic diversity on which future improvement work is based. A number of recent varietal introductions of other cereals, namely, sorghum, teff, and oilseeds such as Ethiopian mustard and sesame, have been reported (Institute of Agricultural Research, Addis Ababa, unpublished data).

Crop replacement

The traditional cereal crop of the Ethiopian highlands, barley, is currently being replaced or complemented by oats (*Avena sativa*) (Jutzi & Gryseels, 1984) and, at slightly lower altitudes, by wheat. Maize is currently becoming a menace to sorghum in most traditional sorghum growing areas. Moreover, while sorghum is being replaced by bulrush millet (*Pennisetum americanum*), which is early maturing and drought-resistant (Michael, 1983), pearl millet is also becoming popular at the expense of finger millet in northern Ethiopia.

Change in cropping patterns

Owing to the soaring prices of some agricultural crops in recent years, the acreage of crops such as teff has increased at the expense of other crops, such as wheat and millet. This subsequently reduces the chance of maintaining the landraces of other traditional crops, as local farmers favour the more profitable crops.

Change in land uses

Currently it is not difficult to see the deforestation process that is taking place in the country. The semi-wild coffee is seriously threatened by this process and also because it is being replaced by crops, such as maize. Furthermore, large-scale agriculture is expanding, new roads are opening up remote areas, bushland is being taken into cultivation and several other developments are taking place at the expense of the wild and primitive crop germplasm.

Exploration and collecting activities

PGRC/E embarked on a systematic field collecting operation in the 1977 crop season. Since then, many collecting expeditions have been undertaken in all administrative regions of the country, covering a wide range of ecological zones.

Based on relevant and available information, collecting activities were initiated according to well defined priorities for both crops and areas. The priorities were based on criteria such as the economic and social importance of the crops and their respective degree of genetic erosion. Revisions of the priority list are made during regular workshops with the plant breeders on the basis of factual and up-to-date data. Based on the criteria mentioned, the following list was formulated at the start of the collecting activities (Ebba, 1978):

1. Wheat
2. Barley
3. Teff
4. Sorghum
5. Legume and oil crops
6. Root crops
7. Vegetable crops
8. Forage crops
9. Coffee
10. Medicinal plants
11. Forest trees
12. Others

In the first 10 years of its operation, the Centre has conducted more than 76 collecting expeditions. As a result, some 15 000 accessions of about 75 species or groups of species of mainly crop plants have been collected (Table 1).

Table 1. *Crop samples collected by PGRC/E from the various administrative regions of Ethiopia*

Species	1[a]	2	3	4	5	6	7	8	9	10	11	12	13	14	15	Total
Brassica spp.	42	34	1	13	72	90	36	25	22	108	24	4	123	74	0	668
Capsicum spp.	6	5	0	10	15	17	4	6	4	8	3	0	18	12	4	112
Cicer arietinum	16	11	1	6	106	110	19	0	0	203	22	15	5	30	184	728
Coffea arabica	0	0	0	0	0	0	0	0	39	0	0	0	101	0	0	140
Eleusine coracana	0	0	5	12	161	107	4	8	3	4	4	16	88	3	8	423
Eragrostis tef	16	20	8	10	204	223	11	19	44	81	34	59	97	197	44	1067
Guizotia abyssinica	12	8	25	7	172	123	1	7	5	102	1	5	80	78	12	638
Hordeum vulgare	462	142	78	152	278	584	215	5	37	689	120	81	123	252	70	3288
Lathyrus sativus	2	0	0	0	60	20	0	0	22	0	0	12	5	20	13	154
Lens culinaris	21	13	0	6	25	1	16	1	0	52	1	14	3	71	9	233
Linum usitatissimum	51	29	10	7	48	116	19	3	6	67	5	27	19	109	22	538
Phaseolus spp.	1	7	1	32	32	4	61	6	3	9	32	2	50	6	10	256
Phytolacca dodecandra	0	0	0	0	0	1	0	0	0	143	0	0	0	0	0	144
Pisum sativum	51	44	5	23	56	148	20	0	14	93	13	17	55	129	34	702
Ricinus communis	9	13	4	9	27	10	49	3	1	27	12	0	14	17	15	210
Sesamum indicum	1	0	1	9	13	1	5	1	4	11	1	0	27	96	48	218
Sorghum bicolor	17	10	52	179	47	119	70	176	50	44	48	31	45	67	189	1144
Trigonella foenum-graecum	9	9	1	2	59	55	10	0	1	17	3	6	8	36	8	224
Triticum spp.	225	121	31	51	429	204	107	1	5	672	37	114	57	344	42	2440
Vicia faba	30	21	5	16	70	152	29	2	15	121	12	16	44	179	13	725
Other crops	21	23	1	110	100	130	84	26	17	74	41	8	110	41	46	832
Grand total	992	510	229	654	1974	2215	760	289	292	2525	413	427	1072	1761	771	14884

[a] Administrative regions: 1, Arsi; 2, Bale; 3, Eritrea; 4, Gamo Gofa; 5, Gojam; 6, Gondar; 7, Harerge; 8, Ilubabor; 9, Kefa; 10, Shewa; 11, Sidamo; 12, Tigray; 13, Welega; 14, Welo; 15, Unknown.

Collection strategy and sampling technique

Collection strategy

The collection strategy was based largely on broad-based or non-crop specific missions rather than on crop specific or so-called pointed collecting missions. During a non-crop specific mission the crops that mature at more or less the same time in a given region are collected in the farmer's field or in the markets. This strategy was favoured particularly in drought-stricken areas where the ultimate objective was to rescue whatever germplasm was still available. However, in order to make better use of the foreign crop-specialist collectors or to concentrate on the species of interest to any collaborating collectors (e.g. national plant breeders or scientists), pointed collection missions have been undertaken. In this way, a closer and more detailed view of a given crop can be obtained and thus exploration missions, diversity studies and preliminary surveys of Ethiopian oil crops (Demissie, 1984; Seegeler, 1986), as well as of other important crops such as chickpea, sorghum and finger millet, have been carried out in collaboration with interested international agencies. Furthermore, several missions have been launched to collect spices and lesser known but potentially valuable plant species such as *Moringa stenopetala*, *Amorphophallus* spp. and *Sauromatum* sp.

Sampling technique

The optimum sample size per collecting site is the number of plants required to obtain, with 95 per cent certainty, all the alleles in a population that occur in 5 per cent frequency or more (Marshall & Brown, 1975). Hawkes (1976) suggested that bulked seed samples from up to 50 individual plants, and certainly not more than 100, should be collected non-selectively to obtain optimum sample size. These plants are taken at random by walking backward and forward through the field taking a sample every so many paces until the 50 plants have been sampled. Whenever rare types, i.e. plants which show characters not included in the random sampling, are noticed a selective sampling technique is adopted. The sample should be given a different collection number (Hawkes, 1985).

The sampling of vegetatively propagated material such as sweet potato, yams, taro, ensete, etc. requires distinct sampling techniques since such crops do not occur as large populations, but are highly selected individual genotypes. These crops are sampled on the basis of information obtained from local farmers on the type and number of varieties they grow. In general, it should be pointed out that

PGRC/E COLLECTION RECORD SHEET

COLLECTION NO.________
DATE________
COUNTRY________
ADM. REGION________
AWRAJA________
WEREDA________
VILLAGE/SITE________

MAP NO.________
LAT.______ LONG.______
ALTITUDE______(m)

TOPOGRAPHY:
1 Swamp
2 Flood plain
3 Plain level
4 Undulating
5 Hilly
6 Hilly dissected
7 Steeply dissected
8 Mountainous
9 Other (specify)

SITE:
1 Level
2 Slope
3 Summit
4 Depression

SOIL TEXTURE:
1 Sand
2 Sandy loam
3 Loam
4 Clay loam
5 Clay
6 Silt
7 Highly organic

SOIL COLOUR:
1 Black
2 Brown
3 Red
4 Orange
5 Yellow
6 Other (specify)

STONINESS:
0 None
1 Low
2 Medium

DRAINAGE:
1 Poor
2 Moderate
3 Well drained
4 Excessive

SOIL PH:________

ACC. NO.________
CROP________
GENUS______ SPECIES______
LOCAL/VARIETY NAME________
ETHNIC GROUP________
LANGUAGE________

SAMPLE TYPE:
1 Single plant
2 Pure line/clone
3 Population/mixture
4 Other (specify)

GENETIC STATUS:
1 Wild
2 Weed
3 Primitive cultivar/landrace
4 Breeders line
5 Advanced cultivar

SOURCE OF COLLECTION:
1 Field
2 Backyard
3 Farm store/threshing place
4 Market
5 Agricultural institute
6 Natural vegetation
7 Other (specify)

MATERIAL:
1 Seed
2 Spike
3 Pods
4 Other (specify)

HERBARIUM SPECIMEN: Yes No
PHOTOGRAPH: Yes No
SOWING MONTH: 1 2 3 4 5 6 7 8 9 10 11 12
Early/Mid/Late
HARVEST MONTH: 1 2 3 4 5 6 7 8 9 10 11 12
Early/Mid/Late

ORIGIN OF SEED: Local/Elsewhere________

USAGE (specify)________

DISEASES AND PESTS________

NOTES: (Associated wild-weedy crop species, local flora, disturbance factors, morphological variation, husbandry, etc.)

Fig. 1. PGRC/E's collecting form.

vegetatively propagated material is often encountered in isolated conditions and sampling is often determined by the availability of material. During collecting, relevant data are collected on the sampling site (ecology, soil, vegetation, etc.) and on the germplasm itself. The data format in use has been adapted from the International Board for Plant Genetic Resources (IBPGR) in order to meet the specific local conditions (Fig. 1).

International cooperation in collecting activities

In 1981 and 1984, PGRC/E and the International Crops Research Institute for the Semi-Arid Tropics (ICRISAT) organized joint collecting expeditions which resulted in the acquisition of a wide range of material. The first expedition focused on collecting Zera-zera sorghum types which were identified as good in terms of their agronomic desirability and their tolerance to diseases and drought (Prasada Rao & Mengesha, 1981). The second mission was mainly geared to collecting chickpeas (Pundir & Mengesha, 1982).

IBPGR has provided partial funding to a number of collecting expeditions that have been carried out in the last 10 years. In July 1980, IBPGR assisted PGRC/E to collect 'belg' or small rainy season crops (Toll, 1980). A similar type of assistance was offered by IBPGR to collect the 'meher' or main rainy season crops which are associated with the June–August rains of the Ethiopian highlands (Toll, 1982). Furthermore, IBPGR co-sponsored a *Brassica* spp. collecting expedition in Ethiopia (Astley, Mahteme & Toll, 1982) as well as the visit of a barley expert.

The German Agency for Technical Cooperation (GTZ) financed a consultant-collector to assess the degree of genetic erosion in Ethiopian oil crops and to advise on priorities of collection (Seegeler, 1986).

Germplasm collected

In view of the importance of cereals as a staple food and the degree of genetic erosion, PGRC/E gave top priority to the collection of cereal germplasm. Well over 56 per cent of the material collected consisted of the major cereals, i.e. wheat, sorghum, teff and barley. The results of these cereal collection missions are discussed below, together with facts and figures on the collection of some other traditional Ethiopian crops. A summary, by administrative region, of all crops collected by PGRC/E is presented in Table 1.

In order to show, in a general way, the intensity of collection for some of the major crops, Table 2 gives the calculated hectarage per

Table 2. *Intensity of collecting of some of the major crops by administrative regions of Ethiopia, expressed in hectares of production per collected accession (July 1986)*

	Cereals				Pulses			Oil crops	
Administrative region	Teff	Sorghum	Barley	Wheat	Faba bean	Field pea	Chickpea	Noog	Linseed
Arsi	1631	1270	309	716	1073	296	83	42	122
Bale	380	80	461	296	119	250	9	625	172
Eritrea	3275	1325	369	526	520	860	5800	336	840
Gamo Gofa	1350	31	163	35	87	113	250	14	14
Gojam	1198	392	300	64	464	234	73	265	321
Gondar	950	515	167	160	309	128	505	385	23
Harerge	364	2564	9	73	7	20	11	400	21
Ilubabor	2574	80	740	3500	950	–	–	143	67
Kefa	2216	578	514	1920	887	607	–	300	17
Shewa	3927	2996	246	284	891	295	266	83	148
Sidamo	482	151	343	268	242	69	186	100	20
Tigray	1107	1429	349	353	600	153	1087	920	119
Welega	1760	1509	152	84	175	60	280	339	68
Welo	279	1182	24	67	131	27	87	3	8
Overall mean for PGRC/E collections	1224	638	222	232	393	166	208	236	100
Overall mean for total Ethiopian collections	575	111	99	89	238	104	208	163	62
Total hectarage of major crops in Ethiopia ('000 ha)	1305.6	766.1	728.3	565.0	297.8	116.8	157.6	136.3	53.3

collected sample for each of the administrative regions. Since the total PGRC/E holdings per crop are generally much higher, due to donations and selections, the hectarage per accession of some crops in the genebank has also been calculated. In both cases, the actual number of hectares under a given crop per administrative region, as far as is known, is used to calculate the intensity.

The figures in Table 2 do not allow any conclusions to be drawn on the genetic value of the collected (or non-collected) samples. Aspects such as the genetic diversity within and between accessions, and the coverage of any given area, are important as well. If the collection sites are plotted on a map of Ethiopia, it can easily be seen that by far the majority of accessions have been collected along accessible roads and only rarely have collections been made far from the roadsides.

Wheat (*Triticum* spp.)

In Ethiopia, it is common for a farmer to grow durum and poulard wheat in the same fields. More recently, bread and durum wheat have frequently been found together in the same field and even a combination of these with *T. polonicum* or *T. compactum* can occasionally be found.

Wheat is one of the most important cereals in Ethiopia, both in terms of production and of acreage. According to the latest statistics, a total of *ca.* 565 000 hectares are under wheat production (Anonymous, 1984). From the genetic resources point of view also, wheat is an important crop since several species have a secondary centre of diversity in Ethiopia and the majority of the fields are still planted with landrace populations. Wheat collections on a regional basis ranged from one sample in Ilubabor to 675 in Shewa. The number of hectares per sample ranged from one sample for every 35 hectares in Gamo Gofa to one for every 3500 hectares in Ilubabor. Overall, when considering the total holdings of PGRC/E, including donations and selections, this figure comes to 89 hectares per accession, which is close to the figure reported by Chapman (1985) who probably considered other Ethiopian material not yet in PGRC/E's possession. This figure shows that wheat is one of the best collected crops in Ethiopia in terms of hectarage per accession. The other important aspect is the genetic diversity between accessions, but this has not yet been assessed in a systematic way.

The altitudinal range for wheat in the areas covered varies from 1200 to 3300 m above sea level with the majority of the accessions collected from altitudes above 2500 m (Table 3). A general estimate of diversity of various crops was provided in a study by Mengesha (1975). A detailed analysis of the regional pattern of phenotypic diversity in a limited number of Ethiopian durum and bread wheat accessions was carried out by Bekele (1984a). He reported that the total variation differed from character to character and that the total phenotypic variation was highest within populations followed by differences among populations within a region, and the least between regions.

Sorghum (*Sorghum bicolor*)

Sorghum is a major cereal in Ethiopia and is the third most important in terms of production (Anonymous, 1984). Apart from being a staple food crop, sorghum has several other uses, e.g. the stalks are used for fuel and for house construction.

The first major effort to collect sorghum germplasm was launched in 1967 after the formation of the Ethiopian Sorghum Improvement Project. Since then, numerous collecting expeditions have been undertaken by PGRC/E, partly in collaboration with ICRISAT (Prasada Rao & Mengesha, 1981). Emphasis was mainly on areas that had been less extensively collected.

The number of hectares per sample is *ca.* 638, when considering the PGRC/E collection alone. The overall figure, considering all Ethiopian accessions held by PGRC/E, shows a much more favourable figure, namely 111 hectares per accession (Table 2).

The altitudinal range for sorghum in the areas covered is from 400 to 2940 m (Table 3), which shows the wide ecological amplitude of sorghum in Ethiopia.

Teff (*Eragrostis tef*)

Teff is the most important cereal crop and stands first in terms of acreage, with 1 305 600 hectares of land under cultivation (Anonymous, 1984). Its cultivation as a cereal crop is confined almost entirely to Ethiopia and it is mainly used to make 'injera', a local bread and an important part of the national dish.

Teff is known to have been domesticated in Ethiopia and a wealth of diversity exists. Efforts have been made to collect landraces systematically from each of the ecological zones. The germplasm collections made by PGRC/E in the last 10 years exceed 1100 accessions. In this case also, the majority of the accessions are landraces.

The altitudinal range of the collected teff varies from 1120 to 2950 m. The intensity of collections from the different administrative regions and the overall mean are shown in Table 2.

Table 3. *Altitudinal range and frequency of occurrence of the various crops*

Niger		*Brassica* spp.		Chickpea		Field pea		Faba bean	
Altitude (m)	%	Altitude (m)	%	Altitude (m)	%	Altitude (m)	%	Altitude (m)	%
≤1300	1.11	≤1300	1.07	≤1300	0.28	≤1600	0.30	≤1800	0.79
1301–1500	3.54	1301–1500	1.70	1301–1500	3.54	1601–1800	0.60	1801–2000	5.24
1501–1700	9.50	1501–1700	7.86	1501–1700	7.88	1801–2000	3.29	2001–2200	9.43
1701–1900	19.87	1701–1900	17.20	1701–1900	22.01	2001–2200	6.87	2201–2400	15.71
1901–2100	21.42	1901–2100	16.78	1901–2100	24.19	2201–2400	17.62	2401–2600	27.23
2101–2300	16.12	2101–2300	15.50	2101–2300	11.69	2401–2600	25.97	2601–2800	24.87
2301–2500	17.22	2301–2500	19.96	2301–2500	16.85	2601–2800	22.09	2801–3000	13.62
2501–2700	10.60	2501–2700	14.87	2501–2700	12.50	2801–3000	15.53	≥3001	3.15
≥2701	0.67	2701–2900	3.87	≥2701	1.09	3001–3200	5.68		
		≥2901	1.28			≥3201	2.09		
Altitude range (m)									
1100–2950		1050–3170		1200–2880		1560–3380		1300–3150	

Table 3 (*cont.*)

Wheat		Barley		Teff		Sorghum		Linseed	
Altitude (m)	%	Altitude (m)	%	Altitude (m)	%	Altitude (m)	%	Altitude (m)	%
≤1500	0.19	≤1700	0.63	≤1200	0.33	≤950	5.07	≤1700	1.06
1501–1700	0.49	1701–1900	2.60	1201–1400	1.48	951–1150	1.77	1701–1900	2.46
1701–1900	3.27	1901–2100	6.91	1401–1600	7.38	1151–1350	8.81	1901–2100	9.13
1901–2100	5.70	2101–2300	8.67	1601–1800	12.63	1351–1550	20.27	2101–2300	12.28
2101–2300	11.82	2301–2500	15.58	1801–2000	20.50	1551–1750	21.81	2301–2500	20.00
2301–2500	25.50	2501–2700	21.34	2001–2200	13.28	1751–1950	17.85	2501–2700	26.32
2501–2700	30.41	2701–2900	20.41	2201–2400	15.74	1951–2150	9.26	2701–2900	14.39
2701–2900	16.60	2901–3100	15.27	2401–2600	20.66	2151–2350	8.59	2901–3100	8.78
2901–3100	5.21	3101–3300	6.39	2601–2800	6.73	2351–2550	3.75	3101–3300	3.86
≥3101	0.85	3301–3500	1.93	≥2801	1.32	2551–2750	1.55	≥3301	1.76
		≥3501	0.32			≥2751	1.33		
Altitude range (m)									
1200–3300		1500–3750		1120–2950		400–2940		1470–3430	

Barley *(Hordeum vulgare)*

Barley was given high priority since it is an important crop in Ethiopia, the third in terms of acreage. It possesses a high genetic diversity (Ethiopia is a secondary gene centre for cultivated barley) and considerable genetic erosion is being observed. The number of accessions collected, as well as the size of the total barley collection maintained at PGRC/E, reflect this priority (Tables 1 and 2).

The potentialities of the barley collections have been demonstrated by Qualset (1975), who listed an impressive number of accessions which possess resistance genes against one or more important diseases, as well as some important quality characters, e.g. high protein and lysine content. Bekele (1983a,b, 1984b) investigated the diversity existing in Ethiopian barley for the allozyme genotypic composition, the genetic distances between populations based on morphological characters and flavonoids. Significant differences between regions were found and conclusions for future collection and conservation were drawn. Engels (1987) compared the diversity indices for the administrative regions and for the country as a whole, for a large collection, with the results of earlier diversity studies and concluded that the diversity indices for the different regions are generally not significant and that the overall index for Ethiopia is high.

Pulses

A number of important food crops belong to this section. Some species (e.g. *Vicia faba, Pisum sativum* and *Cicer arietinum*) have built up a significant diversity in Ethiopia and form an important source for the local and international breeding programmes. Therefore, considerable efforts have been put into the collection of these crops and a total of almost 3000 accessions have been collected throughout Ethiopia (Table 1). The collection intensity of some of the major pulses is comparable with the other major crop species (Table 2). The altitudinal range for some of the pulse crops can be observed in Table 3.

Root and tuber crops

At present, only a modest collection of root and tuber crops is maintained by PGRC/E. A total of some 100 accessions of various species has been collected so far, including a sweet potato collection of 42 accessions, as well as yams (*Dioscorea* spp.), *Coccinia abyssinica* and *Colocasia esculenta*). This collection includes species which are of regional importance in times of drought and food shortage (e.g.

Table 4. *Some of the minor native species with potential value collected by PGRC/E*

Scientific name	Common name	Uses	Distribution
Abelmoschus spp.	Okra (incl. wild species)	Vegetable	Ilubabor, Welega
Arisaema sp.	Burie	Root crop	Gamo Gofa, Sidamo
Amorphophallus abyssinicus	Bagana	Root crop	Gamo Gofa
Oryza longistaminata	Wild rice	Food grain	Gojam, Ilubabor
Brassica oleracea	Gurage gomen	Leaf vegetable	Shewa, Sidamo
Coccinia abyssinica	Anchote	Tuber crop	Welega, Shewa
Plectranthus edulis	Oromo dinich	Tuber crop	Shewa, Welega
Ensete ventricosum	Ensete	Edible pseudostem	Shewa, Kefa, Gamo Gofa, Sidamo
Ipomoea batatas	Sweet potato	Root crop	Harerge, Gamo Gofa, Sidamo, Shewa
Moringa stenopetala	Cabbage tree, Shiferaw, Haleko	Leaf vegetable	Gamo Gofa
Sauromatum nubicum	Banshalla	Root crop	Gamo Gofa, Sidamo
Trigonella foenum-graecum	Fenugreek	Baby food	Spread over the country

Amorphophallus abyssinicus and *Sauromatum nubicum*). The sweet potato collection is being systematically screened for reaction to some prevalent diseases and pests and increasing use is being made of traditional varieties, especially for their adaptability to local stress conditions.

Oil crops

Oil crops are another important group of food plants which are grown extensively in the country. Some species are native to Ethiopia and were first taken into domestication in the Ethiopian highlands, e.g. niger seed or noog (*Guizotia abyssinica*) and Ethiopian mustard or gomen (*Brassica carinata*). Other oil crops such as linseed (*Linum usitatissimum*), sesame (*Sesamum indicum*) and safflower (*Carthamus tinctorius*), have been given due priority (Table 1).

The intensity of collecting for noog ranges from one sample for every 3 hectares in Welo to one sample per 625 hectares in Bale. The overall acreage per collected accession for the whole of Ethiopia is 163 hectares and this is comparable to the intensity for other crops. The considerable altitudinal range, from 1100 to 2950 m, is worth mentioning and shows the broad adaptability of the crops in question.

Miscellaneous and under-utilized crop plants

During the last 10 years, PGRC/E has also explored for and collected some lesser known but potentially valuable indigenous crop plants. These include some regionally important crops which are utilized by local people in times of food shortages (Table 4).

References

Anonymous (1981). Some technical information on seeds. Ethiopian Seed Corporation, Addis Ababa (mimeographed).

Anonymous (1984). General agricultural survey. Preliminary report 1983–4, vol. I. Planning and Programming Department, Ministry of Agriculture, Addis Ababa (mimeographed).

Astley, D., Mahteme, H. G. & Toll, J. (1982). Collecting brassicas in Ethiopia. *Plant Genetic Resources Newsletter*, **51**, 15–20.

Bekele, E. (1983a). Allozyme genotypic composition and genetic distance between the Ethiopian landrace populations of barley. *Hereditas*, **98**, 259–67.

Bekele. E. (1983b). A differential rate of regional distribution of barley flavonoid patterns in Ethiopia and a view on the centre of origin of barley. *Hereditas*, **98**, 269–80.

Bekele, E. (1984a). Analysis of regional pattern of phenotypic diversity in the Ethiopian tetraploid and hexaploid wheats. *Hereditas*, **100**, 131–54.

Bekele, E. (1984b). Relationships between morphological variance, gene diversity and flavonoid patterns in the landrace populations of Ethiopian barley. *Hereditas*, **100**, 271–94.

Chapman, C. D. G. (1985). *The Genetic Resources of Wheat: a survey and strategies for collecting*. IBPGR, Rome.

Demissie, A. (1984). Oilcrops exploration and collection in Ethiopia. PGRC/E, Addis Ababa (mimeographed).

Ebba, T. (1978). Plant Genetic Resources Centre/Ethiopia activities and programs. Crop genetic resources in Africa. *Proceedings of a workshop jointly organized by AAASA and IITA*. International Institute of Tropical Agriculture, Ibadan, pp. 25–30.

Engels, J. M. M. (1987). A diversity study in Ethiopian barley. *In*: J. M. M. Engels (ed.), The conservation and utilization of Ethiopian germplasm. Proceedings of an international symposium, Addis Ababa, Ethiopia, 13–16 October 1986, pp. 124–32 (mimeographed).

Gentry, H. S. (1971). Pea picking in Ethiopia. *Plant Genetic Resources Newsletter*, **26**, 20–4.

Hawkes, J. G. (1976). Sampling gene pools. *Proceedings of Nato conference on conservation of threatened plants. Series 1, Ecology*. Plenum, London.

Hawkes, J. G. (1985). Report on a consultancy mission to Ethiopia for GTZ to advise PGRC/E on germplasm exploration, conservation, multiplication and evaluation. Birmingham (mimeographed).

Jutzi, S. & Gryseels, G. (1984). Oats, a new crop in the Ethiopian highlands. *PGRC/E–ILCA Germplasm Newsletter*, **5**, 22–4.

Marshall, D. R. & Brown, A. H. D. (1975). Optimum sampling strategies in genetic conservation. *In*: O. H. Frankel and J. G. Hawkes (eds), *Crop Genetic Resources for Today and Tomorrow*. Cambridge University Press, Cambridge, pp. 53–80.

Mengesha, M. H. (1975). Crop germplasm diversity and resources in Ethiopia. *In*: O. H. Frankel and J. G. Hawkes (eds), *Crop Genetic Resources for Today and Tomorrow*. Cambridge University Press, Cambridge, pp. 449–53.

Michael, T. (1983). Germplasm conservation in Eritrea. *PGRC/E–ILCA Germplasm Newsletter*, **4**, 7–9.

Prasada Rao, K. E. & Mengesha, M. H. (1981). A pointed collection of 'Zera-Zera' sorghum in the Gambella area of Ethiopia. *Genetic resources progress report 33*, ICRISAT, Patancheru.

Pundir, R. P. S. & Mengesha, M. H. (1982). Collection of chickpea germplasm in Ethiopia. *Genetic resources progress report 44*, ICRISAT, Patancheru.

Qualset, C. D. (1975). Sampling germplasm in a centre of diversity: an example of disease resistance in Ethiopian barley. *In*: O. H. Frankel and J. G. Hawkes (eds), *Crop Genetic Resources for Today and Tomorrow*. Cambridge University Press, Cambridge, pp. 81–94.

Seegeler, C. J. P. (1986). Genetic variability of oil crops in Ethiopia. Consultancy report for GTZ by PGRC/E. Oosterbeek (mimeographed).

Toll, J. (1980). Collecting in Ethiopia. *Plant Genetic Resources Newsletter*, **43**, 36–9.

Toll, J. (1982). Collecting in Ethiopia. *Plant Genetic Resources Newsletter*, **48**, 18–22.

Vavilov, N. I. (1951). The origin, variation, immunity and breeding of cultivated plants. *Chronica Botanica*, **13**, 1–366.

Worede, M. (1983). Crop genetic resources in Ethiopia. *In*: J. C. Holmes and W. M. Tahir (eds), *More Food from Better Technology*. FAO, Rome, pp. 143–7.

Zohary, D. (1970). Centres of diversity and centres of origin. *In*: O. H. Frankel and E. Bennett (eds), *Genetic Resources in Plants, their Exploration and Conservation*. Blackwell, Oxford, pp. 33–42.

15

Collection of Ethiopian forage germplasm at the International Livestock Centre for Africa

JEAN HANSON AND SOLOMON MENGISTU

Introduction

Ethiopia is an area rich in germplasm of many plant species and was considered as a primary centre of crop diversity by Vavilov (1951). Among the most important plant genetic resources of the East African region, and indeed of all of Africa, are forages and especially forage grasses (Zeven & Zhukovsky, 1975). Large areas of Africa are covered with tropical savannah with a great diversity of grasses which are vigorous and polymorphic (Clayton, 1983). Tropical forage legumes and browse species are also endemic. In particular, Africa has been described as the centre of diversity of the browse shrubs of the subfamily Caesalpinioideae (Williams, 1983).

The genetic resources of forages are usually found in wild populations since they have only been cultivated on a commercial scale for about 50 years and no landraces are available (Williams, 1983). This is very different from crop species and therefore the collection strategies for forages differ to accommodate the population structures found in the wild. Marshall & Brown (1983) have defined the objective of forage plant exploration as the collection of material with the maximum amount of useful genetic variability within a strictly limited number of samples. The strategy of the International Livestock Centre for Africa (ILCA) is to collect representative population samples from the wild, although in some cases only a few plants can be found growing together as a population and sampling is therefore limited (Lazier, 1984). Collection is always a compromise between capturing the greatest amount of variation using theoretical collection procedures and practical constraints imposed by field conditions. Reid & Strickland (1983) support the concept of Harlan (1975) which considers that, given the restrictions and practical constraints found in the field, it is more important to sample the maximum number of sites than to collect the theoretically ideal number of plants per site, because more variation is likely to be captured by sampling in different ecological zones. Whenever possible, large seed samples are preferred because these are more representative of the total variation in the wild and more likely to capture genes which occur with low frequency in the population (Marshall & Brown, 1983). Also the seed sample can be placed into the genebank without a regeneration cycle prior to storage.

Potential of Ethiopian forage species

Large numbers of grass and legume species found in Ethiopia have the potential to be developed as forages. Many species have been collected whose true value is still unknown and these must be evaluated before they can be utilized. When collecting forages for genetic resources, all plants which could prove useful as forage crops should be collected in the field. Plants which may not seem directly useful may have unseen characteristics such as disease, insect or drought tolerance. However, in addition, there are certain target species which have proven forage potential and are therefore given priority during collecting missions.

A large number of grass genera which are endemic to Africa are of special interest during collecting missions. Bogdan (1977) has identified 45 grasses which he considers to be the most important for forage. Of these, 27 are endemic to Africa, indicating the importance of collecting in this region (Clayton, 1983). Zeven & Zhukovsky (1975) have considered that *Brachiaria brizantha*, *B. decumbens*, *B. mutica*, *Chloris gayana*, *Melinis minutiflora*, *Pennisetum clandestinum* and species of *Cynodon*, *Digitaria* and *Setaria* are endemic grasses whose centre of origin lies within the Ethiopian centre of diversity.

Several important genera of forage legumes are native to Ethiopia and have been identified as targets for collecting. Zeven & Zhukovsky (1975) consider *Acacia*, *Crotalaria*, *Indigofera*, *Lablab purpureus* and *Stylosanthes fruticosa* to have their centres of diversity in Africa. The variation found within these genera and species in Ethiopia supports this. Other important legumes showing considerable variation belong to the genera *Aeschynomene*, *Alysicarpus*, *Medicago*, *Neonotonia* (i.e. *N. wightii*) and *Trifolium*. Browse species, including

Acacia, Cordeauxia, Cassia, Sesbania and *Erythrina*, are also native to East Africa and show considerable potential for development as forages, especially in drier areas where their deep roots provide some tolerance to periods of low rainfall.

Forage legumes are especially important in the native agricultural system because they both provide fodder and increase soil fertility by nitrogen fixation in association with *Rhizobia*. In Ethiopia, the native genus *Trifolium* is of interest due to its abundance and observed potential in highland areas. The *Trifolium* species collected by ILCA are being evaluated to identify suitable genotypes for use in the diverse environment and soil types found in the Ethiopian highlands (Kahurananga, 1982; Kahurananga & Tsehay, 1984).

ILCA collecting missions

The forage agronomy programme of ILCA has made 22 collecting missions in Ethiopia in the last six years. The target genera for these missions have been for the most part those which have shown potential in work under way at ILCA or in other forage research organizations. Initially, forage collecting was concentrated largely in accessible parts of the Ethiopian highlands because the potential of native African *Trifolium* species had already been recognized during some preliminary screening in Kenya (Strange, 1958; Bogdan, 1977) and in Ethiopia (Chilalo Agricultural Development Unit, 1972). Although these earlier ILCA missions were primarily for *Trifolium*, other highland legumes of potential were also collected. Later, the collecting missions were extended to the lowlands for more general collecting and also for browse species, an area of interest and research at ILCA. These missions covered a large area of the central and southern parts of the country. The major genera collected are listed in Table 1.

Most of these ILCA collecting missions were funded by the International Board for Plant Genetic Resources (IBPGR) who recognized the need to collect forages in Ethiopia. ILCA collections in 1985 were also carried out in cooperation with staff of the Centro Internacional de Agricultura Tropical (CIAT) with a special emphasis being given to the grass genus *Brachiaria* which has shown potential in Latin America.

Highland collections

There are about 64 endemic legume species in Ethiopia and most of these can be found in the highlands. The most abundant

Table 1. *Number of collections of major forage genera from the administrative regions of Ethiopia*

Genus	Administrative region[a]												Total
	Ar	Ba	Ga	Gj	Gn	Ha	Il	Ke	Sh	Si	We	Wl	
Alysicarpus							6	4	8	4	8	1	31
Argyrolobium										1			1
Cassia							3	3	1	3	1		11
Crotalaria									3	1		2	6
Desmodium				1			2	1		2		2	8
Eriosema								1		2			3
Heteropogon	1												1
Indigofera										2		1	3
Lablab		1		1						2	1		5
Lotus									1				1
Lupinus				12	2								14
Medicago	1	3							6			1	11
Neonotonia	2			4			3	6	1	10	1	2	29
Rhynchosia		1		2				2	1	6		1	13
Stylosanthes		3	1						1	95			100
Tephrosia			3				1	1	30	15	1	5	56
Teramnus	1	1		1			2	2	6			7	20
Trifolium	108	105	14	217	33		14	40	409	38	41	44	1063
Vicia	4	2	1	2			1	1	11	4	5	2	33
Vigna	2	1	2	2			1		7	3	1	7	26
Zornia									2	17			19
Andropogon	1			1									2
Brachiaria		3	5	6	3	2	4	1	2	41	25		92
Cenchrus	1									10	1		12
Chloris									1	8	2		11
Digitaria				1						4			5
Echinochloa										1		1	2
Festuca										2			2
Panicum										4		1	5
Pennisetum				2				1	2	1		3	9
Setaria	1	1							2	4	3	2	13
Acacia	1		3						28	13	1		46
Aeschynomene			1	2					2	1			6
Albizia										1			1
Dichrostachys			1	1					1				3
Entada									1				1
Erythrina		1	1			4		1		6			13
Pseudarthria							1	1		2	1		5
Sesbania				4				1	6	4			15
Total	123	122	32	259	38	6	38	66	532	307	92	82	1697

[a] Ar, Arsi; Ba, Bale; Ga, Gamo Gofa; Gj, Gojam; Gn, Gondar; Ha, Harerge; Il, Ilubabor; Ke, Kefa; Sh, Shewa;, Si, Sidamo; We, Welega; Wl, Welo.

Table 2. *Number of accessions of* Trifolium *collected in Ethiopia from 1982 to 1986 showing the range of altitudes*

	Altitude range (m)					
Species	<1500	1500–2000	2001–2500	2501–3000	<3000	Total[a]
Trifolium acaule			1	3		4
T. baccarinii	2	11	21	16		50
T. bilineatum		17	25	3		45
T. burchellianum		2	11	24	16	53
T. calocephalum			3	14	6	23
T. crytopodium			10	43	13	66
T. decorum	1	5	43	17	1	67
T. mattirolianum	2	33	12	1		48
T. multinerve			1	12		13
T. pichisermollii		1	8	12		21
T. polystachyum		9	24	10		43
T. quartinianum	1	14	15			30
T. rueppellianum		18	53	34		105
T. schimperi		5	18	4		27
T. semipilosum		5	56	71	10	142
T. simense			27	37	2	66
T. steudneri		20	43	10		73
T. tembense		5	57	75	12	149
Total						1025

[a] Thirty-eight of the 1063 accessions of *Trifolium* do not have altitude data.

genus found in the cool highlands is *Trifolium*, and 28 of the 40 species of African *Trifolium* are found in Ethiopia of which nine are endemic (Gillett, Polhill & Verdcourt, 1971; Thulin, 1983).

Collecting began in the highlands in 1980 when the first mission collected *Trifolium* species from Debre Zeit for immediate evaluation (Kahurananga, 1982). Collecting was intensified in 1982–4 when 17 missions covered all accessible areas of the central highlands (Kahurananga & Mengistu, 1983, 1984). A large number of accessions were collected during these missions, the majority belonging to 18 species of the genus *Trifolium* from a wide range of altitudes (Table 2). Accessions from other highland genera including *Vicia*, *Medicago* and *Lotus* were also collected.

Lowland collections

Two general collecting missions were made in 1984 and 1985 for lowland grasses and legumes. The mid-highland and lowland savannah and bushland of the Harerge, Bale, Sidamo, Gamo Gofa,

Welega and Bale administrative regions were a good source of forage grasses including *Brachiaria, Cenchrus, Chloris, Andropogon, Panicum* and *Cynodon*. Forage legumes were also abundant, including *Vigna, Desmodium, Alysicarpus, Macrotyloma, Zornia, Lablab, Stylosanthes* and *Neonotonia*. *Neonotonia* is very widespread and was found at a wide range of altitudes, growing in diverse vegetation types. Both *Neonotonia* and *Stylosanthes fruticosa* are vigorous and abundant in the Rift Valley and are showing considerable promise in the preliminary forage evaluation plots in that area.

Three collecting missions were undertaken in 1984 to collect browse tree and shrub germplasm and to identify browse species which could be used as fodder. These missions covered the lowlands of Harerge, Sidamo, Gamo Gofa, Kefa and Ilubabor. The collection routes covered more than 7000 km across varied vegetation types, differing terrain and a wide range of altitudes and reached as far as the Kenyan frontier in the south and the Sudan frontier in the south-west. In all, 103 accessions were collected, most of which belonged to the genera *Acacia, Aeschynomene, Albizia, Erythrina, Sesbania* and *Tamarindus* (Mengistu, 1985). Data on palatability, growth season and traditional feeding method of native forage species were collected from the rural people. This information, together with visual observations in the field, has been extremely useful to determine which genera and species are of particular interest for collecting and which are likely to show potential for development as forages.

Priorities for future collecting

Additional collecting of forages in Ethiopia is necessary to broaden further the genetic base of the forage crops available and identify promising genera and species for further study. Forage collecting at ILCA must fill gaps in current collections, salvage endangered germplasm, fulfil the needs of ILCA, national and international forage programmes and, to some extent, concentrate on the species of known forage potential.

Collection objectives are closely linked to the needs and goals of the ILCA forage programme and the need to complement the available germplasm with species of potential forage value. Native Ethiopian legumes of value include, among others, *Macrotyloma axillare, Neonotonia wightii* and *Stylosanthes fruticosa*. Ecotypes with high dry matter yield, high nitrogren fixing capacity, vigorous growth in minimal rainfall areas and wide adaptation beyond their normal environmental range will be sought. Greater emphasis will be placed on the collecting of associated *Rhizobia* with the different legumes, so that maximal productivity is achieved in the field. Other priority genera for collecting include *Acacia, Trifolium, Erythrina, Eriosema, Galactia* and *Lotus*.

Collecting will continue in specific environments where potentially important genetic variation can exist and in areas not yet fully sampled. Certain genera and species have not yet been extensively collected by ILCA. This is especially true of the grasses, where only a few collections have been made, but a large amount of variation occurs in the field. Grasses, legumes and browse are all of interest from environments which are swampy, seasonally inundated, arid or semi-arid.

ILCA will continue to collect forages in areas where germplasm is endangered and genetic erosion is occurring due to such factors as drought, land clearing, intensive cultivation and heavy grazing. Collecting around the periphery of such areas can yield very promising types because of the extreme selection pressures. Collecting can both salvage germplasm and provide a gene pool of ecologically adapted material for further development for use in similar areas.

References

Bogdan, A. V. (1977). *Tropical Pasture and Fodder Plants*. Longman, London.

Chilalo Agricultural Development Unit (1972). *Report on Surveys and Experiments Carried Out in 1972*. Publication No. 80, CADU, Ethiopia.

Clayton, W. D. (1983). Tropical grasses. *In*: J. G. McIvor and R. A. Bray (eds), *Genetic Resources of Forage Plants*. CSIRO, Australia, pp. 39–46.

Gillett, J. B., Polhill, R. M. & Verdcourt, R. (1971). Leguminoseae (part 3). *In*: E. Milne-Redhead and R. M. Polhill (eds), *Flora of Tropical East Africa*. Crown Agents, London, pp. 1016–36.

Harlan, J. (1975). Seed crops. *In*: O. H. Frankel and J. G. Hawkes (eds), *Crop Genetic Resources for Today and Tomorrow*. Cambridge University Press, Cambridge, pp. 111–15.

Kahurananga, J. (1982). ILCA's forage legume work in the Ethiopian Highlands. *ILCA Newsletter*, **1**, 5–6.

Kahurananga, J. & Mengistu, S. (1983). ILCA native forage germplasm collection in Ethiopia for 1982–1983. *PGRC/E–ILCA Germplasm Newsletter*, **3**, 6–9.

Kahurananga, J. & Mengistu, S. (1984). ILCA native forage germplasm collection in Ethiopia during 1983. *PGRC/E–ILCA Germplasm Newsletter*, **5**, 8–18.

Kahurananga, J. & Tsehay, A. (1984). Preliminary assessment of some annual Ethiopian *Trifolium* species for hay production. *Tropical Grasslands*, **18**, 215–17.

Lazier, J. R. (1984). Theory and practice in forage germplasm collection. Paper presented at PANESA Workshop on Pasture Improvement Research in Eastern and Southern Africa, Harare, Zimbabwe, 17–21 September 1984. IDRC proceedings series 237-e.

Marshall, D. R. & Brown, A. H. D. (1983). Theory of forage plant collection. *In*: J. G. McIvor and R. A. Bray (eds), *Genetic Resources of Forage Plants*. CSIRO, Australia, pp. 135–48.

Mengistu, S. (1985). ILCA browse collection in Ethiopia 1984. *PGRC/E–ILCA Germplasm Newsletter*, **8**, 19–23.

Reid, R. & Strickland, R. W. (1983). Forage plant collection in practice. *In*: J. G. McIvor and R. A. Bray (eds), *Genetic Resources of Forage Plants*. CSIRO, Australia, pp. 149–56.

Strange, R. (1958). Preliminary trials with grasses and legumes under grazing. *East African Agricultural Journal*, **24**, 92–102.

Thulin, M. (1983). Leguminoseae of Ethiopia. *Opera Botanica*, **68**.

Vavilov, N. I. (1951). The origin, variation, immunity and breeding of cultivated crops. *Chronica Botanica*, **13**, 1–366.

Williams, R. J. (1983). Tropical legumes. *In*: J. G. McIvor and R. A. Bray (eds), *Genetic Resources of Forage Plants*. CSIRO, Australia, pp. 17–37.

Zeven, A. C. & Zhukovsky, P. M. (1975). *Dictionary of Cultivated Plants and their Centres of Diversity*. PUDOC, Wageningen.

16

Germplasm conservation at PGRC/E

REGASSA FEYISSA

Introduction

Genetic conservation has arisen as a solution to some of the problems caused by Man in his social and agricultural relationship with the environment (Simmonds, 1979). Unwise exploitation of nature has caused an irreversible loss of variability and has become the major cause of worldwide genetic erosion. The seriousness and rapid expansion of the problem has created a universal need to collect and conserve genotypes that would no longer be available if not conserved today. This can best be achieved by maintaining a wide range of plant materials covering the maximum variability existing at present.

Taking into account these needs, and being aware of the enormous diversity of crops in Ethiopia, the Plant Genetic Resources Centre (PGRC/E), is currently working on the conservation of both orthodox and recalcitrant crops. At present, the centre holds 40 000 accessions of 78 different species, including the germplasm material preserved in field genebanks.

Facilities, personnel and organization of the Conservation Division

The longevity of any conserved material depends upon the system of conservation used and this, in turn, is affected by the facilities existing at any given genebank and the quality of technical knowledge available. The inadequacy of the infrastructures for the maintenance and utilization of plant genetic resources remains the major limiting factor in the establishment of a genebank in a developing country. Storage facilities require large inputs in terms of construction, equipment and maintenance costs, as well as capable technicians and a reliable electricity supply. The size of accessions to be stored, their safety, frequency of rejuvenation and flow of samples in evaluation and exchange are dependent on the quality and size of the storage facilities. Similarly, the equipment required by a genebank varies according to the number of samples and species to be stored.

The diversity of species maintained at PGRC/E requires an assortment of equipment. At present the Conservation Division is equipped with germination room, maintained at about +20 °C with 80 per cent relative humidity (RH), and a germinator with alternating temperature. The drying room is fitted with a dehumidifier and an air conditioner which allow operation at 18–20 °C with 15–18 per cent RH. The seed processing and germination test laboratories are equipped with seed blowers, several types of scales, a vacuum system with counting heads used to place seeds in petri dishes, a metal can sealer, and a sealer for aluminium foil bags. The three cold stores possess a total volume of about 225 m^3 of which 175 m^3 is used for storage at −10 °C. The remaining 50 m^3 store is maintained at +4 °C and 35 per cent RH for temporary storage of accessions with insufficient seeds for long-term storage. The cold rooms are fitted with mobile shelves which give convenient working conditions and economize on space. With the help of a computerized numbering system each of the accessions can be located without problems. Electricity is supplied by two independent lines and a standby generator automatically switches on if the power supply fails or the voltage varies too much. Personnel in the Conservation Division includes two physiologists, five trained technical assistants, eight laboratory assistants and eight supporting staff.

Efficiency and caution must be given high priority during the processing and storage of the seeds and during the compilation of information. This can be achieved by organizing the activities in an orderly and logical manner. Figure 1 shows the organization and distribution of activities between units in the Germplasm Conservation Division at PGRC/E.

The process of seed deterioration begins immediately after the seeds have matured on the plant itself. Thus, seeds collected from the farmers' fields, stores and marketplaces, as well as from seed increase, have to be processed and stored at the genebank as quickly as possible to avoid further deterioration. A general scheme developed by PGRC/E, which helps to perform these activities efficiently, is presented in Fig. 2.

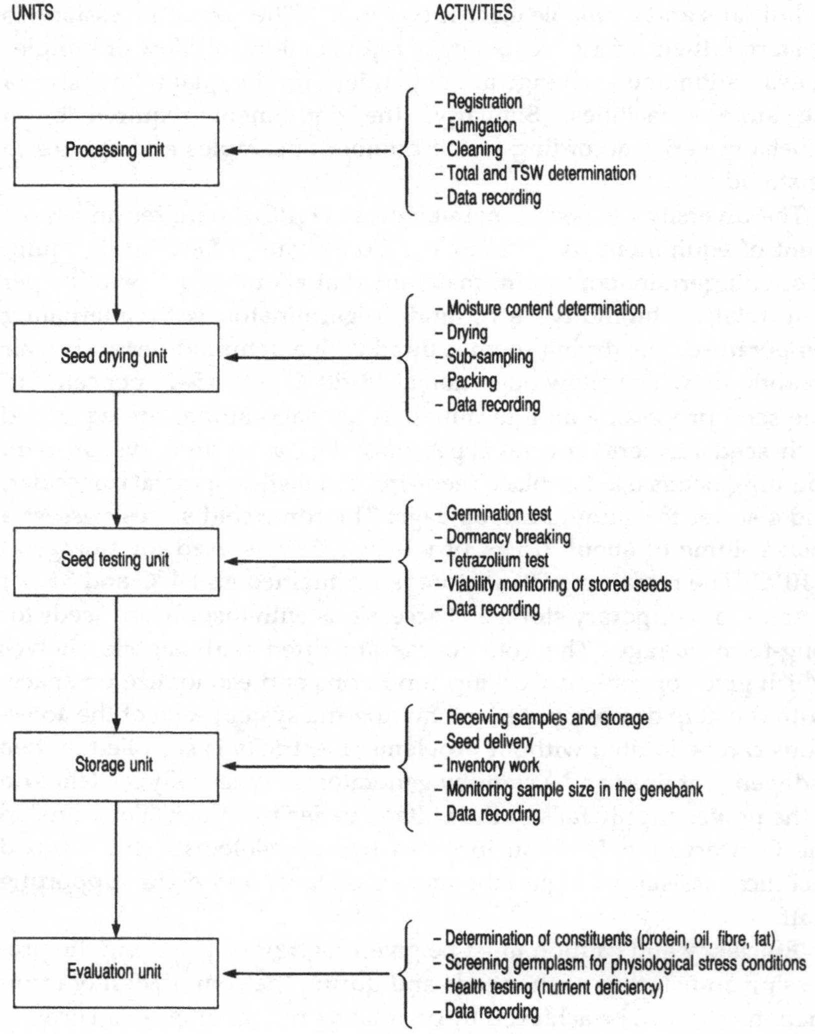

Fig. 1. Organization of PGRC/E's conservation units and their respective activities.

Sources of germplasm

Collection, selection, donation and repatriation are the main sources of our germplasm material (Table 1). Details of the major crop species are presented in Table 2 and in a paper on Documentation at PGRC/E (Sendek & Engels, 1988, Chapter 17).

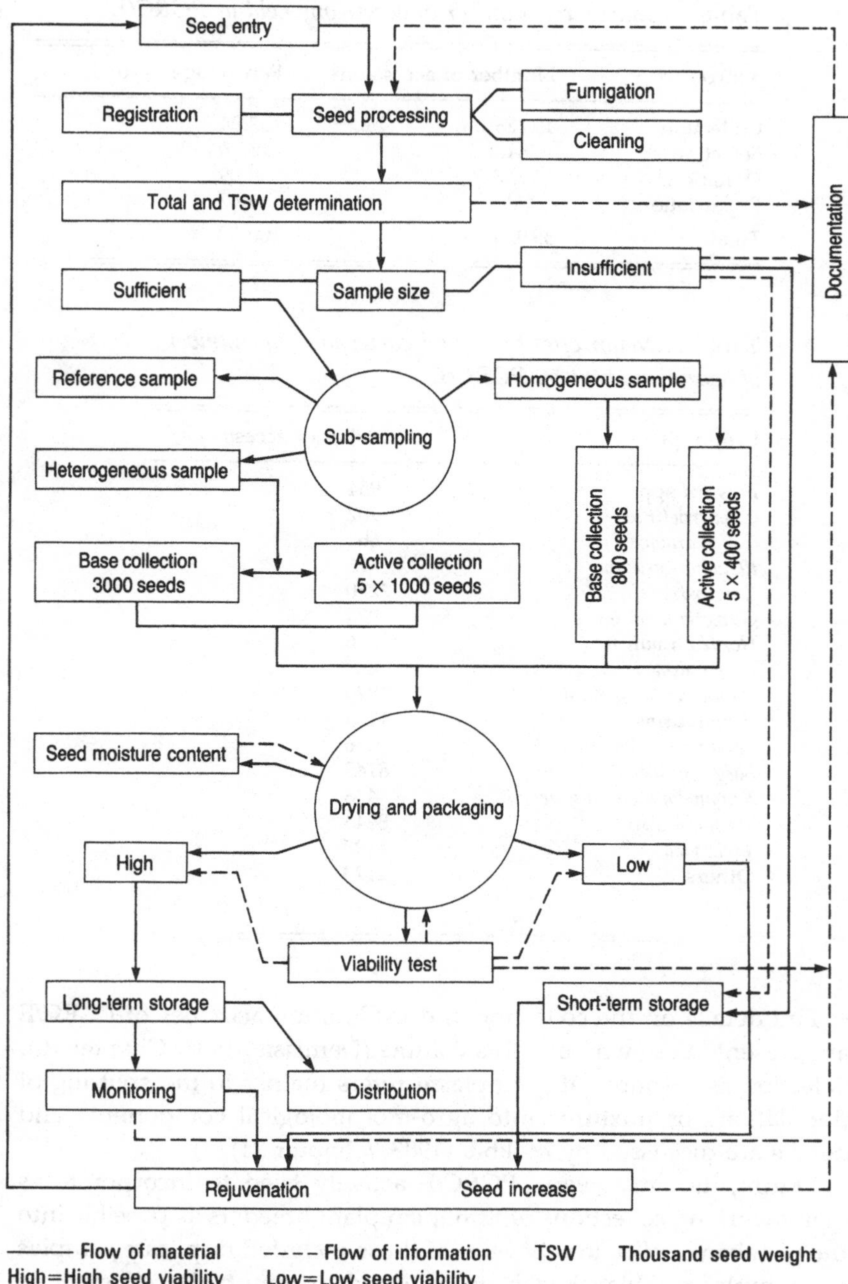

Fig. 2. Organizational chart of the seed conservation division at PGRC/E.

Table 1. *Source and number of accessions held at PGRC/E*

Source	Number of accessions	Percentage of total
Collection	15126	37.84
Selection	5840	14.60
Donations	15227	38.09
Repatriation	3785	9.47
Total	39978	100.00

Table 2. *Major crop types and corresponding number of accessions kept by PGRC/E*

Crop type	Number of accessions
Brassica spp.	954
Cicer arietinum	902
Coffea arabica	662
Eleusine coracana	796
Eragrostis tef	2270
Guizotia abyssinica	924
Hordeum vulgare	9316
Lens culinaris	427
Linum usitatissimum	1820
Pisum sativum	1133
Sesamum indicum	376
Sorghum bicolor	8145
Trigonella foenum-graecum	438
Triticum spp.	8444
Vicia faba	1198
Others	2173
Total	39978

Full details on the collection and exploration activities of PGRC/E are presented elsewhere in this volume (Demissie, 1988, Chapter 16). Selection as a source of germplasm refers mainly to the splitting of populations or mixtures into agro-morphological components and details are discussed by Mekbib (1988, Chapter 21).

During its early years, PGRC/E actively tried to incorporate as many working collections of Ethiopian plant breeders as possible into the genebank collection. Several of these included duplicate samples of germplasm which had been collected by reputable collectors or institutes, mainly during the 1960s.

Since the International Board for Plant Genetic Resources (IBPGR)

Table 3. *Sample size required for long-term storage*

Sample type	TSW 5–200 g	TSW >200 g
Heterogeneous		
Total	8000 seeds	4000 seeds
B.C.	3000 seeds	1500 seeds
A.C.	5 × 1000 seeds	5 × 500 seeds
Homogeneous		
Total	3200 seeds	1600 seeds
B.C.	800 seeds	400 seeds
A.C.	6 × 400 seeds	6 × 200 seeds

B.C., base collection; A.C., active collection; TSW, thousand seed weight.

assigned global or regional responsibilities to PGRC/E for a number of crops, duplicate samples are regularly received for conservation purposes. Major collections in several countries, e.g. USA, USSR, Japan, Italy and the Netherlands, form an important source of Ethiopian germplasm. This germplasm was collected in the past and the duplicate samples left behind in Ethiopia have sometimes been lost, partly because of inadequate storage.

Preparing seeds for long-term storage

Registration and cleaning

The first step in seed preparation is the registration of samples. Identification data related to each sample must be recorded and care taken in order to avoid duplications and/or errors. As a phytosanitary measure, seeds coming into the genebank are fumigated for 72 hours with phosphine to control further infestation and damage. Due to the heterogeneity of some of the samples, mechanical selection of seeds is likely to occur if seed cleaning equipment is used. To avoid this, any debris or seeds from other species are cleaned away by hand.

Sample size and thousand seed weight determination

Thousand seed weight (TSW) and the total available number of seeds are two of the important characters which need recording for each accession held by a genebank. TSW is determined electronically; data are recorded in a crop specific 'seed processing file' and later transferred to the computerized documentation system. The required sample size for long-term storage is 8000 seeds for genetically heterogeneous and 3200 seeds for genetically homogeneous material. For pragmatic and economic reasons, the final sample size of species with a TSW greater than 200 g is reduced to a smaller number of seeds. Samples fulfilling the minimum requirements are subdivided according to their storage status (Table 3).

Seed drying

Seed drying is a complex process and its impact varies according to the nature of the seed and the drying conditions applied. The purpose of drying the seeds is to minimize the rate of seed deterioration during storage.

The viability of seeds can be significantly affected by overdrying or exposure to high temperatures. The equilibrium seed moisture content at a given drying temperature depends on the relative humidity of the ambient air, air flow rate, and the oil content of the seeds. The higher the oil content the lower the equilibrium seed moisture content due to the hydrophobic nature of lipids (Cromarty, Ellis & Roberts, 1982). Nevertheless, seeds of various species have different equilibrium hygroscopic relationships and thus, when exposed to a given relative humidity, will have different moisture contents after the equilibrium is reached. By extending the drying period and lowering the relative humidity for those species which have the highest equilibrium seed moisture content, it is possible to achieve identical moisture content for seeds of different species (Cromarty *et al.*, 1982). When seeds are dried, it is important that the uniformity of the equilibrium moisture content within a seed lot is kept and this can be achieved by thin-layer drying or through adequate ventilation.

At PGRC/E seeds are dried under forced ventilation in a drying room maintained at 18–20 °C and 18 per cent RH. Samples are dried in cloth bags or plastic net bags in thin layers on shelves made of wire mesh in which the air can circulate freely. Initial and final moisture content is determined in compliance with the International Seed Testing Association (ISTA, 1985) rules, by oven method and, more recently, with a near-infrared reflectance analyser.

The desired moisture content for cereals and pulses is 4–7 per cent of the seed weight and for oil crops 3–5 per cent. Dried seeds are packed and sealed in moisture-proof aluminium foil bags and all relevant data are entered into the corresponding data file.

Viability test

Viability is the most important attribute of any seed stored in a genebank. The initial viability test is made after drying just before long-term storage and this viability is monitored at regular intervals during storage. ISTA rules (ISTA, 1985) and IBPGR recommendations (Ellis, Hong & Roberts, 1985) are followed as a guide for selecting viability assessment procedures for the various species.

The most commonly used method at PGRC/E is the standard germination test. This is conducted either in a room maintained at a temperature of about ±20 °C or in an incubator with an alternating temperature facility. The result of the test is taken as a percentage of 4 × 50 seeds that are able to produce normal seedlings. In cases where viability is thought to be affected by dormancy, a tetrazolium test is conducted.

Viability during the course of storage is monitored by using the sequential germination test method (Ellis, Roberts & Whitehead, 1980) for an 85 per cent regeneration standard using groups of 40 seeds.

Seed storage

The longevity of seeds under storage depends mainly on the moisture content of the seed and the storage temperature. According to Harrington's rule of thumb, longevity of seed is doubled for every 1 per cent reduction in moisture content and 5 °C in temperature (Harrington, 1963). Various genebanks maintain their germplasm under different storage conditions for a number of reasons. At PGRC/E, seeds meant for both base and active collections are dried to 3–7 per cent moisture content and are kept at −10 °C in laminated aluminium foil bags. Germplasm accessions which are too small to fulfil the sample size required for long-term storage are kept in paper bags at ±4 °C and 35 per cent RH. According to the established criteria these accessions are increased as soon as possible in order to meet the minimum sample size to allow long-term storage.

Seed distribution

Seeds held in the genebank are distributed mainly for research work, increase and rejuvenation. Full details on PGRC/E's germplasm flow policy are given by Kebebew (1988).

At PGRC/E, germplasm is distributed only from active collections and the number of seeds distributed for research purposes depends on several factors such as seed size, homogeneity, precise purpose, etc. For seed increase and rejuvenation, the number of seeds to be sown depends on the genetic composition of the sample, the amount of seed available and the required sample size for long-term storage.

Too frequent exposure of germplasm to either rejuvenation or multiplication is also avoided.

References

Cromarty, A. S., Ellis, R. H. & Roberts, E. H. (1982). *The Design of Seed Storage Facilities for Genetic Conservation*. IBPGR, Rome.

Demissie, A. (1988). A decade of germplasm exploration and collection activities by PGRC/E. *In*: J. M. M. Engels (ed.), The conservation and utilization of Ethiopian germplasm. Proceedings of an international symposium, Addis Ababa, 13–16 October 1986, pp. 28–41 (mimeographed).

Ellis, R. H., Hong, T. D. & Roberts, E. H. (1985). *Handbook of Seed Technology for Genebanks, vol. I, Principles and methodology*. IBPGR, Rome.

Ellis, R. H., Roberts, E. H. & Whitehead, J. (1980). A new, more economic and accurate approach to monitoring the viability of accessions during storage in seed banks. *Plant Genetic Resources Newsletter*, **41**, 3–15.

Harrington, J. F. (1963). Practical instructions and advice on seed storage. *Proceedings of the International Seed Testing Association*, **28**, 289–94.

International Seed Testing Association (1985). International rules for seed testing. *Seed Science and Technology*, **13**, 432–63.

Kebebew, F. (1988). Germplasm exchange and distribution by PGRC/E. *In*: J. M. M. Engels (ed.), The conservation and utilization of Ethiopian germplasm. Proceedings of an international symposium, Addis Ababa, 13–16 October 1986, pp. 276–84 (mimeographed).

Mekbib, H. (1988). Crop germplasm multiplication, characterization, evaluation and utilization by PGRC/E. *In*: J. M. M. Engels (ed.), The conservation and utilization of Ethiopian germplasm. Proceedings of an international symposium, Addis Ababa, 13–16 October 1986, pp. 170–78 (mimeographed).

Sendek, E. & Engels, J. M. M. (1988). Documentation at PGRC/E. *In*: J. M. M. Engels (ed.), The conservation and utilization of Ethiopian germplasm. Proceedings of an international symposium, Addis Ababa, 13–16 October 1986, pp. 87–96 (mimeographed).

Simmonds, N. W. (1979). *Genetic Conservation: an Introductory Discussion of Needs and Principles. Seed technology for genebanks*. IBPGR, Rome, pp. 1–2.

17

Documentation at PGRC/E

ENYAT SENDEK AND J. M. M. ENGELS

Introduction

The quantity and complexity of the information acquired by the Plant Genetic Resources Centre/Ethiopia (PGRC/E) through active collecting, donation and repatriation of germplasm require comprehensive and efficient data management systems. Plant genetic resources can only be successfully utilized if detailed and reliable data on each accession are available to genebank users, breeders, research workers and policy makers.

In 1979, a proposal was made to base such a documentation system on edge-punched cards (Engels, 1979) and this manual system was used until 1982, by which time the amount of data had reached such proportions that the system was overcharged and it became necessary to computerize the documentation activities (Engels, 1985).

At present the documentation system at PGRC/E is based on electronic data processing technology and this has greatly facilitated the handling of the enormous amount of data currently being generated at the centre.

Information sources and descriptor development

The major sources of information in PGRC/E are:

- exploration and collection;
- germplasm introduction and accessioning;
- temporary storage (at +4 °C and 30–40 per cent RH);
- field genebanks;
- multiplication and rejuvenation;
- characterization and preliminary evaluation (in field and laboratory);
- seed processing;

Table 1. *Sources of information, recording method, number of descriptors involved and relationship with the documentation division*

Source	Method of recording original data	Number of descriptors	Relationship with documentation and remarks
Exploration/collection	Form	24	Passport data[a]
Germplasm introduction and accessioning	Master book	13	Reference[b]
	Form	10	Internal use for redistribution
Temporary storage (+4 °C, 40% RH)	Form	8	Short-term storage data[a]
Field genebanks	To be developed	Crop specific	Field management data[a]
Multiplication/rejuvenation	Form	36	Climatic, cultivation and soil data[a]
Characterization and preliminary evaluation (field)	Forms	Crop specific	Direct entry[a]
Characterization and preliminary evaluation (laboratory)	Forms	Crop specific	Direct entry[a]
Seed processing:			
(a) 1000 grain weight and total weight	Form	7	–
(b) seed drying and packing	Form	–	Internal use only
Seed testing:			
(a) germination	Form	23	Direct entry[a]
(b) seed moisture content	Form	7	Direct entry[a]
Seed storage of active and base collection at −10 °C	Form	7	Long-term storage data[a]
	Form	6	
Germplasm dispatch	Forms (3)	7	Partly registration, partly internal management[b]

Further evaluation:			
(a) cytogenetic studies	Form	1	Direct entry[a]
(b) Near-Infra-Red analyses	Computerized	2	Computers not compatible; re-entry[a]
(c) disease resistance screening	Report	Open	Extraction[b]
(d) environmental stress resistance screening	Lists	2	Direct entry[a]
(e) taxonomic studies	Reports	Open	Extraction[b]
(f) literature, reports	Various	Open	External information[b]
Utilization	Reports	Open	External information
Computer services to users	Computer file	—	Internal use only[a]
Library indexing	—	—	Planned
Miscellaneous	Lists, reports	—	Mainly internal services

[a] Computerized.
[b] Partly computerized.

Source: Engels, 1986.

- seed testing;
- seed storage (base and active collection at −10 °C);
- germplasm flow;
- further evaluation;
- utilization;
- miscellaneous activities.

Further details are presented in Table 1 and can also be found elsewhere in this volume (Chapters 14, 16, 19).

Data from all these sources are forwarded to the documentation division, generally on specially designed forms to facilitate further handling. The information is transformed and handled in the form of descriptors and their corresponding states. Each descriptor is properly defined and its possible states are chosen in such a way that no overlapping occurs (International Board for Plant Genetic Resources, 1984). As far as possible, the international norms and recommendations are followed to define and compile descriptors for the various activities. The descriptors and their definitions for characterization, preliminary and further evaluation are developed in consultation with the scientists who utilize germplasm, mainly plant breeders.

The amount of data generated at each stage, i.e. from collection, through processing and multiplication/characterization, to final storage, is considerable. To give an example, one accession of wheat will have 88 items of information or descriptors (32 for passport data, 35 for seed storage and processing and 21 for characterization). This means that for wheat alone, with a collection of 8500 accessions, a total of 748 000 data items will be generated. For the complete germplasm holding at PGRC/E it will amount to some 4 000 000 data items. To get useful information out of such a vast amount of data, it is necessary to handle the data systematically in order to meet the needs of germplasm users as well as the different sections within the genebank.

Activities of the documentation division

The day-to-day activities of the documentation division are organized in the following units:

- germplasm accessioning and data acquisition;
- data compilation, preparation, data entry and correction;
- data processing, retrieval and research.

The first unit deals with material newly arrived at the genebank. It is the responsibility of the germplasm collector or donor to supply proper and complete data sets to the genebank. The filled data form-sheets are then sorted by genus and species as well as by collection or donation number before the assignment of a unique number in the master book. A copy of the information will be forwarded to the conservation division for the necessary follow-up action. Since the establishment of PGRC/E, some 40 000 accessions have been registered. They comprise new collections and selections out of the populations, as well as donations from national and international institutions (Table 2).

The main task of the second unit is the compilation of the received and/or actively collected data. These data are converted, combined, completed, etc., according to the standards established at PGRC/E, by defining each of the descriptors and their respective states. Furthermore, reference files for the Latin names of the genera and species and another for the administrative units (e.g. regions, awrajas and woredas) have been developed for easy checking. After the data set is completed it is entered into the computer and a first printout is made, to be corrected as necessary by the division concerned.

The third and most important component of the documentation division is the data management system. Processing and retrieval activities as well as provision of data requested by plant breeders are handled and assistance for research operations is given. This unit is also responsible for the production of seed lists and catalogues in order to disseminate information to potential users. At the same time, it greatly facilitates the monitoring function of the conservation division and supports the collection division in planning new collecting activities. The research activities include, among others, diversity studies in the various crops based on the evaluation data, as well as services provided to scientists for the analysis of experimental data.

In general, it can be said that the information management system supplies the genebank administration and management with the necessary facts and details for optimal operation.

Data management system

The computers in use at PGRC/E are two HP 125 micro-computers each with 64 Kbytes of memory. Accessories include double disc drives with 5¼-inch floppy diskettes of 248 Kbytes each, a hard disc winchester drive, a T-switch and a daisy wheel printer. The software comprises DBASE II (Ashton-Tate, 1982), WORD/125 (Hewlett Packard, 1982), MICROSTAT (Anonymous, 1981a), WORDSTAR/125 (Anonymous, 1981b), BASIC/125 (Hewlett Packard, 1981) and STATPAK (Anonymous, 1982).

Table 2. *Germplasm accessions donated to or collected and/or selected by PGRC/E as at 30 June 1986*

Species	PGRC/E collected accessions	Donated accessions	Selected accessions	Total
Abelmoschus esculentus	9	0	0	9
Alframomum korarima	16	0	0	16
Allium spp.	28	0	0	28
Amaranthus spp.	30	0	0	30
Amorphophallus sp.	8	0	0	8
Arachis hypogaea	15	4	0	19
Arisaema sp.	1	0	0	1
Avena spp.	23	0	0	23
Brassica spp.	668	286	0	954
Cajanus cajan	27	0	0	27
Capsicum spp.	112	14	0	126
Carthamus tinctorius	75	56	0	131
Carum copticum	16	0	0	16
Celosia sp.	1	0	0	1
Cicer arietinum	728	12	162	902
Coccinia abyssinica	5	0	0	5
Coffea arabica	140	522	0	662
Colocasia sp.	21	0	0	21
Corchorus olitorius	1	0	0	1
Coriandrum sativum	42	1	0	43
Crambe abyssinica	1	0	0	1
Cucurbita spp.	54	12	0	66
Cuminum cyminum	6	0	0	6
Curcuma longa	1	0	0	1
Cyphomandra betacea	1	0	0	1
Datura stramonium	1	0	0	1
Dioscorea sp.	13	0	0	13
Eleusine africana	1	6	0	7
Eleusine coracana	423	373	0	796
Embelia schimperi	8	1	0	9
Eragrostis tef	1067	1203	0	2270
Fagopyrum esculentum	1	0	0	1
Gossypium spp.	5	0	0	5
Guizotia abyssinica	638	286	0	924
Guizotia scabra	6	0	1	7
Helianthus annuus	20	0	0	20
Hordeum vulgare	3290	4088	1938	9316
Ipomoea batatas	46	0	0	46
Lablab purpureus	37	0	0	37
Lagenaria spp.	11	0	0	11
Lathyrus sativus	156	61	0	217
Lens culinaris	293	134	0	427
Lepidium sativum	57	0	0	57
Linum usitatissimum	538	1250	32	1820
Lupinus spp.	25	0	0	25

Table 2 (*cont.*)

Species	PGRC/E collected accessions	Donated accessions	Selected accessions	Total
Lycopersicon spp.	4	15	0	19
Medicago sativa	2	3	0	5
Moringa stenopetala	2	0	0	2
Myrsine africana	1	0	0	1
Nicotiana tabacum	0	24	0	24
Nigella sativa	28	0	0	28
Ocimum spp.	13	0	0	13
Oryza spp.	24	0	0	24
Oxytenanthera abyssinica	1	0	0	1
Pennisetum typhoides	5	10	0	15
Phaseolus spp.	256	7	0	263
Phytolacca dodecandra	144	0	0	144
Pimpinella anisum	3	0	0	3
Piper longum	3	0	0	3
Pisum sativum	702	420	11	1133
Plectranthus edulis	6	0	0	6
Raphanus sativus	5	0	0	5
Ricinus communis	210	62	0	272
Rumex abyssinica	1	0	0	1
Ruta chalepensis	1	0	0	1
Sesamum indicum	218	158	0	376
Solanum incanum	8	0	0	8
Sorghum bicolor	1144	5419	1582	8145
Tamarindus indica	0	1	0	1
Trigonella foenum-graecum	224	214	0	438
Triticum spp.	2376	3890	2114	8380
Vernonia spp.	22	0	0	22
Vicia faba	725	473	0	1198
Vigna unguiculata	35	0	0	35
Voandzeia subterranea	1	0	0	1
Zea mays	179	7	0	186
Zingiber officinale	48	0	0	48
Unknown	6	0	0	6
Grand total	15 062	19 012	5839	39 913

In order to increase efficiency in the data management, a general data base system has been designed in which each description has its own fixed place and which allows extension of the number of descriptors and the number of accessions cropwise all the time. The main characteristics of the general data base are:

- the accessions are grouped by crop;
- the descriptors are grouped according to main activities (e.g. passport data, conservation and evaluation);

Table 3. *Files and types of data presently kept by the documentation division of PGRC/E*

File name	Number of files	Type of data	Number of descriptors	Number of records
PGRCE	1	Passport	24	39 978
SEED:STR	1	Seed storage and testing	25	27 755
SEED:CHR	30	Characterization and evaluation Crop specific	9–29	14 110
PGRCCOLN	1	Summary of PGRC/E holdings	6	78
SEED:REF	1	Crop reference	5	119
LOC:REF	1	Locality reference	3	552
SEED:DES	1	Seed dispatch	12	41
SEED:REC	1	Seed acquisition	11	28
DESC:DEFN	1	Descriptor definitions	8	233

- new accessions can be added without restrictions (vertical increase);
- new descriptors can be added without limitations (horizontal increase).

The files at present in use within the PGRC/E documentation division are listed in Table 3. Each contains a specific type of data.

The passport data base at present contains data on almost 35 000 accessions of approximately 75 different crops or groups of crop types. For each accession, 24 descriptors are recorded: accession number, crop name, genus, species, subspecies, local name/cultivar, country of origin, administrative region/state, awraja/district, woreda/area, village/locality, latitude, longitude, altitude, collecting institute, collection team/collector, collector's number, donor number, collection date (day, month, year), collection source, sample type and genetic status.

Some applications of the data management system

The importance of the documentation system for the genebank can be readily illustrated by listing its uses:

- seed inventory, monitoring and handling (seed storage and testing file);
- assisting the exploration and collection division to plan collecting missions by area, crop and time (passport file);
- assisting the respective divisions with planning and decision

making in germplasm rejuvenation, multiplication and/or characterization (seed storage and characterization files);
- facilitating exchange of germplasm through the publications of germplasm lists (passport file);
- supporting plant breeders' use of germplasm by analysing the data and publishing the available information in crop catalogues (characterization/evaluation files);
- helping the users of germplasm to make a first (rough) selection from the available germplasm based on specific criteria (characterization/evaluation files);
- analysing the available information on accessions and collection sites to predict possible areas where germplasm with specific traits can be found (passport and characterization/evaluation files);
- carrying out taxonomic analyses and classification using the characterization data;
- identifying of duplicate accessions, unknown accessions, etc.

Publications and library

The PGRC/E–ILCA Germplasm Newsletter is published jointly with the Forage Legume Agronomy Group (FLAG) of the International Livestock Centre for Africa (ILCA) and produced on the PGRC/E computing system. Its first issue appeared in December 1982. The newsletter is circulated both nationally and internationally to more than 2000 addresses and has already shown its value as a link between the germplasm centres and the users of these genetic resources worldwide.

The PGRC/E library has a considerable number of specialized books in stock, as well as a collection of reprints in the field of plant genetic resources.

References

Anonymous (1981a). *MICROSTAT User's Manual*. Lifeboat Associates, New York.

Anonymous (1981b). *WORDSTAR/125 Reference Manual*. Micropro International Corporation, San Rafael, California.

Anonymous (1982). *The NWA STATPAK, Version 2.1 Preliminary Manual*. Northwest Analytical Incorporated, Portland, Oregon.

Ashton-Tate (1982). *dBASE II Version 2.3B Assembly-Language Relational Database Management System*. Ashton-Tate, Culver City, California.

Engels, J. M. M. (1979). Proposal for a documentation system for the Plant Genetic Resources Centre at Addis Ababa, Ethiopia. GTZ, Eschborn (mimeographed).

Engels, J. M. M. (1985). Documentation and information management at PGRC/E. *PGRC/E–ILCA Germplasm Newsletter*, **9**, 20–7.

Engels, J. M. M. (1986). The documentation at the Plant Genetic Resources Centre/Ethiopia. *Acta Horticulturae*, **182**, 387–92.

Hewlett Packard (1981). *BASIC/125*. Hewlett Packard, Sunnyvale, California.

Hewlett Packard (1982). *WORD/125*. Hewlett Packard, Cupertino, California.

International Board for Plant Genetic Resources (1984). *Annual Report*. IBPGR, Rome.

Part IV

Evaluation and utilization of Ethiopian genetic resources

18

Germplasm evaluation with special reference to the role of taxonomy in genebanks

J. G. HAWKES

Introduction

Of all the varied activities of genetic resources centres, that of evaluation is probably the most neglected. Genetic resources centres in general carry out excellently their tasks of exploration, accession of samples and all the complex processes of seed storage and data management. They perhaps do not pay enough attention to surveys and many only undertake the initial stages of evaluation, thus to some extent hindering the subsequent processes in the management chain (Fig. 1). This means that the important function of a genetic resources centre may be deflected, that is to say, the material in it may not be made completely available to the breeders and thus may run the risk of not being incorporated into their new selections and varieties. In this way much of the money and effort expended in exploration and conservation is in danger of being wasted and the very existence of the genebank itself may be threatened (Hawkes, 1985a).

Evaluation

To understand the above assertion we should look more closely at the activities of germplasm evaluation in the broad sense. The International Board for Plant Genetic Resources (IBPGR) has usefully divided the process of evaluation into three main categories (Erskine & Williams, 1980).

- Characterization. This includes the scoring of morphological and agronomic characters of high heritability, not likely to

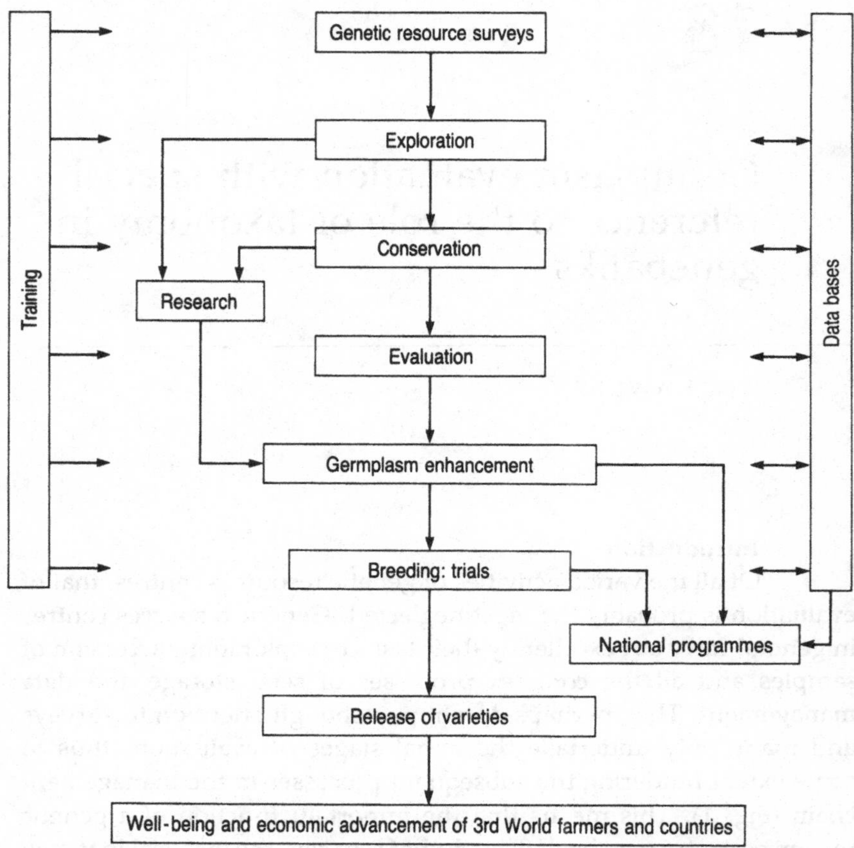

Fig. 1. Genetic resources impact chain. In the central sequence each activity exerts a direct impact on the ones following it. Thus the development of the lower levels of activity is influenced positively or negatively in accordance with the efficient or non-efficient development of those above them (Hawkes, 1985a).

change very much under different environmental conditions.

- Preliminary evaluation. This category comprises a limited number of agronomic traits thought to be desirable by a consensus of users of the crop in question.
- Full evaluation (also termed secondary, in-depth or further evaluation). This concerns the scoring of characters conferring tolerance, resistance or immunity to the pests and diseases that attack a particular crop. It also includes characters of adaptation and resistance to stress conditions such as

frost, cold, heat and drought, and to adverse soil conditions such as high acidity, salinity, aluminium, sulphates, etc.

The work in each category needs to be discussed in detail and certain difficulties noted.

Under the first category – characterization – a series of committees convened by IBPGR has advised on the publication of a set of crop descriptor lists. These enable standardized characterization of crop after crop to be made in different parts of the world in a uniform manner by the use of the descriptor and descriptor states listed. Such standardization is clearly to be welcomed, but some words of warning should be given.

Many of the descriptor lists are very long and genebank staff may perhaps spend too much time on these activities at the expense of others that ought to be given a higher priority. Genebank staff may feel uneasy about omitting some of the descriptors. However, IBPGR itself has reduced the number of recommended descriptors in its second edition lists. Thus the first wheat and *Aegilops* list of 1978 included 26 morphological and 29 agronomic descriptors – a total of 55. The second edition (1981) reduced the list to 12 (6 plus 6), less than a quarter of the original descriptor number. The change has been brought about largely by the breeders themselves, who became aware that many of the characters scored were of very little value to them in selecting initial material for their breeding programmes. This point will be referred to later.

It has also been said that a long list of morphological characters will help the genebank staff to recognize a particular sample if it loses its label or if it is suspected that it has become mixed with, or been replaced by, another sample. In general, however, a series of voucher herbarium specimens, cereal spikes and tubes of seeds or grains will be more useful than any written descriptions to identify possible errors and mixtures.

In addition, it has been argued that a large morpho-botanical descriptor list will help genebank staff identify duplicates with the aid of a computer-generated study. The International Potato Centre (CIP) used this method with great success to identify identical morphotypes in their clonal potato collection. Fifty-four morpho-botanical and 14 agronomic characters were used to reduce a collection of some 15 000 entries to about one-third. It must be stressed that the method would be difficult to apply to outbreeding seed crops but would probably be useful in certain cases for inbreeders. It could not easily be applied to population samples.

In the preceding paragraphs it has been assumed that both the morphological and agronomic characters were of high heritability. This cannot always be assumed, since the differences between high and low heritability characters are not very clearly defined. A decision 'can only be made on the basis of a profound knowledge of a particular crop' (Erskine & Williams, 1980). Evidence of the difficulties encountered may be quoted from Tyler (1985) working with outbreeding forage grasses. He states that because of the quantitative nature of inheritance, the characters are often not easily seen, and most of them are influenced, often quite strongly, by environmental factors. He goes on to say that the majority of characters used for the registration of new varieties have 'little relevance to agricultural performance or indeed to breeding objectives'. Traits such as inflorescence emergence date, vegetative habit and habit at flowering have more relevance, for instance, than the length and width of the flag leaf, ear length and other characters. Differences in the expression of many agronomic characters due to season and locality may also be strongly marked. Such high genotype–environment interactions are inevitable for many characters. To overcome this to some extent, all screening must bear full site data as well as seasonal or weather data wherever possible.

The second category – preliminary evaluation – may include other characters that are not used in characterization, but the difficulties encountered in the selection of characters of high heritability still remain.

Full, secondary or in-depth evaluation is considered by IBPGR to be the task of the breeders, the two previous activities being undertaken by genebank staff.

In an ideal world, the breeders might be able to undertake secondary evaluation. In the real world, this is not often attainable because of pressure of work from existing programmes or through lack of interest in genebank material. A happy situation exists in Ethiopia where breeders are encouraged to examine grown-out genebank material in the experimental field and to make notes of promising lines for crossing and performance trials. They take advantage of this opportunity in a most satisfactory manner. Nevertheless, it has to be said quite frankly that this is an exception, and in any case it refers only to field inspections and not to laboratory or glasshouse screening.

Normally, the genebank staff do not possess the equipment or the time to undertake secondary evaluation. Nevertheless, it should be one of the clear responsibilities of a genebank manager to arrange for secondary evaluation to be carried out. Admittedly, this is not always easy, but where national facilities exist in universities, sister institutes or even in commercial companies, screening for resistance to pests, diseases and environmental stress can be arranged free of charge. The collaborators receive publishable data and the genebank manager obtains results to add to his inventories.

A more difficult problem arises when materials need to be sent abroad for screening because of lack of national facilities. Political pressures may have prevented this in the past but with the new Food and Agriculture Organization (FAO) convention on the free international exchange of materials, few or no difficulties should arise in transmitting materials to other countries for screening. Most laboratories in developed countries would probably charge a fee for this work, but it should not be too difficult to obtain bilateral or multilateral aid funding to cover the costs of screening and postage.

When the screening results are sent back to the genebank they must be entered into the computerized data base and inventories with such results should be published with the least possible delay. Some authorities question the need for the publication of inventories, stating that if a breeder wants material with certain characters he can always ask for the appropriate information from the data base. In theory this could be so, but in practice the genebank manager is the one who needs to take the initiative. After contacts have been made and the breeder has received useful material he will probably ask for more, but the genebank manager should make the first move by printing and distributing inventories. Simple accession lists will not do; the hard facts of the screening results must also be available. These points are stressed by Roelofsen (1985), Hawkes (1985b) and van Soest (1985) who, incidentally, all took part in a recent symposium entitled 'Evaluation for the Better Use of Genetic Resources Materials' held in Prague, Czechoslovakia (Rogalewicz, 1985), and all stress the importance of secondary evaluation as seen by European breeders and genetic resources personnel. It is of interest to note that Peeters & Williams (1984) also stress the need for secondary evaluation, based on data provided by breeders.

Pre-breeding

When a breeder is convinced that a certain genebank accession contains characters of value to his programme, he will have no hesitation in using it, providing, however, that its agronomic characters are on or near the same level as the materials he is already using. If the valuable characters are to be found only in unproductive or poor quality landraces, or in related wild species, he may hesitate to use them because the high-level features of his breeding stocks will be diminished by contamination with the unacceptable low-level characters which will be introduced from the non-elite line together with the single feature in which he is interested. Such contamination would set back his programme for several years and he will, understandably, be reluctant to use this material. It would then be the responsibility of the genebank personnel to look for the same useful feature in other genebank accessions that might at the same time possess better agronomic characters. If this is not possible, then a programme of pre-breeding, or 'germplasm enhancement' as it is often called, will be needed. By this is meant the transference of the useful gene or genes into good agronomic lines that would be more acceptable to the breeder. But who is to do this pre-breeding? It depends very much on the circumstances of the country concerned. Often a university department or laboratory can be asked to cooperate and the work can be undertaken partly by research students. Sometimes a breeding institute is interested or the breeders themselves are enthusiastic enough to carry out the work. On a world scale, however, but with notable exceptions, it must be admitted that secondary evaluation and pre-breeding constitute 'bottlenecks', thus preventing the adequate utilization of genetic materials.

Taxonomy

A taxonomic approach, in the wide sense, to genetic resources work is essential. Beginning with the survey work (Fig. 1), it is essential to know the nature and names of the materials for which the genebank is responsible. This perhaps hardly needs saying. Secondly, it is essential to know where they are to be found and whether cultivated or wild. Thirdly, we must know the taxonomic system of a crop. Is it a single species, as in *Secale;* is it a group of species at different ploidy levels, as in *Triticum;* does it possess a series of wild or weedy forms related to it, evolving in a parallel manner and exchanging genes from time to time through natural hybridization, as in *Sorghum;* does it have no closely related wild relatives, as in *Vicia faba;* or does it have something of all these situations? To answer some, at least, of these questions the crop plant taxonomist must possess not only a solid background of classical taxonomy but also experience and/or an interest in crop plant taxonomy and in the related disciplines of cytogenetics, phytochemistry, numerical taxonomy, reproductive biology and ecology. In short, taxonomy functions as a filing system in which all other data on resistance, adaptation, crossability, cytogenetics, etc. can be stored. For such a system to be reasonably useful it should be simple, logical and stable (Hawkes, 1980).

On an immediately practical level the taxonomist must identify the materials accessed into the genebank and if necessary also classify them at the infraspecific level. This may be done by using the categories of formal taxonomy (Parker, 1978) or by confining these to the species and subspecies only and using an informal group method for the lower categories at infraspecific level (Hawkes, 1986). Whatever infraspecific classification is used for cultivated plants it is certain that a large amount of genetic diversity at this level cannot be encompassed in such categories. For this we must use the descriptor and descriptor state methods as mentioned above.

Formal taxonomy is obviously necessary, but we must go beyond it to an understanding of reproductive biology which develops from the methods of what is generally known as 'experimental taxonomy'. For instance, breeders need to know about the possibility of gene transfer and the presence, if any, of genetic incompatibility or 'incongruity' barriers between species. Such barriers exist, for instance, between maize and *Tripsacum*, which can cross with difficulty but have many problems in their hybrids. On the other hand, maize and teosinte cross readily and produce viable offspring – a fairly predictable result if we assume (as many people now do) that they belong to the same genus (*Zea mays* and *Z. mexicana*) and that the latter may even be the wild prototype of the former. We can see a similar situation in diploid wheats. The cultivated species, *Triticum monococcum*, crosses very readily with *T. boeoticum*, its wild progenitor, and the hybrids suffer no loss of fertility. In fact, some authorities class them as two subspecies of the same species.

It will be of value at this point to mention Harlan & de Wet's (1971) gene pool hypothesis (see also Harlan & de Wet, 1986). Their idea is that for any species of cultivated plant the primary gene pool (GP-1) represents the concept of the biological species (*Triticum* in the example mentioned in the previous paragraph). The secondary gene pool (GP-2) includes individuals that can be crossed with GP-1 although there are distinct partial genetic barriers present (maize and *Tripsacum*, quoted above). Tertiary gene pool species (GP-3) could only yield hybrids with GP-1 that were completely sterile. Related species with different genome compositions from those of GP-1 might fit into this category.

It would be most interesting to investigate some of the lesser known Ethiopian species and genera from this point of view and, in particular, the assumed wild prototypes. In this way the materials in GP-2 and GP-3 might be made more available to breeders.

Chemotaxonomic studies are also of great interest and could form some of the research projects of genebank personnel. Relationships between species might be investigated by comparative serology and immuno-electrophoresis and by two-way electrophoresis of leaf phenolics. Isozyme analyses may prove illuminating by showing areas where the greatest amount of isozyme diversity occurs and linking these with results for disease and pest resistance screening. Results from all aspects of this kind of work can be assessed by means of the usual computer-aided techniques (see also Yndgaard & Hoskuldsson, 1985).

Two words of warning are necessary here. First, although the diversity of isozymes may be intensive in certain areas, does this mean that useful genes for resistance to pests and diseases may also be found with greater intensity in these same areas? No-one has ever looked into this problem and indeed few have even asked the question. Yet the question must be asked and answered. Otherwise, isozyme studies, useful as they are from a theoretical viewpoint, may turn out to be of no practical value whatsoever.

Secondly, no matter how interesting scientifically these results may be, the acid test is to show whether gene transfer from one species to another can be accomplished. Thus, if by principal components analysis of isozyme results we can show that a certain accession of a possibly related wild species seems to be closely similar to the cultigen, the question still remains. Will the two taxa hybridize and if so can the transfer of useful genes be effected from the wild relative to the crop itself? For this answer we must always go back to crossability studies, the cytogenetical analysis of the progenies and, if necessary, pre-breeding work.

Apart from this extremely important function of taxonomy the morpho-geographical approach is still essential. If we find useful resistance in one area, as Qualset (1975) describes for barley yellow dwarf virus disease resistance in Ethiopia, then this is the area to which we should return to collect further samples. It seems highly likely that the evolution of this type of resistance took place in such an area, and samples should be made at points where the proportion of resistant to susceptible plants is greatest – in the case of barley yellow dwarf virus, apparently at elevations higher than 3600 m above sea

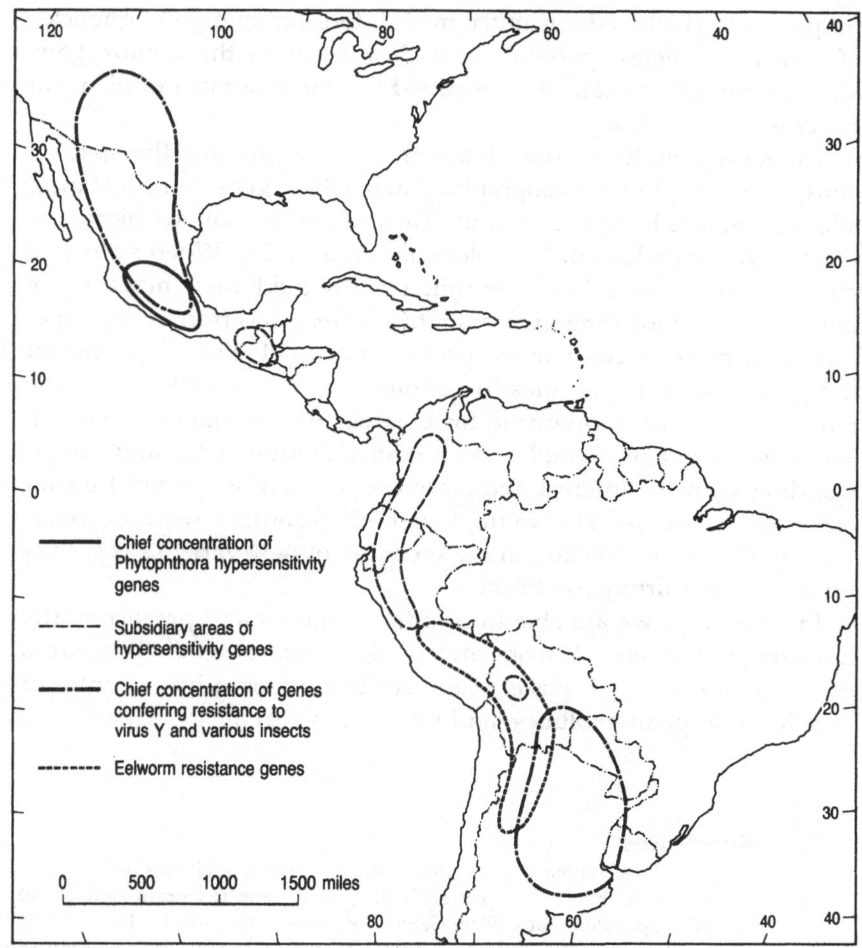

Fig. 2. Gene mapping in potatoes. This map represents an early attempt at mapping some of the genes conferring resistance to certain potato diseases and pests (Hawkes, 1958).

level, according to Qualset. Evolution of character complexes linked to the resistance gene also points to materials on which maximum efforts should be concentrated.

A similar example can be quoted for the resistance of wild potatoes to various biotypes of the potato round cyst nematode. Since potatoes are outbreeders, we are here concerned with population sampling techniques. Populations from some sites may possess low frequencies of resistance genes, others medium and yet others quite high frequencies. The breeder is much more interested in high frequencies of resistance alleles, particularly if they occur in the homozygous state, because these can be transmitted to a large number of plants in his crossing progenies.

The lesson we learn from this is to identify carefully the species, subspecies, etc. and the geographical area where genes for resistance, adaptation, etc. have been found. This, of course, follows along the lines of Vavilov's Law of Homologous Series (1926). When such species and areas have been identified we should then make really detailed collections (fine-grid) from those areas and from related species in the same places, if such species exist. Vavilov's Law points out the geographical regularities to be found in such situations and the author has found it to be true for potatoes in the Americas (Fig. 2). Thus, we can begin to build up an edifice of knowledge and understanding with the interest and close cooperation of specialist evaluators and breeders. The value of genetic resources work becomes clear to all and its function in the breeding of new varieties is greatly enhanced and firmly established.

In such ways we are able to coordinate theory and practice to the advantage of genebank work and to show that a genetic resources centre is not merely a place where seeds are stored but a centre of excellence in plant evolution and breeding research.

References

Erskine, W. & Williams, J. T. (1980). The principles, problems and responsibilities of the preliminary evaluation of genetic resources samples of seed-propagated crops. *Plant Genetic Resources Newsletter*, **41**, 19–33.

Harlan, J. R. & de Wet, J. M. J. (1971). Toward a rational classification of cultivated plants. *Taxon*, **20**, 509–17.

Harlan, J. R. & de Wet, J. M. J. (1986). Problems in merging populations and counterfeit hybrids. *In*: B. T. Styles (ed.), *Infraspecific Classification of Wild and Cultivated Plants*. Oxford University Press, pp. 71–6.

Hawkes, J. G. (1958). Significance of wild species and primitive forms for potato breeding. *Euphytica*, **7**, 257–70.

Hawkes, J. G. (1980). The taxonomy of cultivated plants and its importance in plant breeding research. *In*: *Perspectives in World Agriculture*. Commonwealth Agricultural Bureaux, Farnham Royal, UK, pp. 49–66.

Hawkes, J. G. (1985a). Plant genetic resources. The impact of the international agricultural research centres. *CGIAR Study Paper No. 3*. World Bank, Washington DC.

Hawkes, J. G. (1985b). Genetic resources evaluation. An overview. *In*: V. Rogalewicz (ed.), *Evaluation for the Better Use of Genetic Resources Materials*. Research Institute of Plant Production, Prague.

Hawkes, J. G. (1986). Problems of taxonomy and nomenclature in cultivated plants. *In*: L. J. G. van der Maesen (ed.), *First International Symposium on the Taxonomy of Cultivated Plants. Acta Horticulturae*, **182**, 41–52.

Parker, P. F. (1978). The classification of crop plants. *In*: H. E. Street (ed.), *Essays in Plant Taxonomy*. Academic Press, London, pp. 97–124.

Peeters, J. P. & Williams, J. T. (1984). Towards better use of genebanks with special reference to information. *Plant Genetic Resources Newsletter*, **60**, 22–32.

Qualset, C. O. (1975). Sampling germplasm in a centre of diversity: an example of disease resistance in Ethiopian barley. *In*: O. H. Frankel and J. G. Hawkes (eds), *Crop Genetic Resources for Today and Tomorrow*. Cambridge University Press, Cambridge, pp. 81–96.

Roelofsen, H. (1985). Using evaluation data: an information problem. *In*: V. Rogalewicz (ed.), *Evaluation for the Better Use of Genetic Resources Materials*. Research Institute of Plant Production, Prague, pp. 167–73.

Rogalewicz, V. (ed.) (1985). *Evaluation for the better use of genetic resources materials. Proceedings of the Eucarpia Genetic Resources Section international symposium*. Research Institute of Plant Production, Prague.

Tyler, B. F. (1985). Evaluation of forage grass genetic resources for characterization and breeding potential. *In*: V. Rogalewicz (ed.), *Evaluation for the Better Use of Genetic Resources Materials*. Research Institute of Plant Production, Prague, pp. 215–24.

van Soest, L. J. M. (1985). Some aspects concerning the better use of germplasm collections. *In*: V. Rogalewicz (ed.), *Evaluation for the Better Use of Genetic Resources Materials*. Research Institute of Plant Production, Prague, pp. 189–206.

Vavilov, N. I. (1926). Studies on the origin of cultivated plants. *Bulletin of Applied Botany, Genetics and Plant Breeding*, **16**, 1–248.

Yndgaard, F. & Hoskuldsson, A. (1985). Electrophoresis: a tool for genebanks. *Plant Genetic Resources Newsletter*, **63**, 34–40.

19

Crop germplasm multiplication, characterization, evaluation and utilization at PGRC/E

HAILU MEKBIB

Introduction

The majority of the germplasm accessions maintained by the Plant Genetic Resources Centre/Ethiopia (PGRC/E) are landraces which have evolved under local conditions in the farmers' fields since time immemorial. Such gene pools are the reservoirs of variation which provide the raw material for crop improvement. Samples in the form of seeds or whole plants, representing the spectrum of genetic variation within cultivated species and their wild relatives, are currently being collected and maintained in seedbanks and field genebanks throughout the world (Frankel & Hawkes, 1975; Williams, 1984). Of fundamental importance in the management of these resources is the determination of the variation they represent. To this end, characterization of the various crop germplasm collections is undertaken by the multiplication, characterization and evaluation division of PGRC/E in close collaboration with the plant breeders.

In the past, characterization activities were limited in scope and greater attention was given to collection and conservation activities. During the last three or four years, however, the priorities have changed and now include the extension and intensification of characterization and evaluation work, as well as support for the utilization of germplasm. Highest priority has been given to the major economic crops (e.g. cereals, pulses and oil crops) with the aim of providing useful materials for the breeding programmes.

Crop germplasm multiplication and rejuvenation

Because of the earlier priorities, a systematic increase of the collected germplasm accessions did not start until 1982. In order to cope with the considerable backlog, the division has had to handle more than 8000 accessions annually up to now. Since the size of the collected samples is generally not large enough to meet the minimum requirements for long-term storage, seed increase is a necessity. Furthermore, during the course of storage, a considerable number of seeds will be used for viability monitoring, research and evaluation work, etc. and this will lead to the next cycle of increase. By knowing the multiplication factor for the various crop species, as well as the initial amount of seed and the genetic structure of a given accession, the actual required sample size for multiplication can be determined in order to fulfil the need for safe and adequate conservation. The larger the original sample, the easier and the safer its multiplication and evaluation will be (Frankel, 1970).

Where germplasm accessions need only to be multiplied or rejuvenated, only a few characters, such as flower colour, flowering date and growth habit, are recorded to ensure the identity of the accession with the original sample and to avoid or trace possible mistakes.

Rejuvenation of germplasm accessions is needed when there is a significant decrease in the germination percentage. This activity is conducted under the same conditions used for seed increase, i.e. growing conditions, with maximum care being taken so that no, or as little as possible, natural selection can take place to alter the identity of the accession. Multiplication methods appropriate to the species and cultural practices that maximize the yield of qualitative seeds are being used.

Germplasm multiplication and characterization sites

The increase and rejuvenation of germplasm should take place principally where the accessions were collected, or under similar conditions (Hawkes, 1985). The gene frequencies of each character should remain unchanged from generation to generation, i.e. selection pressure should be minimized.

PGRC/E is in a favourable situation to deal with germplasm which is almost entirely of Ethiopian origin. Therefore, and because of the Ethiopian topography, no real problems exist in the multiplication and characterization of crop germplasm in its natural habitat or under very similar conditions (Table 1). The accessions of a given crop are divided into groups according to the altitude obtained from the passport data and are evaluated accordingly at sub-stations in corresponding altitudinal zones. For example, highland sorghum is evaluated at

Table 1. *Multiplication sites of PGRC/E with altitude, average daily temperature and average annual precipitation*

Sites	Altitude (m)	Temperature (°C)	Precipitation (mm)
Addis Ababa	2450	16.0	1163
Arsi Negele	1960	18.4	1763
Asmara	2325	16.6	525
Awassa	1750	19.2	961
Bekoji	2850	12.9	1203
Debre Zeit and Dembi	1900	18.7	866
Fitche	2800	12.9	1283
Ginchi	2240	17.0	1075
Holetta and Kuyu	2390	13.8	1097
Jima	1577	18.8	1469
Kulumsa	2200	16.3	938
Melkassa	1558	21.5	806
Melka Werer	737	26.3	471
Mieso	1320	22.7	694

1960 m above sea level at Arsi Negele, intermediate sorghum types at 1580 m at Melkassa and lowland types at 1320 m in Mieso.

Procedures followed and difficulties encountered

The most important goal of any seed increase activity is to produce healthy seeds without changing the genetic composition of an accession. Therefore, different approaches have to be followed according to the flower biology of a species, the genetic composition of an accession, the total number of accessions to be multiplied, the initial status of an accession (sample size, health condition), etc.

The biological characteristics of self-pollinated crops make the task of handling germplasm relatively easy. In general, the plants of a given accession do not require any additional input and the seeds can be harvested from the whole plot. However, some of the self-pollinated crops have shown a low percentage of cross-pollination. Therefore, plans have been made to adjust the procedures for self-pollinated crops in such a way that the chance of occasional outcrossing will be reduced to nearly zero. A second group of species comprises those crops which are predominantly cross-pollinators. These species require special isolation techniques and, in addition, extra attention has to be paid to the (artificial) pollination of sufficient plants per accession (Demarly, 1981). Some of the problematic crops are presented below as an illustration.

Niger seed or noog (Guizotia abyssinica)

This annual oil crop belongs to the Compositae family. The species is pollinated predominantly by insects and is predominantly genetically self-incompatible. This is mainly because the receptive part of the stigma rarely touches the pollen of the floret (Seegeler, 1983). Since selfing of this species would cause serious inbreeding depression and might expose the accessions to unwanted natural selection, a form of sibbing within each accession is applied. A sufficiently large number of plants are covered with cheesecloth bags and at regular intervals hand pollination between plants within the accession takes place. The use of pollinating insects within a bag is under study and could increase the degree of seed setting.

Brassica *species*

Brassica carinata, B. nigra and *B. oleracea* are species either indigenous to Ethiopia or widely grown. All are predominantly cross-pollinated and therefore require special attention to avoid gene flow between accessions. In this case also, cheesecloth bags are used to isolate a group of plants of the same accession and natural pollination within a bag is used to obtain the required seed amount. Problems with fungi and low seed setting are still hampering an adequate increase procedure.

Faba bean (Vicia faba)

This generally self-pollinating species shows, under ideal conditions, up to 50 per cent outcrossing, caused mainly by bees. The isolation method presently in use is based on a *Brassica* fence around each accession. The assumption is that the pollinating bees stay mainly within a plot and before approaching another plot (=another accession) they will be attracted by the *Brassica* plants where the faba bean pollen will be 'brushed off'. Although this method does not ensure the avoidance of gene flow between accessions it is relatively simple and cheap and thus large numbers of accessions can be increased.

Castor bean (Ricinus communis)

This monoecious species is a wind pollinator. Although selfing might cause inbreeding depression to a certain extent, individual inflorescences are bagged to avoid cross-pollination. The same procedure is followed with sorghum.

Characterization and evaluation

Since the value of the conserved germplasm will depend greatly upon the information available on each accession, high priority is given to the systematic characterization work. More than 20 descriptor lists for the different crop types have been discussed and agreed upon by the respective plant breeders and are presently in use. The descriptor lists issued by the International Board for Plant Genetic Resources (IBPGR) are followed, as far as they are relevant to Ethiopian conditions, to record morphological and preliminary evaluation data on various crop types.

By 1986 a total of about 35 000 accessions of more than 25 crop types had been multiplied and characterized. Details of the number of descriptors used for each of the crops as well as the respective multiplication sites are presented in Table 2. Priorities for the crops to be multiplied were based on various criteria which were defined ahead of time. Major aspects taken into consideration were the degree of dependency of the plant breeder on local germplasm, the economic importance of the crop in Ethiopia, the amount of diversity found or expected in the crop, the sample size and viability condition of the individual accessions, etc.

The genetic composition of an accession has far-reaching consequences for the characterization and evaluation of germplasm. Genetically homogeneous accessions of self-pollinating crops are easy to handle and are not any different from medium uniform varieties. However, genetically heterogeneous accessions cause many complications to a genebank and they frequently require an extra treatment. To overcome such complications and difficulties as far as possible, PGRC/E has started to separate heterogeneous accessions or landraces of self-pollinated crops, such as wheat and barley, into agromorphological components. Each component has some obvious and important agronomic and taxonomic characters in common although the individuals may vary in other characters. Such components are treated as pure lines following the procedures described below.

During the first year of increase, spikes with highly heritable characters in common within an accession are combined to form a component. The next season, five ears from each component are planted ear to row. Rows and components are compared with each other and where there is no difference they will be lumped. Distinguishable components will receive new accession numbers and be treated as independent samples.

The characters of highest interest to the plant breeders are

Table 2. *Crop germplasm multiplication and characterization by PGRC/E in the period 1982–6*

Crop type	Total number of accessions planted and characterized					Total	Number of descriptors employed	Multiplication/ characterization sites
	1982	1983	1984	1985	1986			
Cereals								
Wheat	1644	225	1777	2898	2595	9139	20	Asmara, Holetta, Debre Zeit, Combolcha
Barley	424	716	4201	2943	2510	10794	20	Holetta, Kulumsa
Sorghum	2036	2991	500	953	1777	8257	21	Arsi Negele, Melkassa
Teff	–	350	2400	36	472	3258	20	Debre Zeit
Finger-millet	–	–	–	210	316	526	16	Melkassa
Pearl millet	–	–	–	11	8	19	27	Melkassa
Oil crops								
Rapeseed	303	264	438	280	191	1476	27	Holetta
Noog	243	184	344	376	300	1447	23	Holetta
Linseed	193	300	177	298	83	1051	20	Holetta
Safflower	75	103	35	7	–	220	21	Debre Zeit, Melka Werer
Sunflower	–	11	8	2	–	21	4	Awasa, Arsi Negele
Sesame	90	104	81	120	92	487	28	Melka Werer
Castor bean	–	–	200	49	–	249	26	Arsi Negele, Awassa
Legumes								
Faba bean	102	143	176	80	262	763	13	Debre Zeit, Bekoji
Field pea	153	34	125	415	296	1023	14	Bekoji
Lentils	168	–	145	214	130	657	13	Debre Zeit
Chickpea	321	–	335	224	150	1030	14	Debre Zeit, Ginchi
Fenugreek	–	24	–	99	–	123	13	Debre Zeit
Lathyrus	–	–	–	74	–	74	15	Debre Zeit
Phaseolus	–	–	–	157	93	250	19	Melkassa
Cowpea	–	–	–	23	20	43	19	Melkassa
Lablab	–	–	–	30	25	55	19	Melkassa
Total	5752	5449	10942	9499	9320	40962		

generally less heritable or their determination needs special experimental designs or complicated equipment. The activity to record such characters is called further evaluation and is the principal responsibility of the plant breeder. However, PGRC/E has initiated several further evaluation activities, in close contact with the breeder concerned, in order to increase the value of its germplasm. Examples include the evaluation of some 1700 sorghum accessions for their agronomic performance and for possible resistance or tolerance to bacterial leaf streak and stalk borer. These activities were conducted in collaboration with scientists from the International Crops Research Institute for the Semi-Arid Tropics (ICRISAT) and Addis Ababa University.

In 1983, PGRC/E and the Scientific Phytopathological Laboratory in Ambo started a pilot project to screen durum wheat accessions for resistance to stem rust, stripe rust and leaf rust. Of 502 genotypes, only 110 were found to be susceptible to all three rust diseases. The other 392 accessions showed resistance to one, two or all three of the rust populations under investigation. For further screening and field verification, the genotypes showing resistance to stem rust were selected. Of the 101 accessions, 63 were also stripe rust resistant, 23 were both stripe rust and leaf rust resistant and 15 were stem rust resistant only. They were planted at the Addis Ababa University's experimental site at Debre Zeit in cooperation with the durum wheat breeders. This site is known as a 'hot spot' for stem rust and has a high disease pressure.

In order to allow mass screening of germplasm under abiotic stress conditions (e.g. drought and aluminium toxicity tolerance) laboratory screening methods are now being developed with encouraging results so far. The installation of a near-infrared-reflectance analyser allows easy and precise determination of chemical constituents, such as protein, oil, water and fibre content, of seeds and other organic parts.

Utilization activities

The ultimate goal of any germplasm evaluation is its utilization. The national plant breeders form the main target group and PGRC/E tries to involve the breeders as much as possible in its routine activities. Examples of close cooperation include the workshops held with plant breeders and other potential users to discuss and agree on priorities regarding germplasm collection and evaluation, the participation of breeders in the collection missions and the cooperation that

Table 3. *Results of an evaluation of the sesame collection. The best 11 accessions are compared with the standard*

	1	2	3	4	5	6	7	8	5 × 7 × 8:1000
Accession number	Days to 50% flowering	Days to pod-setting	Days to maturity	Plant height (cm)	Number of capsules per plant	Capsule length (mm)	Seeds per capsule	1000 seed weight (g)	Theoretical yield[a] (g)
111505	79	91	145	224	149.0	17.0	70.2	2.28	23.8
111518	40	79	104	119	58.2	23.5	75.3	3.91	17.1
111809	36	44	104	98	183.2	24.9	69.1	3.47	43.9
111815	36	41	99	104	65.2	27.4	84.1	3.71	20.4
111823	34	38	98	121	80.4	20.9	57.6	3.70	17.1
111824	34	39	98	118	93.8	23.8	64.4	3.87	23.4
111829	33	40	98	111	161.8	24.5	70.9	4.27	49.0
111833	21	63	107	155	108.8	26.8	68.8	3.73	27.9
111834	15	50	111	132	76.4	23.8	77.6	3.00	17.8
111859	31	39	112	181	119.4	26.1	79.4	3.31	31.4
111861	47	57	107	132	112.2	24.7	62.0	2.97	20.7
Standard T85	47	53	101	156	87.8	22.6	67.8	3.52	21.0
Experimental average	47	55	120	148	74.9	22.1	64.2	3.23	15.5

[a] Calculated as a product of number of capsules per plant × number of seeds per capsule × thousand seed weight:1000.

Table 4. *Number of accessions included in the Pre-national (PNYT) and National Yield Trials (NYT) as well as in observation trials (OT) in the period 1983–6*

Crop type	Number of accessions used	Type of trial
Faba bean	16	PNYT/NYT
Field pea	1	NYT
Chickpea	4	PNYT/NYT
Linum	1	PNYT
Niger	4	PNYT
Brassica	3	PNYT/NYT
Sesame	4	PNYT
Niger	7	NYT
Brassica	2	NYT
Linseed	3	NYT
Sesame	4	NYT
Groundnut	3	NYT
Sorghum	12	OT
Barley	50	OT
Niger	7	OT
Lentil	13	OT
Chickpea	9	OT
Faba bean	2	OT
Field pea	5	OT
Castor bean	14	OT

occurs during the characterization and multiplication of germplasm. Furthermore, PGRC/E presents as many results as possible from evaluation activities in the PGRC/E–ILCA Germplasm Newsletter or during meetings with potential users. (An example of the characterization work on sesame is presented in Table 3.) This approach has led to intensive utilization of germplasm by the breeders and some of the results can be observed from Table 4. A more detailed account of the procedures followed is given by Engels & Mekbib (1987).

References

Demarly, Y. (1981). Theoretical problems of seed regeneration in cross-pollinated plants. *In*: *Seed regeneration in cross-pollinated species. Proceedings of the CEC/Eucarpia Seminar, Nyborg, Denmark, 15–17 July 1981*. Balkema, Rotterdam, pp. 7–31.

Engels, J. M. M. & Mekbib, H. (1987). The utilization of germplasm in Ethiopia and role of PGRC/E. *PGRC/E–ILCA Germplasm Newsletter*, **13**, 21–5.

Frankel, O. H. (1970). Evaluation and utilization – introductory remarks. *In*:

O. H. Frankel and E. Bennett (eds), *Genetic Resources in Plants, their Exploration and Conservation*. Blackwell, Oxford, pp. 395–401.

Frankel, O. H. & Hawkes, J. G. (eds) (1975). *Crop Genetic Resources for Today and Tomorrow*. Cambridge University Press, Cambridge.

Hawkes, J. G. (1985). Report on a consultancy mission to Ethiopia for GTZ to advise PGRC/E on germplasm exploration, conservation, multiplication and evaluation. Birmingham (mimeographed).

Seegeler, C. J. P. (1983). *Oil Plants in Ethiopia, their Taxonomy and Agricultural Significance*. PUDOC, Wageningen, pp. 122–46.

Williams, J. T. (1984). A decade of crop genetic resources research. *In*: J. H. W. Holden and J. T. Williams (eds), *Crop Genetic Resources: Conservation and Evaluation*. Allen and Unwin, London, pp. 1–16.

20

Evaluation methods and utilization of germplasm of annual crop species

J. B. SMITHSON

Introduction

Proper evaluation of the very large germplasm collections now assembled for many crop species presents major problems. These arise principally from the effects of the environment on the expression of plant characteristics. For qualitative characteristics there is little difficulty as their expression is usually affected little by environment. Examples are seed coat and flower colour and colour pattern. A single evaluation is all that is required to characterize a set of materials for such characteristics.

It is, however, the quantitative characteristics, and these are of most interest to the breeder, that are especially intransigent as their expressions are always modified by environment to some degree, so that the separation of the contributions of genotype and environment to the phenotype requires special techniques.

Environment and genotype × environment interaction

The modification of plant characteristics by environment takes two forms. First, there is a general reduction or increase in expression of a character across all genotypes. Environmental features such as soil fertility, moisture availability, temperature and pathogens, pests and weeds may all affect plant characters in this way. The result is what is often termed 'field variability' and this will always occur in a single evaluation at a single location in a single season. It also occurs across locations and seasons.

Secondly, there is the situation where all genotypes are not affected equally by differences in environment, normally described as 'genotype × environment (g × e) interaction'. Such interaction is usually associated with a set of materials being grown in more than one location or season and, indeed, is only detectable in this way. It also occurs in a set of materials at a single location and season but is in that case not separable from general environmental effects.

The two features of the relationship between plants and their environments pose serious problems in the evaluation of large numbers of materials, whether germplasm or breeding lines, and in the interpretation of the collected data. Two examples from chickpeas illustrate the importance of environmental effects in evaluation and the misinterpretations that can arise as a result.

The first concerns seed protein percentages (International Crops Research Institute for the Semi-Arid Tropics, unpublished data). Seed protein percentages were routinely determined on seeds from successive evaluations of different germplasm accessions over a period of five years. The seeds of 100 accessions representing the largest and smallest seed protein percentages in each of the five years were grown in a replicated trial at ICRISAT in 1982–3 and 1983–4 and the seed protein percentages of their produce determined.

Examination of the data revealed very poor correlations between the seed protein percentages from the germplasm evaluation and those from the replicated trials in either of the two years. Large and small values from the germplasm evaluation tended towards the mean and many of the differences that had been demonstrated earlier disappeared. There were, however, good correlations between the protein percentages obtained from the trials in the two years, indicating that the discrepancies arose from environmental differences within or between seasons or both but illustrating that repeatable results can be obtained with appropriate methods.

The second example concerns the number of days to flowering (ICRISAT, 1980). Based on germplasm evaluation data, five groups of lines, flowering respectively in 45, 45–56, 57–68, 69–80 and more than 80 days after sowing, were included in five replicated trials, again at ICRISAT, to examine their adaptability to sowing one month earlier than normal. Mean flowering times for the trials were roughly according to expectation. The ranges of flowering times were, however, 33–80, 34–66, 41–80, 40–106 and 44–124, overlapping completely, and with very poor correlations with the earlier records, and this is exactly what would be expected from the effects of different day lengths and temperatures on the flowering times of sets of materials differing in their photoperiod and temperature responses.

As usual, there is no complete solution. None the less, methods of measuring and controlling environmental variation are available and some of these will be described and discussed in this chapter.

The questions of the need for evaluation of genetic resources and of the characters to be evaluated are not considered here. It is assumed that accessions are maintained discretely in order to retain character combinations and that some kind of evaluation is required for the purposes of utilization. In the text, materials being evaluated will be termed test entries or materials. Performance refers to the value assigned to any quantitative character, be it size, concentration or number of any plant component, or rate of any plant process. It is also assumed that performance is being measured in field nurseries and with the necessary precision and accuracy. Disease and pest reactions are not considered since they require specialized techniques.

Field variability

First, we will consider means of handling a set of test entries in a single location and season. Where numbers are relatively small (say 500 or less), orthodox statistical designs such as randomized blocks or lattices may be employed, utilizing techniques such as randomization, replication and sub-grouping within replicates to measure and/or control environmental effects.

But germplasm collections and early generation breeding materials are usually too numerous to allow replication and, were replication even feasible, and with the most uniform field environment conceivable, the replicates would be too large to cope with field variability by orthodox analysis of variance. What then can be done to measure and/or control environmental effects in such situations?

Regular check entries

The simplest means of obtaining a measure of environmental variability in an unreplicated set of entries is the inclusion of check entries. This is common practice in germplasm evaluation, comprising the inclusion of two or three different checks in the nursery at the rate of one check for every 10 test entries. Unfortunately, the mere inclusion of checks is not sufficient. They must also be used to assess and reduce field variability by some form of adjustment of the performances of the test entries according to the performance of the checks, and this is rarely practised.

The simplest form of adjustment is to express the performance of each test entry as a percentage or proportion of some measure of the performance of the checks in the same sector of the field. Where several different checks are included, the best measure is likely to be the mean of the performances of the nearest full set or sets of checks.

Alternatively, the performance of each test line may be adjusted by subtracting the deviation of the mean of the performances of the nearest full set of checks from the mean of the performance of all the checks. For example, if the mean number of seeds per pod of the nearest full set of checks is 9.5 and the mean for all checks is 9.0 seeds per pod, the number of seeds per pod of each test entry in the same sector of the field is adjusted downwards by $9.5 - 9.0 = 0.5$.

Conceptually, the latter adjustment is more desirable than the first in that the actual and the adjusted data are of the same units. Both have the disadvantage of providing no estimate of error for a comparison of the differences among the test lines.

Augmented designs

An extension of the regular check system is the augmented design (Federer, 1956). The test materials are again unreplicated but are randomized and grouped in blocks of convenient size (say 20–50) for the number of materials, size of field and number of checks. An appropriate number of different check entries (2–5 according to block size) is then randomized within each block. Check performance can then be analysed in the form of a complete randomized block and the estimate of error so derived used to compare the performances of the unreplicated test lines.

The performances of the test lines may also be adjusted according to the deviation of the mean of the checks in the block in which they occur from the mean of all the checks, so that $\bar{B}_j - \bar{y}..$ is the adjustment for the performance of each of the test lines in the jth block, where $\bar{B}_j$ is the mean of the checks in the jth block and $\bar{y}..$ is the overall check mean. Note that the variance of the difference of two test entries in the same block will be twice the error mean square, while that of two test entries in different blocks will contain an additional quantity for block differences.

The method assumes that the random components associated with the checks and the test lines are similar. This may not be so, but is more likely if the checks are chosen to represent the range of variability in the collection. However, it does provide a measure of environmental variation and a means of adjusting the performances of the test lines to remove some of the variability, and thus is an improvement on other methods.

One further point should be mentioned. It is common practice to include test lines in order of origin, or to group them in some other way, so that similar materials are compared more accurately. But in evaluating a set of materials we are also interested in the relative performances of dissimilar materials, so in the absence of very compelling reasons for grouping, less biased comparisons are obtained by randomizing the test lines.

Nearest neighbour analysis

A third and less often used method of handling field variability is adjustment according to the performances of neighbouring plots, first proposed by Papadakis (1937) and described with a worked example in Pearce (1983). Adjustment by neighbouring plots may be applied to any replicated field layout and is especially useful for large sets of materials. It adjusts the performance of each plot according to the mean performance of its neighbours. In most cases, the four plots adjoining the ends and sides of each plot are used for the adjustment. Where plots are long and narrow, it is more appropriate to use only the plots along each side. In the case of end plots there are only three neighbours and corner plots have only two. It has been found that the first cycle of adjustment is often erratic, but it is possible to iterate (i.e. repeat the calculation with the adjusted values until the adjustments remain similar) as is usually done when estimating values for more than a single missing plot.

The analysis proceeds as follows:

- compute the deviations of each plot from the mean of all plots of that treatment;
- compute the mean deviations (X) of the neighbours of each plot;
- compute treatment totals and means for the X values;
- compute an analysis of covariance of the actual values (Y) on X;
- compute the regression (b) of Y on X;
- adjust the Y value for each plot subtracting $b(X-x)$;
- iterate the above steps until the adjustments are the same;
- compute the analysis of variance of the adjusted values.

Since the test entries must be replicated the area required will be large but the method allows for adjustment for patchy field variability, which is not possible in an orthodox analysis of variance, in addition to providing an estimate of error.

Summary

It should be noted that these methods of analysis are not mutually exclusive. Common checks ought always to be included to enable adjustment across seasons and, provided there is replication and randomization, both types of adjustment are theoretically possible. Augmented designs and nearest neighbour analysis both provide estimates of error for comparison of differences among entries; nearest neighbour analysis takes up more land because at least two replicates are required, but this may be accommodated to some extent by reducing plot size. Computer facilities are desirable for all because of the large volume of material to be examined. Nearest neighbour analysis can be expected to produce more accurate results because of replication and the opportunity to adjust for patchy field variation and (see next section) to assess the magnitude of $g \times e$ interaction.

Evaluation across sites and seasons

Because of $g \times e$ interaction, the relative performance of a set of materials in one environment is unlikely to reflect its relative performance in other locations and seasons. Therefore the value of a set of data obtained in one environment is doubtful, especially so in a country as varied as Ethiopia.

Furthermore, the very large numbers now present in germplasm collections virtually preclude the possibility of evaluating all materials at a single time. The whole available Centro Internacional de Agricultura Tropical (CIAT) bean collection (more than 17 000) was evaluated for resistance to angular leaf spot in hill plots this year (CIAT, 1987) but this is a special case. There is also the case of continuing collection and the need to evaluate newly assembled groups of materials. For these reasons, the evaluation of germplasm collections in successive seasons and/or at more than a single location is inevitable.

Care should be taken to ensure that environments are as uniform as possible. For example, in the evaluation of the International Institute of Tropical Agriculture (IITA) cowpea collection in two successive years, sowing was on the same date and exactly the same cultural practices were used, with sufficient fertilizer and irrigation to reasonably eliminate any soil nutrient or moisture stress (IITA, 1974). Nevertheless, it must be accepted that the removal of all environmental variability likely to affect performance is impossible. General environmental effects can be accommodated to some degree by the inclusion of common checks and by using augmented designs to adjust performance across seasons in the same manner as within seasons.

G × e interactions cannot be handled in this way. Several methods of assessing their magnitude and of characterizing them have been developed. They include: combined analysis across locations and seasons; regression of individual entry performance on environment mean performance or some other environmental measure; and, more recently, multivariate analysis. A comprehensive account of these techniques is given by Hill (1975). Such methods have helped in the understanding of g × e interaction but are not appropriate to the evaluation of large numbers of materials. There appears, therefore, to be no escape from the need to evaluate the same test materials in more than a single environment. Seasons are unsatisfactory as they are unpredictable, so this means testing at a number of locations. Multivariate analysis can be used to choose locations that represent the range of environments in which the test entries are likely to be utilized. The greater the number of locations, the more complete will be our characterization, but practically, three or four (say, two extremes and two intermediates) may be all that is feasible or, for some characters, even necessary. For example, based on growth cabinet studies with chickpeas and lentils, Roberts, Hadley & Summerfield (1985) and Summerfield *et al.* (1985) concluded that evaluation in field nurseries at three properly chosen locations is sufficient to characterize accessions of these species for their flowering responses to photoperiod and temperature. Alternatively, the number of environments may be increased and the number of materials reduced by selecting representatives of the total variability by some form of multivariate analysis.

Finally, environmental data are every bit as important as plant character data in any evaluation if we are to understand variation in performance across environments. It is vital, therefore, to characterize the physical and biological environments in which evaluations are conducted as thoroughly as we characterize our plants. Physical factors should include, at least: latitude; altitude; maximum and minimum temperatures and rainfall on 10-day mean bases during the growing season; physical and chemical properties of soils. The biological environment will include diseases, insects and weeds.

Suggested procedures

Based on the above considerations it is possible to suggest optimal procedures for germplasm evaluation.

Whole collections should be evaluated at three or four locations selected to represent a range of situations. Operationally, this will have to proceed in groups of around 2000 entries. These should be selected at random from those available for evaluation. They should be deliberately grouped only if absolutely necessary, e.g. bush and climbing types. They should be sown in an augmented design with a set of frequent, common checks, chosen to cover the total variation in the collection as far as possible. If possible, two replicates should be sown at each location to allow a nearest neighbour analysis. Plant and environment data should be collected on each evaluation.

Multivariate analysis of the data should be conducted to select sets of different sizes representing the variability in the collection, as was done with the IITA cowpea collection in the late 1970s (Rawal, Kaltenhauser & Snyder, 1977). These sets should be of different sizes (say, 100, 500, 1000 and 2000) to accommodate different capacities. The evaluation of these sets by breeders in other environments and the return of the data for continuous updating of information should be vigorously encouraged.

Finally, a note of caution should be added. The techniques described are merely tools to aid evaluation. In most circumstances they can be expected to be useful and we have a duty to use them. But there is always the possibility that they may distort differences rather than reduce environmental variation. There is, therefore, no substitute for knowledge of the crop, careful observation of the test materials in the field, careful examination of the actual and adjusted data and the application of common sense in their interpretation.

Dissemination of information

A further area requiring thought is the method of presentation of data. Pulse germplasm catalogues include those for cowpea (IITA, 1974), beans (CIAT, 1983) and chickpea (Singh, Malhotra & Witcombe, 1983). All comprise long lists of accessions and characters recorded. For cowpea, there are 46 characters for 4224 accessions; for chickpea, there are 29 characters for about 3300 accessions.

Such forms of presentation are very difficult for an aspiring breeder to assimilate and are therefore not very useful. The information is important for the institutions conducting the evaluation but can be on computer file for manipulation as it is unlikely that a hard copy of the complete information is ever going to be required.

For the breeder, it would be more useful to have a summary for each character, perhaps in the form of a histogram showing its frequency distribution together with an estimate of the total variation, such as the coefficient of variation. This should be accompanied by important characters (e.g. disease resistance) or important combinations of characters. Data on the environments in which the evaluations are carried out and a summary of the check performance should also be included. The intending breeder can then see easily what kind of variation is available and can request whatever number of materials, having the range of characteristics and adaptation in which he is interested, that he is capable of handling.

These procedures presuppose that the intending breeder knows what he requires. But in many cases this may not be so. In other situations, materials adapted to a wide range of environments are needed. These are additional reasons why performance in different environments is important and environmental information is required for every evalution.

In this situation, the different sized sets representing the variation are important. The breeder evaluates a set of the size he can handle in his own environment, identifies the most promising materials, requests additional accessions of similar origin and character from the distributing institution and returns the data for updating the information base.

The development of these kinds of relationships between genetic resource specialists and breeders is vital if we are to make the most effective use of our genetic resources.

References

Centro Internacional de Agricultura Tropical (1983). *Catalogo de Frijoles*. CIAT, Cali.

Centro Internacional de Agricultura Tropical (1987). *Annual Report 1986*. CIAT, Cali.

Federer, W. T. (1956). Augmented (or Hoonuiaku) designs. *Hawaiian Planters Record*, **55**, 191–207.

Hill, J. (1975). Genotype × environment interactions – a challenge for plant breeding. *Journal of Agricultural Science, Cambridge*, **85**, 477–93.

International Crops Research Institute for the Semi-Arid Tropics (1980). *Annual Report 1979*. ICRISAT, Hyderabad.

International Institute for Tropical Agriculture (1974). *Cowpea Germplasm Catalogue no. 1*. IITA, Ibadan, Nigeria.

Papadakis, J. (1937). Méthode statistique pour des expériences sur champ. *Bulletin de l'Institute d'Amélioration des Plantes, Salonika 23*.

Pearce, S. C. (1983). *The Agricultural Field Experiment*. Wiley, Chichester.

Rawal, K. M., Kaltenhauser, J. & Snyder, M. J. (1977). Agronomy abstracts. 69th Annual Meeting, American Society of Agronomy 68.

Roberts, E. H., Hadley, P. & Summerfield, R. J. (1985). Effects of temperature and photoperiod on flowering in chickpeas (*Cicer arietinum* L.). *Annals of Botany*, **55**, 881–92.

Singh, K. B., Malhotra, R. S. & Witcombe, J. (1983). *Kabuli Chickpea Germplasm Catalogue.* ICARDA, Aleppo.

Summerfield, R. J., Roberts, E. H., Erskine, W. & Ellis, R. H. (1985). Effects of temperature and photoperiod on flowering in lentils (*Lens culinaris* Medic.). *Annals of Botany*, **56**, 659–71.

21

Evaluation and utilization of Ethiopian forage species

J. R. LAZIER AND ALEMAYEHU MENGISTU

Introduction

Plants which are utilized as fodder for livestock are confined to those which can provide maximum yields in animal production with minimum management inputs. Such plants are usually members of the families Gramineae and Leguminosae and are mainly herbs or subshrubs. In recent years leguminous shrub and tree species have been receiving increasing attention, particularly for small farmers in developing countries.

The Gramineae and Leguminosae are major sources of human nutrition as cereals and pulses while the by-products of these crops are major sources of nutrition for livestock. This report will be confined, however, to plants which are planted primarily as livestock fodder.

Ethiopia, as part of the African continent, shares many genera and species of grasses and legumes with the rest of the continent. However, its great variations in climate and relief and its heavily dissected landscape have provided the opportunity for further evolution of species and genotypes. About 64 species of legumes, mainly montane (10–11 per cent of the total), have been reported as probably being endemic to Ethiopia, while 30 species of grasses are endemic (Thulin, 1983).

African grasses are the main source of cultivated commercial grass species and cultivars in the tropics and subtropics worldwide and are almost all represented in Ethiopia. While Africa is not the major centre of diversity in legumes it is a major centre of diversity for such genera of fodder potential as *Aeschynomene, Alysicarpus, Indigofera, Lablab, Lotononis, Macrotyloma, Neonotonia, Trifolium* and *Vigna*. Other genera of importance which are represented include *Cajanus, Clitoria, Galactia, Stylosanthes* and *Zornia*. Browse species of potential in Ethiopia include *Acacia, Albizia, Bauhinia, Cassia, Dichorostachys, Eriosema, Erythrina, Flemingia, Lonchocarpus, Millettia, Parkinsonia, Piliostigma, Sesbania* and *Tamarindus*. Considerable diversity is present in *Acacia* and *Sesbania*. Thirteen of 24 important legume genera are represented and 14 of 26 browse genera (Skerman, 1977). Undoubtedly potential exists in other leguminous genera as little work has yet been done on collection and evaluation.

Early work

Early work in the screening of native forage germplasm was done by a Food and Agriculture Organization (FAO) project which included a number of local collections in unreplicated adaptation plots at Adami Tulu and Abernosa Ranch in the Rift Valley. The local germplasm, which was not named, was less successful than exotic germplasm (Ibrahim, 1975).

The Chilalo Agricultural Development Unit (CADU) has screened a number of native lines of commercialized grass species (Carlsson, 1972; Froman, 1975). Eight *Cenchrus ciliaris* ecotypes were found to be more productive than a Kenyan variety in the lowlands of Chilalo but were not sufficiently productive to be of interest. Sixteen *Chloris gayana* lines were screened and found to be very variable but to form a good source of germplasm for development as both creeping, stoloniferous and upright, tufted, many-headed types were identified. Six similar accessions of *Phalaris arundinacea*, which were collected 2200 m above sea level in wet areas, were found to be more productive than exotic lines. Germplasm of *Hyparrhenia* species and *Panicum maximum* was also screened. *H. hirta* was less stemmy than other *Hyparrhenia* species though all reportedly had low palatability. Considerable variation was found in 13 ecotypes of *P. maximum* germplasm which were collected from 1500 to 1700 m; four lines were regarded as promising. Indigenous lines of *Setaria sphacelata* were found to be better adapted to higher altitudes than cv. Nandi. Local lines were also more rhizomatous.

An investigation of the leguminous species of the Arsi region by the same project resulted in the identification of some 90 taxa of which about 20 were considered promising for fodder development. Indigenous material of *Trifolium semipilosum* var. *semipilosum* was more vigorous than the cultivar Safari and one line of *Neonotonia wightii* was earlier flowering, heavier seeding and perhaps slightly lower yielding than commercial lines.

Recent work

Over the past five years the International Livestock Centre for Africa (ILCA) has been involved in the evaluation of native Ethiopian germplasm as part of the normal procedures of the ILCA Genetic Resources Section. Agronomic description and evaluation are done for all lines in the collection at the same time as the germplasm is multiplied. Multilocation evaluation is then done by the Forage Network in Ethiopia (FNE) in a range of environments. In addition, national and ILCA scientists have been pursuing their own evaluation programmes on ILCA germplasm. This report will consider the screening procedures used and the results obtained in the regular screening exercises undertaken by ILCA and FNE.

Highland species

Annual Trifolium

The initial description, agronomic evaluation and seed multiplication of the annual native *Trifolium* species is done at ILCA headquarters in Addis Ababa (9°02′ N, 38°42′ E) at an altitude of 2400 m on a transitional soil lying between nitosols and vertisols which is somewhat more free-draining than the vertisols. Unreplicated strips, 5 m long, at a spacing of 1 m, are planted with a single line of scarified seed sown at 0.6 g per plot. As the strips are unreplicated, control strips are planted in each fifth plot. The soil is slightly mounded along each strip and abundant small drains are provided in the trial to prevent water transporting or burying the seed. Fertilizer is applied at 10 kg P/ha, banded below the planting level beside the line of planting. The amount applied is calculated on a 0.5 × 5 m plot basis. The plots are weeded while the pathways are allowed to accumulate weeds, which are mown periodically.

Regular three-weekly observations are taken on general characteristics, with more frequent observations on germination and date of flowering. All parameters observed are defined. The most promising lines for planting in yield trials are determined by using grouping techniques on the observational data. Principal component analysis has been found to be the most useful technique.

Screening has been undertaken in this manner since 1983 and more than 500 lines have been screened to date. The strips normally provide sufficient seed for long-term and duplicate storage, as well as for the second stage evaluation.

The second level of screening (stage 2) is done in small replicated plots (5 × 2 m) in which selected lines of the annual species are plan-

Table 1. *Agronomic characteristics of selected lines of promising Ethiopian annual* Trifolium *species. 1984 replicated yield trial*

Species	Number of lines	Range of DM yields (kg/ha)	Range of seed yields (kg/ha)	Days to 50% flowering
Trifolium decorum	5	4200–7500	10[a]–40[a]	>122[a]
T. quartinianum	4	5500–7700	860–1590	80–102
T. rueppellianum	5	3700–5300	150–560	72–87
T. steudeneri	5	4400–7000	1100–1480	67–93
T. tembense	5	5000–6100	71[a]–790	85–123

[a] Dried before maturity due to cessation of rain.

Source: Kahurananga & Tsehay (n.d.).

ted in pure stand. Seed and hay yields are recorded. The most recent form of this trial includes a plot which is left unplanted and a plot planted to oats. These plots provide controls in the second year of the trial when oats are planted in all plots. N fertilizer is then applied to half of each plot at 40 kg/ha. The trials are run for two or three years to determine the long-term effects of N fixation by the *Trifolium* lines planted in the first year.

In these replicated trials hay yields have generally been good, with yields varying greatly within species. Generally *T. quartinianum*, *T. decorum* and *T. steudeneri* have provided the best yields, with the most productive lines in years of good rainfall producing 7 t/ha or more of dry matter (DM) and up to 1.5 t/ha of seed (Table 1).

Other studies have indicated that a *T. tembense* content of 20–25 per cent in teff straw increased intake of the mixture by 20–30 per cent compared with the pure teff straw control (Butterworth, Mosi & Preston, 1985). Similarly, the addition of *T. tembense* to a number of cereal straws increased the apparent digestibility of the DM, crude protein and phosphorus of the diets (Mosi & Butterworth, 1985).

Since 1985 FNE has planted multilocation initial evaluation trials of the elite lines from stage 2 screening. The trials are planted as unreplicated strips of 20 genotypes belonging to five species (*T. quartinianum*, *T. decorum*, *T. rueppellianum*, *T. steudeneri* and *T. tembense*) with *T. quartinianum* ILCA 6301 as control in each fifth plot and about the trial perimeter. The 7.5 m strips have three P treatments imposed: 0, 10 and 20 kg P/ha. The trials have been planted by ILCA, the Institute for Agricultural Research (IAR) and the Arsi Rural Development Unit at

Addis Ababa, Debre Berhan (Enwari), Holetta, Kulumsa and Gobe in a range of soil types and altitudes. Standard observations are made (Anonymous, 1985). ILCA has also recently started screening of native germplasm at Agew Midir, in Gojam, on a nitosol.

Screening is currently under way in other countries using ILCA–supplied Ethiopian germplasm. These include six other African countries, five in Central/South America, three each in Europe and Asia, as well as Australia, New Zealand, Canada and the USA.

Stage 2 multilocation screening of native annual highland *Trifolium* species has been under way since 1985 (Anonymous, 1985). Three promising lines of native annual *Trifolium* species (*T. quartinianum* ILCA 6301, *T. tembense* ILCA 5774 and *T. rueppellianum* ILCA 9690) are planted in mixture with the two most productive grasses from earlier FNE multilocation trials, *Festuca arundinacea* cv. Demeter and *Phalaris aquatica* cv. Sirocco. These trials are being undertaken at sites at or above 2400 m by ILCA, the Arsi Rural Development Unit and IAR at Holetta, Addis Ababa, Gobe and Debre Berhan (Enwari).

Perennial Trifolium

The difficulty of maintaining pure lines of the outcrossing perennial highland *Trifolium* species has inhibited seed multiplication and the amount of screening done.

In 1985, at ILCA headquarters in Addis Ababa, 98 accessions of perennial *Trifolium* species (*T. burchellianum* 26, *T. cryptopodium* 13 and *T. semipilosum* 59), mainly from Ethiopia, were planted as single plants transplanted from pots to the field in unreplicated rows with 10 plants per row at spacings of 1 × 1 m. Fertilizer was applied once at 40 kg/ha.

Standard parameters are observed every three weeks as well as such morphological parameters as petiole length, middle leaf length and width, stolon width, internode length and time of flowering. Harvests are taken to obtain dry matter yields.

Vicia *spp.*

Initial screening of 100 accessions of exotic *Vicia sativa* germplasm and eight collections of native *Vicia* species was done at ILCA headquarters in 1984 using the same procedures as previously described for the annual highland *Trifolium* species. The native lines had comparable performance to the introduced lines, with two lines (ILCA 8324, 8047) being grouped among the most promising.

Tropical and subtropical screening

Acid soils, rainfed

Stage 1 screening of mainly perennial species of tropical and subtropical exotic and native germplasm on acid soils in the Sodo area (1850 m, 1100 mm rainfall) has been under way for more than two years and the initial 1984 plantings have thus completed their third growing season.

Screening in Sodo is done using unreplicated microplots (2 × 1.5 m), each with six plants spaced at 0.5 m and established by seed or transplanting from pots. The plots are slightly mounded for uniform drainage and natural vegetation is allowed to establish in the plots and pathways to provide a more natural environment for the growth of the plants. The weeds are cut back from time to time and regular observations are taken at six-week intervals on defined morphological and agronomic parameters on standard ILCA observation sheets. P fertilizer is added at 10 kg/ha at planting and at the same rate annually as a split application after alternate observations.

The lines planted in 1984 for initial screening were mainly single representatives of a wide range of leguminous species. As there had been little collection of native lowland species at that time, few introductions were of native germplasm. An early assessment of the results (Lazier, 1986) indicated that of the six species rated as excellent, three are native to Ethiopia: *Argyrolobium ramosissimum, Macrotyloma axillare* and *Eriosema psoraleoides*. Of these lines only *E. psoraleoides* was collected in Ethiopia. Included in the 11 lines ranked as good were one line each of *Neonotonia wightii, Indigofera arrecta* and a *Crotalaria* species, all of Ethiopian origin.

The 1985 initial evaluation microplot plantings included more than 1000 lines. Major plantings were made of lines of *Stylosanthes fruticosa* and *Neonotonia wightii*, both of which are important, widely represented species in Ethiopia. Of these, the lines of *S. fruticosa* being screened are all native. The germplasm is relatively similar in appearance and thus analyses are being done on details of morphological and agronomic characters in an effort to group accessions which are similar.

Thirty-seven *Neonotonia wightii* lines were screened (36 native and one exotic). Five native lines were outstanding in adaptation and performed better than an earlier planting of a commercial line (cv. Cooper). In limited plantings of grass germplasm, mainly commercial, a local collection of *Melinis minutiflora* performed as well as a commercial line. Ten lines of native *Zornia* species are being screened along with 161 lines of mainly South American origin in a 1985 planting. Only one line of the native Zornias is recorded as having good vigour.

Basic soils, irrigated

All tropical and subtropical germplasm of grasses, herbaceous legumes and browse which can grow reasonably well on basic soils and at medium altitude is multiplied at the 4 hectare ILCA seed multiplication site on the Ministry of State Farms, Horticulture Division's site at Zwai (1650 m, pH 8.0). Here, since 1983, germplasm has been planted and replanted as required to provide seed for the genebank and early stage agronomic evaluations.

Plantings are done in 5 × 5 m plots either by seed or by transplanting pot-raised plants. The size of the plantings varies, from one plant to a 5 × 5 m plot, depending on the amount of seed available. P fertilizer is applied at 50 kg/ha to all plots and N at 20 kg/ha to grass and non-legume plots. The applications are made after alternate six-weekly observations. Observations are made of defined characters on standard ILCA seed multiplication observation forms.

Ideally, all germplasm of one species would be planted at the same time to provide more uniform screening conditions. However, such organized plantings are unlikely to be possible until the large backlog of germplasm awaiting urgent multiplication is reduced. Despite the variations in time of planting, considerable useful data are acquired from the plots as the moisture availability is uniform year-round, due to irrigation, and the plants are normally in the plots for more than one year.

The number of native Ethiopian grass and legume lines which have been and are being multiplied in this manner is very large, approaching 3000 lines, and a huge amount of observational data has been acquired. These data are currently being computerized for analysis.

Mainly vegetative collections of native materials of *Erythrina* have been made and these have been screened under irrigation at Zwai. Six accessions, which are leafy, relatively thornless, vigorous and with a more prostrate growth habit, have been selected as elite lines and are currently being further screened at Sodo and at ILCA headquarters at Addis Ababa. The species involved have not yet been definitely identified as the plants have yet to flower. *E. abyssinica* and *E. brucei* appear to be the species utilized as forage locally.

Feeding trials

Leguminous trees are important in the local farming system as sources of forage and as important agents in maintaining soil fertility and stability (e.g. Lazier, 1985). Their feeding value is currently being studied by the ILCA Nutrition Unit. *Acacia seyal* has been found to be as good a feed for sheep as *Sesbania sesban*, once the animals have become accustomed to it. The animals lost weight on *A. cyanophylla*, an exotic species commonly planted in reafforestation projects. Goats appeared to be better adapted to the digestion of *Acacia albida* pods than sheep (Anonymous, 1986).

Current utilization

While no commercialized or large-scale plantings of native forage germplasm have been carried out in Ethiopia, considerable use is made of native forage germplasm. The valley bottoms of the highlands are commonly used as reserves or common grazing lands, or as sources of hay. Native *Trifolium* species are frequently abundant in these areas and undoubtedly make a considerable contribution to the diet of livestock. A farming system has been discovered in the Gojam administrative region in which dense stands of native *Trifolium decorum* appear in the fallow year after a crop of teff. Teff and the native legume are thus grown in alternate years as the farmers recognize the feed value of the legume and its beneficial effect on the crop the following year (J. Kahurananga, personal communication).

In the Welayita awraja of Sidamo administrative region above 1600 m, human population densities are high with farm sizes averaging about 0.5 ha. Here, native *Stylosanthes fruticosa* stands can be a major component of the vegetation of eroded areas, sometimes being the dominant component, while *Neonotonia wightii*, *Zornia* spp. and *Indigofera* spp. also commonly occur. It has been reported from this region (Anonymous, 1983; Ochang, 1985) that areas of such legumes are not grazed but are hand-picked to feed to productive animals, particularly milking cows. The legumes are commonly cooked before feeding.

Erythrina species are commonly planted as browse hedges in the middle altitude (1400–2000 m) areas of western Ethiopia on acid nitosol soils. Such hedges have been reported as high as 2800 m at Chencha in Gamo Gofa administrative region. Decreasing rainfall and lower human population densities limit the distribution at lower elevations. The leaves are normally fed at the beginning of the dry season. Reports on palatability vary markedly from area to area; however, animals appear not to graze it in the wet season, suggesting either that hunger drives the animals to graze it, or that its palatability improves in the dry season. One farmer claimed that the cuttings used locally were of a selection which was dwarf, leafy and sterile (Lazier & Mengistu, 1984).

Sesbania sesban, a palatable browse plant of considerable potential which is native to Ethiopia, is currently being planted widely in Ethiopia by the Soil and Water Conservation Department of the Ministry of Agriculture. Exotic and native germplasm is being used. It is also commonly used by farmers to shade young coffee plants. The origin of this material is not known.

ILCA's collections of native *Sesbania* germplasm show considerable potential in initial evaluation plots. *S. sesban* (an exotic line) has done well in the ILCA Highland Programme trials in stabilizing terraces and as a green manure for cereal crops (Anonymous, 1986). *S. sesban* germplasm is despatched regularly by ILCA, as part of a basic initial evaluation package, to researchers in the tropics and subtropics worldwide.

References

Anonymous (1983). Natural *Stylosanthes* pastures in Ethiopia. *PGRC/E–ILCA Germplasm Newsletter*, **3**, 10–11.

Anonymous (1985). FNE [Forage Network in Ethiopia] trial protocols 1985. *FNE Newsletter*, **8**, 8–15.

Anonymous (1986). *ILCA Annual Report 1985/6*. ILCA, Addis Ababa.

Butterworth, M. H., Mosi, A. & Preston, T. R. (1985). Molasses/urea and legume hay as supplements to poor quality roughage in Ethiopia. XIII International Congress of Nutrition, Brighton (abstract).

Carlsson, J. (1972). Inventory of indigenous ecotypes of some grass species in Chilalo Awraja, Ethiopia. Asela (mimeographed).

Froman, B. (1975). Pasture management in Ethiopia with special reference to conditions in the Chilalo Awraja. Agricultural College of Sweden. Report and dissertation 32.

Ibrahim, K. M. (1975). Pasture and forage crops research programme report. Project ETH/74/002/a/01/12. FAO, Rome (mimeographed).

Kahurananga, J. & Tsehay, A. (n.d.). Interspecific and intraspecific dry matter and seed yield and flowering variation of five annual Ethiopian *Trifolium* species (mimeographed).

Lazier, J. R. (1985). *Acacia*, an important natural forage plant in Ethiopia. *FNE Newsletter*, **10**, 20–2.

Lazier, J. R. (1986). Forage germplasm introduction in Sodo, Welayita: 1984 plantings. *FNE Newsletter*, **12**, 17–22.

Lazier, J. R. & Mengistu, S. (1984). *Erythrina*, a genus with browse potential. *PGRC/E–ILCA Germplasm Newsletter*, **8**, 20–2.

Mosi, A.K. & Butterworth, M.H. (1985). The voluntary intake and digestibility of diets containing different proportions of teff (*Eragrostis tef*) straw and *Trifolium tembense* hay when fed to sheep. *Tropical Animal Production* (in press).

Ochang, J. (1985). Traditional use of native legumes for supplementary feeding of milking cows in the Welayita region of Ethiopia. *PGRC/E–ILCA Germplasm Newsletter*, **8**, 38–9.

Skerman, P.J. (1977). Tropical forage legumes. *FAO Plant Production and Protection Series No. 2*, Rome.

Thulin, M. (1983). Leguminoseae of Ethiopia. *Opera Botanica*, **68**.

22

Improvement of indigenous durum wheat landraces in Ethiopia

TESFAYE TESEMMA

Importance of wheat in Ethiopia

Wheat has been and continues to be one of the most important cereal crops in Ethiopia in terms of both area under cultivation and production. In 1983, the area under wheat production was estimated at 625 590 ha with an average production of 1065 kg/ha (Central Statistics Office, 1984). The demand for wheat as a staple food grain is increasing, especially in the urban areas, while its utilization will be high even in the rural sector in the near future. At present, consumer demand for wheat as a staple food grain is increasing, especially in the urban areas, while its utilization will be high even in the rural sector in the near future. At present, consumer demand far exceeds domestic production and wheat imports are costing the country millions of dollars in foreign exchange. Wheat constitutes a large portion of the daily diet of the population and contributes significantly to the calorie and protein intake. It is consumed in several different forms such as leavened bread, pancakes, macaroni and spaghetti, biscuits and pastries. The most common of the Ethiopian recipes are dabo (Ethiopian home-made bread), hambasha (home-made bread from northern Ethiopia), kitta (unleavened bread), injera (thin bread, part of the national dish and prepared mainly from teff), nifro (boiled whole grains, sometimes mixed with pulses), kolo (roasted whole grains), dabo-kollo (ground and seasoned dough, shaped and deep fried) and kinche (crushed kernels, cooked with milk or water and mixed with spiced butter).

All wheat in Ethiopia is produced under rainfed conditions and the important production areas are Arsi, Bale, Shewa, Gojam, Gondar, Eritrea, Tigray, Welo and the highlands of Harerge and Sidamo administrative regions. Durum wheat (*Triticum durum*) is by far the most predominant species and occupies 60–70 per cent of the total area under cultivation, while bread wheat (*T. aestivum*) constitutes the remaining 30–40 per cent (Tesemma & Mohammed, 1982).

Durum wheat is a traditional crop in Ethiopia. It is traditionally grown on the heavy black clay soils (vertisols) of the highlands at altitudes between 1800 and 2800 m above sea level. On the other hand, bread wheat, although of recent introduction, has a wider environmental adaptation than durum wheat. At present, however, it is grown primarily in Arsi and Bale administrative regions.

Genetic diversity in Ethiopian wheats

Nearly all the wheat varieties grown at present are landraces consisting of a large number of different genetic lines; even different species are often found grown as mixtures in wheat fields. It is not uncommon to find three or occasionally four species and often as many as 8–15 botanical varieties in the same field.

The variation in the Ethiopian wheats is so great that N. I. Vavilov was misled and classified Ethiopia as one of the centres of wheat origin, before other genetic resources authorities (Harlan, 1971) located the origin in the Near East and confirmed Ethiopia as a centre of diversity because none of the wild relatives had been found in the country.

According to Porceddu, Perrino & Olita (1973), Ethiopian wheats belong to the following species: *Triticum turgidum, T. durum, T. dicoccum, T. aestivum, T. polonicum, T. pyramidale* and *T. abyssinicum*. Of the tetraploid wheat species, durum wheat (*T. durum*) is the most extensively cultivated. It is believed that the bulk of the wheat types grown in Ethiopia have undergone very little change over the past centuries (Nastasi, 1964).

Characteristics of the Ethiopian tetraploid wheats

In general, the Ethiopian durums differ in spike form, spike density, spike, awn and kernel colours (brown, amber and violet) and awn condition (awnless, short-awned and long-awned varieties). Other distinguishing characteristics include hairy or waxy leaves, hairy or glabrous glumes and presence of pigmentation in glumes, awns and kernels of seedlings and mature plants (Tables 1 and 2).

The author has observed that the kernel structure and pigmentation of the glumes and awns are influenced by altitude. The same variety when grown at different elevations may produce seeds that

Table 1. *Phenotypic diversity in Ethiopian drum wheat germplasm accessions for some quantitative characters, expressed in their range, mean and coefficient of variation (N=1120)*

Character	Range		Mean	CV (%)
	Minimum	Maximum		
Number of kernels per spike	11	63	30.8	22.8
Number of spikelets per spike	2.6	40	18.5	16.2
Days to 50% flowering	49	109	74.3	13.1
Days to 75% maturity	86	151	108.1	11.4
Plant height (cm)	15	135	93.2	15.0

Source: PGRC/E (unpublished data).

look different from each other. At higher elevations (e.g. Cheffe Donsa at 2450 m) the pigmentation of the glumes and awns is more intense and kernels are vitreous. At lower elevations (e.g. around Debre Zeit at 1900 m) the pigmentation may be completely absent on glumes and awns, and the kernels may be affected by yellow-berry. At Akaki (2200 m) different degrees of pigmentation and a mixture of kernels with yellow-berry and amber colours can be noticed.

Ethiopian wheats are important because of their rust resistance, long coleoptile, short culm, low tillering, early ripening and drought resistance (Porceddu *et al.*, 1973). However, from the author's personal observations, it can be said that Ethiopian durum wheat cultivars generally have very location-specific adaptabilities. They also have weak straw which is prone to lodging. As a result, they do not respond very well to fertilizer application and hence are adapted to low soil fertility conditions. They lack satisfactory resistance to leaf rust but appear to be highly resistant to stem rust. They have the ability to survive deep seeding, better than exotic varieties, and tend to be rather low-yielding but more reliable for yield under adverse conditions.

Diseases and insect pests of wheats in Ethiopia

Wheats in Ethiopia are attacked by a number of fungal diseases, such as leaf rust (*Puccinia recondita*), stem rust (*P. graminis tritici*) and stripe rust (*P. glumarum*). Leaf blotch (*Septoria tritici*) and glume blotch (*S. nodorum*) cause considerable damage while bunt or stinking smut (*Tilletia foetida* or *T. caries*), *Helminthosporium* spp., *Fusarium* spp. and powdery mildew (*Erysiphe graminis*) are problems

Table 2. *Phenotype diversity in Ethiopian durum wheat germplasm accessions for some qualitative characters, expressed as frequency distribution over the respective classes (N=1123–1173)*

Character	Character states and the frequency distribution (in %)				
Awnedness	Awned 99.0%	Awnless 0.8%	Mixture 0.2%		
Glume hairiness	Absent 87.4%	Low 3.6%	High 9.0%		
Glume colour	White to yellow 84.5%	Red to brown 14.5%	Purple to black 1.0%		
Spike density	Very lax 1.5%	Lax 9.5%	Intermediate 64.3%	Dense 24.6%	Very dense 0.1%
Kernel colour	White to yellow 47.4%	Red to brown 26.2%	Purple to black 26.4%		
Kernel size	<6 mm 32.3%	6–8 mm 64.8%	>8 mm 2.9%		
Kernel vitreousness	Soft and starchy 4.5%	Partly vitreous 3.8%	Vitreous 91.7%		

Source: PGRC/E (unpublished data).

in some areas. Recently bacterial stripe (*Xanthomonas translucens*) has also been reported as causing some damage.

The extent of damage caused by these diseases depends on the weather conditions during the crop season. However, because the cultivars grown are landrace varieties consisting of genotypes which differ in their reaction to diseases and pests, some lines may be resistant or tolerant to certain races of the pathogen and others to different races. Consequently, wheat diseases rarely reach epidemic proportions because of the mixture of resistant and susceptible genotypes in the population which provides a buffer against rapid disease development and helps to extend the life of the resistance genes (Harlan, 1975).

Among the insects that commonly attack wheat is the wheat aphid (*Rhopalosiphum maydis*) which sucks on the leaves. Occasionally, ladybird beetle larvae (*Chnootriba similis*) feed on the leaves, leaving only the fibre, while stem borer (*Sesamia epunotifera*) kills the plants after heading in scattered spots in wheat fields.

Status of indigenous wheat landrace improvement in Ethiopia

Attempts to improve Ethiopian indigenous wheat landraces started in 1949 at the Paradiso Experimental Station near Asmara (Tesemma & Mohammed, 1982). Among several local wheat collections tested for productivity, stem rust and leaf rust resistance, four selections, namely A 10, R 18, P 20 and H 23, were selected and released to farmers in Eritrea in 1952.

In 1956 and 1957, several crosses were made between local and exotic varieties mainly for the purpose of transferring stem rust resistance of A 10 and R 18 to Mindum, a variety of excellent quality from the USA. Two outstanding lines, Mindum × A 10 and Mindum × R 18 were selected but were later found to be susceptible to leaf rust. An attempt to incorporate leaf rust resistance into the above two lines from an exotic variety (Senator Capelli) resulted in a selection with good resistance to stem and leaf rusts but too late in maturity for the Eritrean environment (Nastasi, 1964). For some unknown reason, work at Paradiso has been discontinued.

Although the Debre Zeit Agricultural Experimental Station was established in 1953, emphasis in wheat improvement has, until very recently, been on exotic varieties. Work on indigenous landraces consisted mainly of making collections and maintaining the germplasm. Subsequently, all of the then 1145 wheat accessions at hand were transferred to the Plant Genetic Resources Centre/Ethiopia (PGRC/E) soon after its creation in 1976. Through mass selection, two local varieties, DZ04-118 and Marou (DZ-688), were released to farmers in Yerer–Kereyu Awraja in 1968 (Tesemma & Mohammed, 1982). On the other hand, as a result of introduction and extensive testing, four high-yielding and adapted exotic varieties of durum wheat (Cocorit 71, Gerado, Ld57/CI 8155 and Boohai), have been released to farmers by the experimental station. These varieties are mainly grown in Yerer and Kereyu and Menagesha awrajas where the extension programme of the experimental station is strong. Although these varieties are higher yielders than the local cultivars, they are equally susceptible to diseases, except for the variety Boohai which has good resistance to leaf and stem rust.

Work on durum wheat hybridization at Debre Zeit was initiated in 1974. Since then, numerous crosses have been made within local parents and local–exotic parents with a view mainly to improve yielding capacity, disease resistance and straw strength. So far no superior line has been identified. Nevertheless, the programme continues. Consequently, there are at present quite a large number of genotypes at an advanced stage of yield testing.

Prospects

Genetic erosion is not yet far advanced, except in Arsi administrative region where the adapted local landraces have been virtually replaced by exotic bread wheat varieties. The administrative region of Bale is the next most vulnerable area for genetic erosion because of the rapid expansion of bread wheat production in the state farms. In general, genetic erosion is bound to accelerate with the release of high-yielding varieties from research stations, improvement of the infrastructure, increase in population and an overall advance in agriculture. Cognizant of this, PGRC/E has identified wheat as a priority crop for collection and preservation. According to Worede (1985) about 7400 wheat samples have been collected and are preserved at PGRC/E for future use. Nevertheless, to be properly and effectively utilized by the breeders, these and other samples must be characterized, screened and evaluated and the information obtained must be properly documented and made easily available, together with the seed samples.

Ethiopia has a wide range of climatic and ecological conditions. The indigenous cultivars are the outcome of the evolutionary processes in response to the heterogeneous environment that prevails in the wheat growing areas of the country. They are staunch survivors in the struggle for 'survival of the fittest' in the vagaries of nature where intermittent 'normal' seasons have forced them to undergo natural selection for centuries (Chavan, 1975). Consequently, the genetic variability in Ethiopian wheats has tremendous potential which, if properly exploited, could be a vital and very useful source of germplasm not only for the country but also for the rest of the world.

In a country like Ethiopia, where research in plant breeding is in its infancy because of a lack of adequately trained personnel and inadequate facilities, it is only logical that the first step in breeding should be maximum utilization of indigenous material. The dissemination of newly developed uniform varieties in any region is highly risky. Due to their narrow genetic base, such varieties are potentially vulnerable to diseases and pests that can cause extensive yield losses. Broad genetic variability is believed to have a buffering effect against the diverse production environments prevailing in the country. Therefore, desirable results could be obtained through a distribution and production programme of some of the better varieties selected from indigenous landraces or segregations of their most productive and valuable forms in hybridization programmes.

One way of improving the general productivity of wheat landraces grown by the farmers is through phenotypic selection of the best plant types from genetically mixed populations (positive mass selection) and subsequent bulking for further multiplication and distribution to the farmers. A slight modification of this would be to make pure line selections of the different genetic lines through yield testing and bulking two or more superior genetic lines for further multiplication and distribution to farmers.

Another strategy for the improvement of landrace populations of tetraploid wheats in Ethiopia is based more or less on the multiline concept, using the isogenic lines that exist in nature. In this approach, it is assumed that a genetic line that grows in certain parts of the country may have a gene(s) for resistance to certain races of a pathogen or pest prevailing in that particular area. When seeds of such 'identical genetic lines', each possessing different genes for resistance, are composited, the resulting mixture will form a multiline variety which will provide a partial protection or buffer against a broad spectrum of races of pathogens or pests. This approach takes the uniformity of seed colour into consideration and is now under investigation in cooperation with the Swedish Agency for Research Cooperation (SAREC).

In spite of their outstanding merits, Ethiopian wheats also have shortcomings such as low yield potential, unsatisfactory resistance to leaf rust and weak straw. Therefore, the incorporation of desirable characteristics of the most promising exotic varieties into the adapted best local varieties, keeping the background of the local varieties, should be the long-term objective of the durum wheat varietal development programme. Along with an improvement of the characters mentioned, due consideration should be given to such research as the development of:

- varieties that are responsive to small fertilizer inputs;
- varieties that are tolerant to drought, desiccating wind and frost;
- better cropping systems that would increase soil fertility;
- better tillage practices that will help soil and moisture conservation;
- varieties with improved grain quality.

To accomplish this task, the wheat breeding programme should utilize the concerted efforts of breeders, agronomists, soil scientists, protection experts and others relevant to the field. Such a holistic and integrated team approach, it is hoped, would lead to self-sufficiency in the ever increasing demand for wheat in general, and durum in particular, thereby saving the nation substantial foreign exchange.

Acknowledgement

The writer is grateful to Dr Mesfin Abebe for his useful suggestions on the preparation of this manuscript.

References

Central Statistics Office (1984). *Ethiopian Statistical Abstract.* CSO, Addis Ababa.

Chavan, V. M. (1975). The importance of indigenous varieties in plant breeding. *The Indian Journal of Genetics and Plant Breeding,* **17**(1), 1–6.

Harlan, J. R. (1971). Agricultural origins: centres and noncentres. *Science,* **174**, 468–73.

Harlan, J. R. (1975). Our vanishing genetic resources. *Science,* **188**, 618–21.

Nastasi, V. (1964). Wheat production in Ethiopia. *Information Bulletin on the Near East Wheat and Barley Improvement and Production Project, vol. 1 (13)*, pp. 13–23.

Porceddu, E., Perrino, P. & Olita, G. (1973). Preliminary information on Ethiopian wheat germplasm collection mission. *Proceedings, symposium on genetics and breeding of durum wheat.* University of Bari, Italy, pp. 181–99.

Tesemma, T. & Mohammed, J. (1982). Review of wheat breeding in Ethiopia. *Ethiopian Journal of Agricultural Science,* **1**, 11–24.

Worede, M. (1985). Crop genetic resources activities in Ethiopia. International Symposium on South East Asian Plant Genetic Resources, Jakarta, 20–24 August, 1985.

23

Use of germplasm resources in breeding wheat for disease resistance

HAILU GEBRE-MARIAM

Introduction

Wheat (*Triticum* spp.) covers 12 per cent of the total area of about 6 million hectares of land, producing approximately 7 million tonnes of food grain, in Ethiopia. Both tetraploid and hexaploid wheats are important in the farming systems; the former takes more than 50 per cent of the share. Comprehensive wheat improvement programmes for durum and bread wheat have been developed to contribute to the overall effort of increasing food grain production in the country.

Diseases are among the major constraints limiting wheat production. The wheat rusts (*Puccinia* spp.), *Septoria* diseases and *Fusarium* spp. are the economically important diseases that therefore require resistance breeding programmes. In particular, with the increase of bread wheat acreage under the management of state farms, the potential danger of yellow rust epidemics (caused by *P. striiformis*) in Arsi and Bale highlands in southern Ethiopia warrants a well organized breeding programme.

This paper deals with the approaches in breeding for disease resistance with particular emphasis on the use of indigenous and exotic germplasm resources in the bread wheat breeding programme in Ethiopia.

Guiding principles

Considering the complexity of the biotic and abiotic environments of the Ethiopian agro-ecosystems, three guiding scientific principles have been essential in planning and implementing the wheat breeding programme for disease resistance. These are (a) careful consideration of the genetic variation phase of the breeding programme; (b) critical analysis of the crop ecosystem; (c) proper evaluation of the pathosystem.

As more and more experience has been gained in breeding for disease resistance, the importance of the acquisition, selection, evaluation and introgression of diverse germplasm to create as broad a genetic base as possible has been recognized. The use of resistant cultivars is the most economic and practical approach for the control of cereal diseases, and this can only be attained by developing and employing an effective resistance breeding programme with sufficient genetic variation. Current advances in the techniques of evaluating, selecting and incorporating germplasm into breeding stocks have been essential in successfully developing disease-resistant high-yielding wheat varieties. As the complexities of the host–pathogen system in relation to the agro-ecosystem increase, sources of resistance genes become limited.

Within a given agro-ecological zone of adaptation, crop breeding involves the incorporation of a specific gene or gene complexes governing pest resistance and other essential traits. The Ethiopian crop ecosystems are very diverse, consisting of a multiplicity of interacting environmental factors such as altitude, climate, soils, etc. as well as of a wide range of crop genetic diversity, ranging from primitive but well adapted landraces to modern, highly uniform, high-yielding varieties. The latter require high inputs whereas the primitive landraces frequently have to perform without any inputs. This has led the wheat improvement programme to use different strategies in the disease resistance breeding work to serve this diversity of conditions. Sharp (1973), after investigating the significant influence on yellow rust of the diverse agro-ecosystems of north-western USA, stressed the difficulty of evaluating sources of resistance. In Ethiopia, the complexity of these biotic and environmental factors is compounded by the traditional subsistence farming systems which add a socio-economic dimension to the problem. Because there is no organized system for the transfer of farming technology, varieties released for a specific agro-ecological condition frequently end up in the wrong domain of recommendation. This results in lower yields caused by the different wheat rusts. This problem is mainly due to the unavailability of sufficient seed of the recommended varieties and a lack of essential extension information on existing varieties. Thus it is necessary to consider the totality of the agro-ecosystems and the socio-economic settings of the wheat producing areas in the country.

This requires not only the testing of breeding materials across diverse agro-ecological zones, but also different scales of rating among genetic sources and types of resistance and an assessment of traits that show correlation between disease intensity and yield loss (Roebelen & Sharp, 1978). Investigations such as this help to determine the effectiveness, or buffering capacity, of resistance to minimize yield losses when varieties are grown under different agro-ecological farming conditions.

In the traditional Ethiopian farming setting, both the *Triticum* spp. and their pathogens are ancient and indigenous. Therefore, the relative levels of resistance or susceptibility of landraces and the virulence of the pathogens fluctuate only to a limited extent. This avoids pressure due to a high level of damage or resistance, thus allowing the host–pathogen balance to be maintained (Buddenhagen & de Ponti, 1984). The breeding programme, however, introduces into such a system high-yielding homogeneous wheat varieties that remove the suppressing effect on an epidemic that is provided by the mixtures of genotypes in landraces. In other words, the successful introduction of high-yielding wheat varieties has resulted in increased acreage of extensive state farms which in turn has resulted in the buildup of pathogen populations and an increasing level of inoculum sufficient to cause outbreaks of epidemics. A good example of this is the 1981 yellow rust outbreak in Arsi and Bale highlands. In addition, an epidemic level of yellow rust has been observed in 1986 and two newly recommended varieties became susceptible. Whether this is due to the appearance of a new race(s) or not is under investigation. Hence, the wheat improvement programme is conscious of the fact that a newly released variety will influence and be influenced by a multitude of factors occurring in the local cropping system. Selection for yellow rust resistance is further complicated because different biotypes of *P. striiformis* and both specific and non-specific forms of resistance may occur in Ethiopia.

Germplasm resources and introgression

To supplement the classical taxonomic classification of Linnaeus, Harlan & de Wet (1971) devised a gene pool or biological species concept to group cross-compatible taxa. Of the three gene pool categories (primary, secondary and tertiary) we have so far used only those materials which fall into the primary gene pool. Nevertheless, germplasm of related wild species will be briefly mentioned in this discussion because of its future potential.

The three major germplasm sources for the national bread wheat breeding programme are: modern advanced varieties and breeding stocks which originate from national programmes, international research centres, and sometimes genebanks; landraces which include local primitive/traditional cultivars obtained from direct field collections, the Plant Genetic Resources Centre/Ethiopia (PGRC/E) and other genebanks; and hybridizations.

Within the cultigen sources of germplasm, advanced varieties and various breeding stocks are the economic and ready sources of desirable genes. Since the breeding programme is directed towards developing disease-resistant varieties for use in the short-term and intermediate periods, the focus is on already improved gene sources from national and international programmes. This includes both selection under local conditions for direct use and the introgression of unadapted germplasm into adapted genetic backgrounds to increase useful genetic variability. This is a very valuable practice since favourable genes or gene complexes in unadapted germplasm may be masked by major genes for adaptation to different environments (Stuber, 1978). In this respect, the international research centres have been invaluable sources of germplasm resources. In 1985–6 the programme handled a total of 19 490 entries in this group. These were 150 genotypes in variety trials, 274 lines in advanced observations, 2831 nursery lines and 16 235 in segregating populations. Most of the segregating populations were generated from the local hybridization programme.

During the past decade, interest in slow rusting or tolerant types of resistance has increased. This type of resistance is of great potential value because it may be more stable than that recognized by clear-cut infection types (Andres & Wilcoxson, 1986). As a result of a study carried out at Holetta, several high-yielding lines tolerant to *Septoria tritici* blotch were identified (Gebre-Mariam & Gebeyehu, 1985) and are presented in Table 1.

The bulk of the Ethiopian wheat germplasm resources are mostly *Triticum turgidum* and *T. dicoccum* types. The diversity in hexaploid wheat is limited. Therefore, a large proportion of 2384 local collections evaluated at Kulumsa and Ginchi in 1985 by the wheat programme were durum types. Of these, 24 lines were found to be resistant to leaf rust (caused by *Puccinia recondita*) under field conditions. The major constraints in using these materials directly, however, are their susceptibility to other diseases and undesirable morphological characteristics.

Table 1. *Comparative evaluation of three groups of bread wheat varieties for their response to* Septoria tritici *blotch at Holetta*

Variety	HT	G7(R)	PY	HI	KQ	GY
Group 1	*Resistant lines*					
HAR 499	105	1.3	1.5	0.37	2−	6530
HAR 487	100	1.2	1.2	0.40	1	6140
KCJ 29	105	1.4	1.2	0.38	2+	6000
HAR 503	125	1.6	0.4	0.33	1−	5760
HAR 544	90	1.2	1.2	0.36	1	5340
Group 2	*Tolerant lines*					
HAR 466	135	2.4	1.8	0.31	1+	6440
YAL 564	142	2.3	2.3	0.34	1−	6290
ET 620BI	105	2.2	1.5	0.36	1−	6080
TAS 879	113	2.4	2.3	0.32	1	4900
ET 12D4	88	1.8	1.8	0.34	1−	5230
Group 3	*Susceptible lines*					
HAR 118	100	3.1	4.1	0.33	2−	5040
PAN 504	101	2.3	2.6	0.35	2	5400
EDC 1138	126	2.8	3.4	0.31	2−	5240
Olaf	98	3.4	3.6	0.28	2−	4650
Derese	106	2.7	4.5	0.29	3+	4080

HT, Plant height (cm); G7(R), *Septoria* scoring using 0 (= no infection) to 5 (= high level of flag leaf infection) at growth stage 7, i.e. milk development stage (after Zadoks *et al.*, 1974); PY, pycnidia count/mm^2 of the flag leaf; HI, harvest index; KQ, kernel quality on a 1–3 scale with 9 classes, i.e. 1+ (excellent quality), 1, 1−, 2+, 2, 2−, 3+, 3 and 3− (very poor quality); GY, grain yield (kg/ha).

Currently, there is a programme of crossing and selection under sub-optimum soil fertility conditions involving landraces or old/local bread wheat cultivars and improved varieties. It is a study attempting to answer some questions concerning the possibility of developing wheat varieties with an economic level of grain yield under relatively low soil fertility conditions so that they may be safely recommended for low-input conditions.

Triticum aestivum is a classic example of utilizing wild species for crop improvement via interspecific hybridization and genome building. The genus *Triticum*, containing more than 20 species, includes a polyploid series ranging from diploids (2n = 14) to hexaploids (2n = 42). Resistance to several diseases has been incorporated into bread wheat from related species of *Triticum*, *Aegilops* and *Agropyron* (Riley, 1965; Sears, 1969).

Among several methods developed to transfer genetic material from wild relatives to wheat, induction of homoeologous pairing and crossing-over is by far the easiest (Sears, 1972). Substitution lines have also been successfully used in transferring chromosomes carrying genes for disease resistance from *Agropyron* species to those of wheat genomes (Schulz-Schaeffer and McNeal, 1977).

Although the wheat programme has not so far used wild relatives of cultivated wheat species, it would be valuable to collect such germplasm for future use.

PGRC/E and the wheat improvement programme

A comprehensive and coordinated wheat improvement programme has been developed to assist in the urgent need for increased food grain production in the country. This effort needs the strong support of the genebank in the following areas:

- greater availability of different types of germplasm from various sources;
- conservation of various types of germplasm for future use;
- registration of useful germplasm;
- preliminary evaluation and classification of available landraces;
- collection of wild/weedy relatives of cultivated wheat in the region.

The genebank can act as an important medium for acquiring cultigen germplasm for use by breeders, including advanced varieties and breeding stocks from genebanks throughout the world and from national and international breeding programmes. Secondly, the genebank's role of storing/conserving and registering germplasm introduced and/or developed by breeders could be of great assistance to the national crop improvement efforts. During the last 20 years, research institutions in the country have introduced and developed a large number of germplasm accessions but none of these, with the exception of those still under production, are available any more because no proper long-term storage facilities existed in the country.

In the conservation of landrace germplasm, PGRC/E is doing a commendable job. Nevertheless, acceleration of the classification and preliminary evaluation work would encourage breeders to utilize existing genetic resources more fruitfully. Since the genetic erosion of wild relatives of cultivated species is equally serious, the landrace collecting efforts of PGRC/E should also include the wild and weedy species. As the existing sources of pest resistance in cultigens are exhausted, the wild–weed–cultigen complex will be very essential.

Therefore, the germplasm resource collection, conservation and utilization activities must be looked at holistically in terms of both space and time.

References

Andres, M. V. & Wilcoxson, R. D. (1986). Selection of barley for slow rusting resistance to leaf rust in epidemics of different severity. *Crop Science*, **26**, 511–14.

Buddenhagen, I. W. & de Ponti, O. M. B. (1984). Crop improvement to minimize future losses to diseases and pests in the tropics. *FAO Plant Production and Protection Paper No. 55*, pp. 23–49.

Gebre-Mariam, H. & Gebeyehu, G. (1985). Evaluation of bread wheat genotypes for their resistance to *Septoria tritici* blotch. *In*: A. L. Scharen (ed.). *Proceedings, Workshop on Septoria of Cereals, Bozeman, 2–4 August 1983*. Bozeman, Montana, pp. 27–30.

Harlan, J. R. & de Wet, J. M. J. (1971). Toward a rational classification of cultivated plants. *Taxon*, **20**, 509–17.

Riley, R. (1965). Cytogenetics and the evolution of wheat. *In*: Sir J. B. Hutchinson (ed.), *Essays on Crop Plant Evolution*. Cambridge University Press, London, pp. 103–22.

Roebelen, B. & Sharp, E. L. (1978). Mode of inheritance, interaction and application of genes conditioning resistance to yellow rust. *Fortschritt der Pflanzenzüchtung, Beihefte zur Zeitschrift für Pflanzenzüchtung, Supplement 9*, pp. 1–88.

Schulz-Schaeffer, J. & McNeal, F. H. (1977). Alien chromosome addition in wheat. *Crop Science*, **17**, 891–6.

Sears, E. R. (1969). Wheat cytogenetics. *Annual Review of Genetics*, **3**, 451–68.

Sears, E. R. (1972). Chromosome engineering in wheat. *Stadler Genetics Symposium, University of Missouri, Colombia*, **4**, 23–38.

Sharp, E. C. (1973). Wheat. *In*: R. R. Nelson (ed.), *Breeding Plants for Disease Resistance, concepts and applications*. Pennsylvania State University Press, University Park, pp. 111–31.

Stuber, C. W. (1978). Exotic sources for broadening genetic diversity in corn breeding programs. *In*: H. D. Loden and D. Wilkinson (eds), *Proceedings, 33rd Annual Corn and Sorghum Research Conference, Chicago, Illinois, 12–24 December 1977*. American Seed Trade Association. Washington, DC, pp. 34–47.

Zadoks, J. C., Chang, T. T. & Konzak, C. F. (1974. A decimal code for the growth stages of cereals. *Eucarpia Bull*. **7**, 42–52.

24

Indigenous barley germplasm in the Ethiopian breeding programme

HAILU GEBRE AND FEKADU ALEMAYEHU

Introduction

Barley in Ethiopia is used for human food, home-made beverages and beer. Its straw is used for animal feed and mattresses. It is produced in the highlands at altitudes ranging from 1800 to 3300 m above sea level, where poor soil fertility, frost, waterlogging and moderate soil acidity are major problems. It occupies an area of about 0.85 million hectares out of a total crop land of 6.0 million hectares with a productivity of about 1.2 t/ha (Central Statistics Office, 1984). At high altitudes it may be the only crop grown, with or without oats; and among the small grains it is the earliest to become available for consumption at the end of the rainy season.

Barley is grown in the main rainy season of June–September ('meher') on sloping and better drained clay soils. Some barley is grown in the short rainy season ('belg') on bottom lands in some regions. A few areas also grow barley from October to January ('bega'), sometimes with supplementary irrigation. Double cropping is a common practice in the major barley growing regions: barley–barley in 'belg' and 'meher' seasons in Shewa, Welo, Arsi, and Bale; barley–barley in 'meher' and 'bega' in Gojam; and barley–pulses in Gondar.

The level of management is traditional. Farmers use their own landraces in most cases. Application of inorganic fertilizers and use of insecticides and herbicides are fairly low. The landraces are tolerant to marginal soil conditions. At high altitudes fallow, sometimes accompanied by soil heating, is a common practice to improve soil fertility for barley production.

Table 1. *Characters used and their respective classes* (N *varying from 3592 to 3759 accessions)*

Character	Character states	Frequency distribution %	Mean	Minimum	Maximum
1. Kernel row number	6 rows	0.42			
	2 rows[a]	0.57			
	irregular	0.01			
2. Spike density	lax	0.42			
	intermediate	0.48			
	dense	0.10			
3. Spikelets per spike	<15	0.03	22.0	6	42
	15–20	0.22			
	20–25	0.49			
	25–30	0.24			
	≥30	0.02			
4. Caryopsis	covered	0.97			
	naked	0.03			
5. Kernel colour	white–brown	0.72			
	purple–black	0.28			
6. Thousand grain weight	<25 g	0.08[b]	33.9	14	55
	25–35 g	0.50			
	35–45 g	0.37			
	45–55 g	0.05			
	≥55 g	0.00			
7. Days to maturity	<100 days	0.03	119.4	74	193
	100–115 days	0.33			
	115–130 days	0.42			

	130–145 days	0.21			
	≥145 days	0.01			
8. Plant height	<60 cm	0.01	94.0	45	180
	60–90 cm	0.34			
	90–120 cm	0.59			
	120–150 cm	0.06			
	≥150 cm	0.00			

[a] Both categories, sterile and rudimentary lateral florets are combined.

[b] These data originated from a smaller but representative sample of 2480 accessions of barley.

Some characteristics of Ethiopian barleys

Ethiopia is considered to be a centre of diversity for barley with many genotypes available. It has even been considered to be a centre of origin (Negassa, 1985b). Ethiopian barleys are mostly two-rowed (including deficiens), with lax to intermediate spike density, white to brown or black coloured, covered kernels and with a diversity of time to maturity (Engels, 1988). Six-rowed genotypes are concentrated in Arsi, Bale and Shewa highlands because of their genes conferring frost resistance (Negassa, 1985b). Ethiopian barleys are distinctive in their range of grain colour, short-haired rachilla and the *deficiens* and *irregulare* spike types (Ward, 1962; Qualset, 1975). Some germplasm accessions have useful traits, especially for resistance to diseases such as powdery mildew (Negassa, 1985a), barley yellow dwarf virus, net blotch, scald, loose smut (Qualset, 1975) and high protein quality (Munck, Karlsson & Hagberg, 1971). Evaluation of some collections at the International Center for Agricultural Research in the Dry Areas (ICARDA) indicates moderate resistance to lodging and higher kernel weight when compared with collections from some countries (Somaroo *et al.*, 1984).

Other useful characteristics of Ethiopian barleys include high tillering capacity; tolerance to marginal soil conditions, barley shoot fly, aphids and frost; vigorous seedling establishment; and quick grain filling period. On the other hand, they tend to show sensitivity to lodging due to weak straw, low grain to straw ratio due to tall growth of straw, and, in some, fragile rachis. The variations for some selected characters are presented in Table 1.

Utilization of indigenous germplasm in the breeding programme

The major centre for barley breeding and coordination since 1967 has been the Holetta Research Centre. The early breeding programme at Holetta focused mainly on the introduction and evaluation of exotic lines using selection under optimum management conditions to develop food and malting barleys. A large number of introductions were evaluated at Holetta but most lines were found to be highly susceptible to scald, leaf blotch and barley shoot fly. Some of the better genotypes were further tested in yield trials at various locations and a few cultivars were recommended for commercial production. However, their adoption was slow because under traditional management practices the farmers' landraces performed as well as

Table 2. *Mean grain yield, heading date and disease data of the elite selections from Ethiopian collections at Bedi, 1977*

Variety	Heading date	Scald (0–9)	Yield (kg/ha)
CI 10389	100	6	2342
IAR/H/199	100	7	1883
IAR/H/165	101	8	2093
IAR/H/189	104	8	1993
CI 12954	105	6	2050
CI 2376	96	8	1781
HB-30	97	8	1970
IAR/H/177	100	7	2066
IAR/H/138	100	8	2142
IAR/H/117	102	7	1846
CI 4375-1	106	4	1606
IAR/H/195	108	6	2105
Local check	101	7	1961
Mean	102		1988
SE (M)			174
LSD 5%			353
CV %			12

Table 3. *Mean grain yield, heading date and disease data of the elite selections from Ethiopian collections at Sheno, 1977*

Variety	Heading date	Leaf blotch (0–9)	Leaf rust (%)	Reaction type	Yield (kg/ha)
816/70	83	7	80	MS	1729
751/70	86	7	15	MS	1527
602/70	89	7	80	MS	1831
814/70	85	8	90	S	1250
WGA 44-4	91	8	100	MS	1165
WGA 68-1	77	8	80	S	1345
435/70	83	8	60	MS	1591
CI 9724	87	8	80	MS	1186
Local check	74	7	40	S	1682
Mean	84				1479
SE (M)					445
LSD 5%					NS
CV %					31

MS, moderately susceptible; S, susceptible.

the new cultivars. Based on this experience selection of indigenous germplasm and a hybridization programme were initiated.

Selection of cultivars from landraces

In the 1970s, 3300 entries of local collections which were made available to Holetta by the United States Department of Agriculture (USDA) and others collected by national staff were screened at Holetta for disease resistance, plant vigour, medium maturity, stiff straw, good grain set and high kernel weight. Additional collections from the Plant Genetic Resources Centre/Ethiopia (PGRC/E) were evaluated in the early 1980s. Some high-yielding selections were identified from this activity. The elite material was further evaluated from grain yield at Holetta and some other locations. A few cultivars such as IAR/H/485 and ARDU 12-60B were recommended for commercial production.

About half of the original 3300 collections were also screened at Bedi and Sheno, where barley is the only crop together with oats that gives a dependable grain yield. Most selections were highly susceptible to scald. The best selections from the material were not superior to established farmers' varieties (checks) at both locations as shown in Tables 2 and 3.

A similar performance was observed at Mekele where soil moisture stress is a limiting factor for crop production. The collections were also screened for grain malting quality. Most of the grain samples were found to be coarse, wrinkled and thin, and some were pigmented, with high nitrogen content, hence unsuitable for malting purposes.

Within the collections about 200 hull-less types were isolated and evaluated in comparison with hulled barleys for grain yield and disease resistance. The naked barleys were found to be highly susceptible to scald, leaf blotch and lodging, with fairly low kernel weight and grain yield, the latter being about half that of hulled barleys.

Hybridization and selection

This programme comprised breeding for malting and food barley. The main objective in the malting barley breeding programmes was to incorporate disease resistance from local germplasm into introduced malting barley cultivars, namely Proctor, Beka, Kenya Research and Zephr using the backcross method. Initially these cultivars were crossed to Holetta local which has good resistance to scald and leaf blotch. In later years other satisfactory donor parents

Table 4. *Mean grain yield, heading date (HD) and disease data for seven Holetta hybrid selections in comparison with ARDU 12-9C (local selection) and the local check as tested at Holetta and Bekoji, 1985*

		Holetta			Bekoji		
Variety	Origin	HD	Scald (0–9)	Yield (kg/ha)	HD	Scald (0–9)	Yield (kg/ha)
HB 32	Hybrid	87	7	5262	92	7	5386
HB 78	Hybrid	96	2	4177	103	5	4356
HB 99	Hybrid	83	4	5561	93	6	5062
HB 100	Hybrid	83	7	5005	92	8	5616
HB 98	Hybrid	74	6	4764	98	3	5077
HB 71	Hybrid	86	8	4007	91	8	4237
HB 91	Hybrid	90	3	4475	96	7	4847
ARDU 12-9C	Arsi	85	4	4929	97	6	6066
Local check		86	8	3650	78	8	3226
Mean		86		4648	93		4875
SE (M)				290.4			332.9
LSD 5%				847.7			971.8
CV %				12.5			13.7

from other barley growing regions, as well as introduced lines, were included. Some high-yielding selections such as Holkr, Balkr, HB-16, HB-28 and others now featuring in yield trials were identified from this effort.

In the food barley breeding project the aim was to improve yielding capacity, disease and pest resistance, and nutritional quality of promising selections. Three methods of breeding were applied: backcross for improving nutritional quality; modified pedigree; and composite crosses for grain yield and disease resistance.

The activity on nutritional quality was a part of the FAO/SIDA/SAREC Project. Five high lysine lines – Hiproly, Riso 1508, SV 73608, SV 75240 and SV 791582 – were used as donor parents to which 20 cultivars from Ethiopia were crossed. Since 1978 about 1411 segregating populations have been evaluated at Holetta for plant vigour, grain set, resistance to scald and leaf blotch, and for protein quality. A large percentage of the material was highly susceptible to diseases, together with poor plant vigour and grain set.

In the other approach to food barley breeding, cultivars from major barley growing areas and promising introduced lines were included. In addition, segregating materials from the International Centre for Maize and Wheat Improvement (CIMMYT) were evaluated. The best

Table 5. *Mean grain yield, heading date (HD) and disease data for selections (local and crosses) as tested at three sites, 1986 (fertilized at 57/26 N/P kg/ha)*

		Debre Tabor			Gohatsion			Shambu		
Variety	Origin	HD	Spot blotch (0–9)	Yield (kg/ha)	HD	Spot blotch (0–9)	Yield (kg/ha)	HD	Scald (0–9)	Yield (kg/ha)
HB 42	Hybrid	88	5	4458	80	4	4638	85	5	4875
HB 43	Hybrid	82	7	4000	68	3	4513	71	5	6875
ARDU 12-60B	Arsi	85	7	4725	71	3	4963	54	5	5538
ARDU 12-9B	Arsi	82	8	4313	71	3	4913	71	6	5500
HB 37	Hybrid	85	8	4238	76	3	4100	88	5	4250
A/HOR 880/61		95	6	3138	79	3	4088	86	6	5000
Local check		60	9	2375	80	1	3163	57	6	3520
Mean		82		3890	75		4339	73		3937
SE (M)				2488			1564			363.8
LSD 5%				1034			541.3			1258.8
CV %				10.9			5.09			10.4

Table 6. *Mean grain yield, heading date (HD) and disease data for selections (local and crosses) as tested at three sites, 1986, without fertilizers*

Variety	Origin	Debre Tabor			Gohatsion			Shambu		
		HD	Spot blotch (0–9)	Yield (ka/ha)	HD	Spot blotch (0–9)	Yield (ka/ha)	HD	Scald (0–9)	Yield (kg/ha)
HB 42	Hybrid	90	5	3750	81	3	3725	91	5	2250
HB 43	Hybrid	87	7	4013	81	4	3438	84	5	4500
ARDU 12-60B	Arsi	87	6	3538	80	5	4413	90	5	3713
ARDU 12-9B	Arsi	87	7	3150	80	3	3313	87	6	3000
HB 37	Hybrid	87	8	2938	79	3	3763	90	5	1625
A/HOR 880/61		101	7	2125	80	3	3563	93	4	2750
Local check		76	9	2250	80	2	2738	75	6	1750
Mean		88		3109	80		3593	87		2798
SE (M)				777.2			257.8			963.3
LSD 5%				NS			89.25			3333
CV %				35.4			10.2			48.7

Table 7. *Malting barley and food barley cultivars with their year of recommendation, trial average yield and suitable areas*

No.	Origin	Cultivar	Year	Trial yield (t/ha)	Suitable areas
Malting barley					
1.	France	Beka	1973	3.9	Chilalo
2.	Kenya	Kenya Research	1973	2.7	Wolmera
3.	UK	Proctor	1973	2.4	Togulet & Bulga
4.	ETH	Holkr	1979	2.4–3.8	Wolmera, Chilalo, Togulet & Bulga, and Degem
5.	ETH	Balkr	1979	2.5–3.7	Wolmera, Chilalo, Togulet & Bulga, and Degem
6.	ETH	HB-16	1982	3.5	Wolmera
7.	ETH	HB-28	1982	3.3	Wolmera
Food barley					
1.	ETH	IAR/H/485	1975	3.6	Wolmera, Gondar, Chilalo and Alemaya
2.	USA	Composite 29	1975	4.1	Wolmera, Gondar, Chilalo and Alemaya
3.	ETH (?)	A/Hor 880/61	1978	4.5	Wolmera, Selale, Robe, Shambu
4.	ETH	HB-7	1980	3.9	Debre Tabor, Mota
5.	ETH	HB-15	1980	3.7	Wolmera
6.	ETH	HB-42	1984	4.2	Wolmera, Chilalo
7.	ETH	ARDU-60B	1985	3.8	Chilalo, Wolmera, Robe

selections from this programme were advanced to yield trials and were tested at several locations. The performance of such a group of selections in comparison to local selections and cultivars (checks) is shown in Tables 4, 5 and 6 for some barley growing areas. There was not much difference among selections from crosses and local collections in grain yield; however, some selections from both groups were superior to farmers' cultivars (local checks) which were earlier maturing but gave less response to nitrogen and phosphorus fertilizers.

Current emphasis in the breeding programme

The breeding programme from 1967 to date has generated a number of high-yielding cultivars of malting and food barley (Table 7). So far, the uptake of malting barley releases has been satisfactory, with state farms being the major producers, using a fairly high level of management. On the other hand, the adoption of food barley releases by peasants has been rather poor. As a result of this setback the food barley breeding programme has been restructured. The programme now focuses on improvement of local landraces by mass selection and cultivar mixtures without fertilizer application. A crossing programme between indigenous and exotic lines is under way using the F_2–progeny method with yield testing under low and optimum fertilizer applications. The objective is to generate high-yielding uniform bulk material.

References

Central Statistics Office (1984). *Time series data on area, production, and yield or principal crops by regions 1979/80–1983/84, vol. I.* Central Statistics Office, Addis Ababa.

Engels, J. M. M. (1988). A diversity study in Ethiopian barley. *In*: J. M. M. Engels (ed.), The conservation and utilization of Ethiopian germplasm. Proceedings of an international symposium, Addis Ababa, 13–16 October 1986, pp. 124–32 (mimeographed).

Munck, L. K., Karlsson, E. & Hagberg, A. (1971). Selection and characterization of a high protein, high-lysine variety from the world barley collection. *In*: R. A. Nilan (ed.), *Barley Genetics, vol. II. Proceedings of the 2nd International Barley Genetics Symposium, Pullman, 1969.* Washington State University Press, Pullman, Washington, pp. 544–58.

Negassa, M. (1985a). Geographic distribution and genotypic diversity of resistance to powdery mildew of barley in Ethiopia. *Hereditas*, **102**, 113–21.

Negassa, M. (1985b). Patterns of phenotypic diversity in an Ethiopian barley collection, and the Arsi–Bale Highland as a centre of origin of barley. *Hereditas*, **102**, 139–50.

Qualset, C. O. (1975). Sampling germplasm in a centre of diversity: an example of disease resistance in Ethiopian barley. *In*: O. H. Frankel and J. G.

Hawkes (eds), *Crop Genetic Resources for Today and Tomorrow*. Cambridge University Press, Cambridge, pp. 81–96.

Somaroo, B., Mekni, M., Adham, Y., Humed, B. & Kawas, B. (1984). Evaluation of barley germplasm at ICARDA. *Rachis*, **3**, 12–15.

Ward, D. J. (1962). Some evolutionary aspects of certain morphological characters in a world collection of barley. *USDA Technical Bulletin* 1276.

25

The role of Ethiopian sorghum germplasm resources in the national breeding programme

YILMA KEBEDE

Introduction

Sorghum is one of the crop types for which Ethiopia has been credited as being a Vavilovian centre of origin or diversity (Harlan, 1969). In the different ecological zones of the country, germplasm resources representing the major and intermediate races of sorghum are found. In addition, the existence of wide variation in plant, grain, inflorescence and fruit characteristics in the Ethiopian sorghum germplasm is well documented (Gebrekidan, 1973; Gebrekidan & Kebede, 1977). Among the sorghum growing population in the rural areas, the importance of this crop is exemplified not only by its use as a staple food and for other purposes, but also in the folklore, songs and some of the local names by which the sorghum varieties are known.

As one of the leading traditional food cereals in Ethiopia, in terms of both total production and area, major research efforts have been directed towards the improvement and stabilization of sorghum yields. At a national level, sorghum improvement involves the manipulation of indigenous and introduced germplasm to develop adapted types for the various ecological zones. In crop improvement work the indigenous germplasm has been found invaluable (Gebrekidan, 1981).

Periodic sorghum germplasm collections made throughout the country have provided the sources of breeding material necessary for the sorghum improvement programme. In the high altitude areas the indigenous germplasm has often been the only adapted material suitable for use. From evaluations of germplasm collections, potential varieties have been identified. Other accessions, which were found to be outstanding in certain features, have been used in crossing programmes. To date about 5–7 per cent of the evaluated collection has been used in various breeding programmes (Kebede, Menkir & Deressa, 1985). Development of new and better cultivars with improved yield potential is a continuous process whether it be to meet the needs of consumers or to stay ahead of yield-limiting factors. Through the use of conventional breeding methods – selection and progeny testing and various crossing schemes – progress has been made in utilizing local sorghum germplasm.

The objective of this chapter is to review work on sorghum breeding in Ethiopia with reference to the variation and utilization of indigenous germplasm.

Collection and characterization

The concern that some populations may become extinct because of habitat destruction and other factors, as well as the expressed need for making useful germplasm readily available to the crop improvement programme, led to the development of an organized collection and conservation system.

Over the years, in addition to the Ethiopian sorghum programme, contributions have been made to the sorghum collections by individuals, research organizations, national establishments and the Plant Genetic Resources Centre/Ethiopia (PGRC/E) staff. The current number of indigenous sorghum germplasm accessions stands at more than 8000 (PGRC/E, 1986). These accessions represent a wide array of diversity as well as the major sorghum growing areas in the country.

These collections are routinely characterized and screened for characters useful in crop improvement. In cooperation with PGRC/E, germplasm collections are grown at appropriate adaptation sites for evaluation of some agronomic and taxonomic characteristics (Table 1). The major characteristics evaluated fall into vegetative, inflorescence and fruit and grain categories. These characterization activities have been important in identifying desirable types with useful traits either for direct use or in crossing programmes (Gebrekidan & Kebede, 1977).

Moreover, attempts to classify the Ethiopian sorghum types into recognizable taxonomic working groups or races have resulted in recognition of 46 morphotypes representing four out of the five basic races recognized by Harlan & de Wet (1972). Kafir sorghums are not found in Ethiopia (Stemler, Harlan & de Wet, 1977). Data in Table 2 show that the dominant type of sorghum is the durra race with

Table 1. *Predominant characteristics in the Ethiopian sorghum germplasm collection*

Plant	
Plant height	>2.0 m
Inflorescence and fruit	
Days to flowering	>100 days
Inflorescence	loose to semi-loose erect branches
Glume colour	purple
Grain covering	a quarter covered
Inflorescence exsertion	<10 cm
Grain	
Grain colour	red, white, brown
Grain size	medium
Endosperm texture	mostly starchy
Threshability	freely threshable

Table 2. *Frequencies and distribution of races and sub-races in the Ethiopian sorghum germplasm collection*

Race	Number of sub-races[a]	Percentage[b]	Distribution
Durra	25	66	North-eastern, central and eastern plateau
Caudatum	3	20	South and west of central plateau and Rift Valley
Bicolor	1	8	West of Rift Valley (Metekel)
Guinea	1	1	Konso
Intermediate	16	5	Variable

[a] Based on Abebe & Wech (1982).
[b] Based on Gebrekidan & Kebede (1977).

characteristically large panicles and good grain types. Race caudatum with asymmetrical grain shape is dominant in the Gambela area of south-western Ethiopia. Variations as a result of intercrossing among the basic races have resulted in intermediate races among which the durra-bicolor sub-race is the most dominant.

A selected group of sorghum types, identified by local farmers for attributes such as end use, quality, morphology and other traits, is presented in Table 3. Admittedly most information on the value of accessions must come from studies of experimental plantings in which economically important traits can be observed. However, the

Table 3. *Local names and meaning of some Ethiopian sorghum landraces grouped according to their most striking characteristics*

Characteristic	Ethiopian sorghum number	Local name	Meaning
End use	1347	Fendisha	Pop sorghum
	2861	Tinkish	Sweet stalk
	1771	Yeshet Ehil	Consumed green
Quality	2390	Sinde Lemine	Equals wheat
	3133	Gan Seber	Good fermentation
	2624	Wetet Begunche	Mouthful of milk
	2970	Marchuke	Oozing honey
Morphology	3149	Dirb Keteto	Twin seed
	—	Rejim Genbo	Large sink
	—	Shufun	Large glumes
	3870	Alequay	Faba bean-like seeds
Other	2611	Hafukagne	Always heads
	3252	Wof Aybelash	Bird-tolerant
	4762	Kitgn Ayfere	Unafraid of syphilis (*striga*)

Adapted from Gebrekidan (1982).

first evaluation of an accession takes place in the natural habitat where one can observe variation and other features of the population. In this respect, the farmer's time-tested knowledge, as shown by the names assigned to the various sorghums, becomes invaluable.

Utilization

In sorghum improvement work, some of the high priorities are stand establishment, seed set, grain yield, tolerance to drought, pests, *Striga* and food-type sorghums. Germplasm resources represent a unique potential that could have an impact on these aspects. The vital importance of genetic resources in crop improvement programmes has been amply demonstrated by the successes in plant breeding using such resources. Our dependence on germplasm resources is even more credible when we consider the achievements of the past.

The Ethiopian sorghum germplasm has been found useful locally and elsewhere as a source of cold tolerance, high protein (lysine), good grain quality (zera-zeras), disease resistance and diversity, as indicated by its use in the US Sorghum Conversion Program. Some of the phenotypic diversity is illustrated in Table 4.

Table 4. *Phenotypic diversity expressed in range, mean and* CV *for some quantitative characters of Ethiopian sorghum germplasm accessions*

Character	Range		Mean	CV (%)	*N*
	Minimum	Maximum			
Plant height (cm)	19	475	233.7	22.7	2599
Ear length (cm)	4	50	21.5	36.6	2511
Ear width (cm)	2	30	9.4	34.1	2599
Peduncle extension (cm)	1	44	14.7	53.8	2254
Days to 50% flowering	76	169	116.7	14.3	2603
Crude protein content (%)	5.0	15.3	9.6	13.1	3644
Thousand seed weight (g)	6.0	61.1	28.5	35.5	200

Source: PGRC/E, unpublished data.

In the Ethiopian sorghum improvement programme, the utilization of germplasm focuses on manipulation of genetic stocks, specific searches for useful genes and direct utilization of landraces.

Manipulation of genetic stocks

This includes hybridization, population improvement and backcross breeding. Every season indigenous accessions selected for desirable traits are used as parents in a crossing programme. In any one year these parents comprise 20–30 per cent of the total number of parents used in the crossing programme. The resulting 100–200 F_1 generations having at least one indigenous parent are routed through the pedigree system through planting in areas of adaptation (Kebede *et al.*, 1985).

For population improvement, Ms_3 and other genetic male sterile carrier stocks are generally poorly adapted to the Ethiopian highlands. Thus an alternative approach in the Ethiopian sorghum programme is the use of the indigenous high-lysine, hl (Gebrekidan & Kebede, 1979) and dented seed marker in identifying crossed seed. Open pollination of the hl and dented (recessive) seeds with pollen from regular (plump) seed parents results in a few (<10 per cent) plump F_1 seeds which can be visually identified and picked from the mother panicle. The pollen source is a composite of elite high elevation adapted indigenous lines. The resulting plump F_1 seeds are selfed and segregate into a ratio of 3 plump:1 dented seeds on the panicle. Through visual evaluation the best panicles are selected from which the dented seeds are picked for the next cycle of random mating. At different stages, single plant selections have been evaluated in progeny rows and some promising lines have been advanced to yield trials. This system enables the gathering of genes from several different local sources into the hl germplasm. This gives rise to several genotypes which, by reselection, produce further improvement. It has also opened an avenue for the intermating of tall sorghum as well as the possible accumulation of high lysine genes associated with the dented seed character (Gebrekidan, 1983).

A backcross breeding programme initiated to transfer resistance to elite indigenous sorghums (ETS 2111, ETS 2113, ETS 3235, WB-77) that were found susceptible to anthracnose (*Colletotrichum graminicola*), using sources of known resistance, has resulted in resistant types and a good agronomic performance. Details of this programme have been given by Menkir, Kebede & Gebrekidan (1986).

Specific searches for useful genes

The Ethiopian sorghum germplasm resources have proved to be useful sources for desirable genes. As a result of pointed evaluation of some of the collections or from information based on farmers' own experiences, some useful genes reported to exist in the Ethiopian germplasm are listed below.

Disease and pest resistance

- Stalk borer
- Downy mildew
- Smuts
- Bacterial streak
- Anthracnose
- *Striga*

Adaptation

- Agronomic desirability (height, maturity, panicle size and shape, etc.)
- Yield potential

Kernel traits

- Endosperm texture
- Grain colour
- Threshability
- Injera quality
- Protein quality (hl)

Direct utilization of germplasm

In general, sorghum landraces are a mixture of related pure lines. Testing and seed multiplication and further maintenance require some sort of varietal identity. Thus, panicle selections from promising accessions are put in progeny rows from which elite types are selected and put into the Advanced Sorghum Selections Nursery (Kebede *et al.*, 1985). This nursery serves as an intermediate evaluation stage for entries selected from new accessions before advancing to national and pre-national yield trials. In the past, in any given season, approximately 2 per cent of the accessions were advanced to such a nursery.

Over the years, about a dozen entries derived from the collections have been recommended for release and currently five such entries are on the recommended list. They are Alemaya 70 and ETS 2752 for areas of high elevation, Dedessa 1057 and Asfaw White for intermediate elevation and Gambela 1107 for low elevation (Kebede *et al.*, 1985).

Summary and projections

The remarkable diversity in crop germplasm in Ethiopia is now widely recognized but these resources have only recently started to be tapped. The inability to reconstitute lost germplasm underlines the necessity for conservation before this natural wealth is depleted completely. Evaluation and characterization of available germplasm based on needs and requirements of the crop improvement programme would aid in better utilization of available resources. Some of the future activities should take into account that:

- evaluation and documentation of sorghum genetic resources have to be accelerated in order to achieve the desired goal of utilization;
- collecting missions have to concentrate on hitherto unexplored areas since expansion of production into new environments may require attributes not presently considered important;
- evaluation of germplasm has to emphasize screening for resistance to drought, pests and *Striga*.

The importance of germplasm availability for the continued improvement of sorghum has now been recognized and one of the most significant developments in this field in Ethiopia in the past decade has been the establishment of PGRC/E. The Centre has fulfilled its functions in developing appropriate systems of conservation and characterization and we shall look to PGRC/E to help breeders find useful germplasm as the need arises. The joint efforts of PGRC/E and the crop breeders will enable all to cope with the varied production environments through the development of germplasm that could increase and stabilize production.

References

Abebe, B. & Wech, H. B. (1982). The 1981 activities of the Plant Genetic Resources Centre/Ethiopia. *In*: *Proceedings of the Regional Workshop on Sorghum Improvement in Eastern Africa, 17–22 October, 1982, Addis Ababa*, pp. 31–45.

Gebrekidan, B. (1973). The importance of the Ethiopian sorghum germplasm in the world sorghum collection. *Economic Botany*, **27**, 442–5.

Gebrekidan, B. (1981). Salient features of the sorghum breeding strategies used in Ethiopia. *Ethiopian Journal of Agricultural Science*, **3**, 97–104.

Gebrekidan, B. (1982). Utilization of germplasm in sorghum improvement. *In*: Sorghum in the Eighties. Proceedings of the International Symposium on Sorghum, 2–7 November 1981. ICRISAT, Patancheru, pp. 335–45.

Gebrekidan, B. (1983). New breeding concepts in partially self pollinated crops with special emphasis on sorghum. *In*: J. C. Holmes and W. M. Tahir (eds), *More Food from Better Technology*. FAO, Rome, pp. 186–92.

Gebrekidan, B. & Kebede, Y. (1977). Ethiopian Sorghum Improvement Project. Progress Report No. 5. Addis Ababa University (mimeographed).

Gebrekidan, B. & Kebede, Y. (1979). The traditional culture and yield potentials of the Ethiopia high-lysine sorghums. *Ethiopian Journal of Agricultural Science*, **1**, 29–40.

Harlan, J. R. (1969). Ethiopia: a centre of diversity. *Economic Botany*, **23**, 309–13.

Harlan, J. R. & de Wet, J. M. J. (1972). A simplified classification of cultivated sorghum. *Crop Science*, **12**, 172–6.

Kebede, Y., Menkir, A. & Deressa, A. (1985). A review of sorghum research work in Ethiopia. Paper presented at Workshop on Review of Field Crops Research in Ethiopia, 25 February–1 March 1985, Addis Ababa (mimeographed).

Menkir, A., Kebede, Y. & Gebrekidan, B. (1986). Incorporating anthracnose resistance into indigenous sorghum. *Ethiopian Journal of Agricultural Science*, **8**, 73–84.

Plant Genetic Resources Centre/Ethiopia (1986). Ten years of collection, conservation and utilization, 1976–1986. PGRC/E, Addis Ababa.

Stemler, A. B. L., Harlan, J. R. & de Wet, J. M. J. (1977). The sorghums of Ethiopia. *Economic Botany*, **31**, 446–60.

26

Germplasm evaluation and breeding work on teff (*Eragrostis tef*) in Ethiopia

SEYFU KETEMA

Introduction

Ethiopia is the only country that produces teff as a cereal crop. Teff occupies the largest area of cultivated land under cereal production in Ethiopia, and as such it is the most important crop. According to the statistical information of five years' average from 1979–80 to 1983–4, teff was cultivated each year on 1.385 million hectares, followed by barley 0.851, wheat 0.609, maize 0.780 and sorghum 0.994 million hectares. The national average grain yield of these cereals for the same five-year period was 9.1 quintals per hectare (q/ha) for teff, barley 11.83, wheat 11.26, maize 17.35 and sorghum 14.57 q/ha (Central Statistics Office, 1984).

Teff is mainly cultivated as a single crop. However, in a few areas it is cultivated under a multiple cropping system. In such cases it is usually grown as an intercrop with *Brassica carinata*, *Carthamus tinctorius* or *Helianthus annuus*. It is also relay cropped with *Zea mays* and *Sorghum bicolor*.

Teff is mainly used for making a pancake-like bread called 'injera'. In some cases it is used to make porridge and native alcoholic drinks called 'tella' and 'katikala'. Its straw is highly valued and is used as feed for cattle. In addition, the straw is incorporated with mud to reinforce it and used for plastering walls of houses.

Teff is on average as nutritious as any of the major cereals. According to Rouk and Mengesha (n.d.), the Ethiopian Nutrition Survey reported that four unspecified teff varieties when analysed were found to contain an average of 300 calories, 11.6 g protein, 0.65 g fat

Table 1. *Content of seed grain (proximate analysis) as percentage of grain*

Item	Teff	Wheat	Rice	Maize	Sorghum	Barley	Oats	Rye
Protein	11.0	11.0	9.7	9.4	8.6	8.5	9.5	10.7
Fat	2.6	1.9	1.8	4.4	3.8	1.5	4.8	1.7
Fibre	3.5	1.9	8.8	2.2	1.9	4.5	10.3	1.9
Carbohydrate	73.0	69.3	64.7	69.2	71.3	67.4	58.4	69.8
Minerals (ash)	3.0	1.7	5.0	1.3	2.4	2.6	3.1	2.0

Source: B. Tareke, unpublished data, Alemaya University of Agriculture, Dire Dawa, Ethiopia.

and 70.56 g carbohydrate per 100 g, and that teff supplied an average of two-thirds of the protein in the Ethiopian diet. Table 1 shows the nutritional status of teff compared with some other cereals.

Teff also contains more calcium, copper, zinc, aluminium and boron than winter wheat, winter barley and sorghum (Mengesha, 1966).

Some of the reasons why present-day farmers grow teff are given as follows.

1. It can be grown in areas experiencing moisture stress.
2. It can be grown in waterlogged areas and withstands anaerobic conditions better than many other cereals including maize, wheat and sorghum.
3. It is suitable for use in multiple cropping systems such as double, relay and intercropping.
4. Its straw is a valuable feed during the dry season when there is an acute shortage. It is highly preferred by cattle over the straw of other cereals and demands higher prices in the markets.
5. It has acceptance in the national diet, has high demand and high market value and hence enables farmers to earn more.
6. It is a reliable and low-risk crop.
7. In moisture stress areas farmers use it as a rescue crop. For example, around Kobo and Zeway, which are areas with low and erratic rainfall, farmers first plant maize around April. If this fails after a month or more due to moisture stress or pest problems they plough it under and plant sorghum. If this also fails after a month or more then they sow teff as a last resort, which often survives on the remaining moisture in the soil and yields some grain for human consumption and straw for feed.

8. It is not attacked by weevils and other storage pests and therefore is easily and safely stored under local storage conditions. This results in reduced post-harvest management costs.
9. Compared with any other cereal grown in Ethiopia it has fewer disease and pest problems (Stewart & Degnachew, 1967).

Domestication and diversity

According to N. I. Vavilov, Ethiopia is the centre of origin for teff (Mengesha, 1966; Tadesse, 1975). Its domestication is believed to have taken place first in the northern highlands of Ethiopia (Tadesse, 1975). Although there are no exact records on the history of its domestication, one hypothesis to explain the situation that led to its domestication is as follows.

The word teff is said to have originated from the Amharic word 'teffa' which means lost, because the grain size of teff is so small that if one grain is dropped it is difficult to find (Rouk & Mengesha, n.d.). Other sources say that the word teff was derived from the Arabic word 'tahf', a name given to a similar wild plant (*Eragrostis* sp.) used by Semites in South Arabia in times of food scarcity (Constanza, 1974; Tadesse, 1975; Endeshaw, 1978). According to Endeshaw (1978) Ciferri and Baldrati stated in 1939 that *E. pilosa*, which is believed to be the ancestor of teff, is collected as food by people in many parts of Africa other than Ethiopia in times of famine. This suggests that the domestication of teff in Ethiopia might have started in times of food scarcity (Tadesse, 1975; Endeshaw, 1978).

Teff was introduced to other parts of the world by the Royal Botanic Gardens at Kew, which imported seed from Ethiopia in 1866 and distributed it to India, Australia, California and South Africa. In 1916 Burt Davy introduced teff into California, Malawi, Zaïre, India, Sri Lanka, Australia, New Zealand and Argentina; Skyes introduced it in 1911 into Zimbabwe, Mozambique, Kenya, Uganda and Tanzania; Horuity in 1940 introduced it into Palestine (Tadesse, 1975).

Teff is a sexually propagated, self-pollinating annual grass species (Tadesse, 1975; Tareke, 1975). It is tetraploid with a chromosome number of $2n = 40$ (Tareke, 1975; Endeshaw, 1978) and an allopolyploid (Tareke, 1981). It is cultivated from sea level up to 2800 m on soils with varying physical and chemical properties, in waterlogged and in well drained soils, in moisture stress areas having less than 300 mm of rainfall as well as in areas having 1000 mm seasonal rainfall. This gives an idea of the tremendous ecological diversity under which the crop can be grown.

The Plant Genetic Resources Centre/Ethiopia has itself made 1050 germplasm collections (PGRCE, 1986). More than one thousand collections made by Debre Zeit Agricultural Research Centre have been given to PGRC/E also. However, many of these collections have not been made from representative sites nor were they collected in a systematic way and with adequate passport data. All this is now being corrected and a systematic collection, characterization and utilization programme is in progress, the last named in conjunction with the breeding programme. In Ethiopia 54 *Eragrostis* spp. are listed of which 14 (or 26 per cent) are endemic (Constanza, 1974). However, so far no collection of the wild species has been made. The total number of accessions of teff germplasm at present is 2175. Tadesse (1975) has characterized and recognized 35 cultivars, and Endeshaw (1978) mentioned that minor variations still exist within many of the cultivars. Some of the variations noted so far include maturity period 60–120 days, plant height 45–150 cm, culm thickness 1.2–3.1 cm, panicle varying from very loose to very compact, lemma colour whitish green, purple, olive grey, pink and various types of these colour combinations, seed colour white, yellowish brown, dark brown and intermediate types.

Utilization

Improvement work on teff through plant breeding began in the late 1950s. This was done only through pure line selection from landraces, since it was then not possible to produce hybrids. In 1972, in order to create variability, mutation breeding was started. This line of investigation established that useful mutations could be induced through the use of physical mutagens such as X-rays at 100–130 krads and Gamma rays at 150 krads, and chemical mutagens such as ethylmethyl-sulphonate at 2.5–4.7 per cent concentration. Until 1974 artificial pollination was attempted in the morning after 0800 h with no success. The observation by Tareke that teff flowers open during the early morning (0645–0745 h) and that they have only a brief pollination time enabled him to make the first successful intraspecific crosses towards the end of 1974 (Tareke, 1975). Now it is realized that artificial crosses may be made either early in the morning between 0600 and 0730 h, or any time during the day, provided that the natural pollination time is delayed through the use of low temperature 4–5 °C or dark treatment, which can be achieved by putting potted plants under cold or dark conditions overnight. Artificial hand pollination is time consuming and cumbersome. The latest available technique for such work, which was suggested by Seyfu (1983); is given as follows.

- Grow one or two plants in a pot of about 13 cm diameter.
- Eight to 18 days after anthesis begins on the central tiller (or any other particular tiller), put the seed parent plant and the pollen donor plant into separate light-tight dark boxes at around 1400 h. Keep the boxes away from direct sunlight at a temperature well below 28 °C (lower temperature improves degree of control over flowering). Next day, crossing may be done at any time before the early afternoon.
- Take donor plant out first. As soon as it starts to open its flowers, detach those spikelets with open florets using forceps and attach them to the moist inner wall of a vial. Label vial with code number of plant.
- Take out seed plant, lay it horizontally under a binocular microscope (×15) and begin to emasculate as soon as flowers start to open (and before the anthers dehisce). Only the basal florets should be emasculated, the other florets on the spikelet being removed. This serves to identify the treated flowers.
- Keeping the emasculated floret under observation, remove a spikelet from the vial and, with forceps, detach an individual anther while observing it under the binocular microscope, and gently squeeze it to release pollen directly onto the stigma of the emasculated floret.
- Label plant for crossing record.

Achievements

The major objective of the breeding programme was and still is to develop lodging resistant, high-yielding and stable varieties. In the major teff growing areas no epidemic disease or pest problems exist. Therefore, the development of disease and pest-resistant varieties or ones which have high nutritive value (e.g. high protein content) is not the major focal point at the moment. The national average grain yield of teff for landraces is 9.1 q/ha. Improved varieties that outyield the landraces have been developed through pure line selection from germplasm as well as through hybridization following the pedigree method of selection. These varieties give a grain yield of

17–22 q/ha on the farmer's field. Some of these released and recommended varieties are DZ-01-354, DZ-01-99, DZ-01-196, DZ-cross-44, DZ-cross-82 and DZ-01-787.

References

Central Statistics Office (1984). Time series data on area, production and yield of principal crops by regions 1979/80–1983/84. Central Statistics Office, Addis Ababa, 219 pp.

Constanza, S. H. (1974). Literature and numerical taxonomy of teff (*Eragrostis tef*). MSc thesis, Cornell University, Urbana, Illinois, USA.

Endeshaw, B. (1978). Biochemical and morphological studies of the relationships of *Eragrostis tef* and some other *Eragrostis* species. MSc thesis, University of Birmingham, UK.

Mengesha, M. H. (1966). Chemical composition of teff (*Eragrostis tef*) compared with that of wheat, barley and grain sorghum. *Economic Botany*, **20**, 268–73.

Plant Genetic Resources Centre/Ethiopia (1986). Ten years of collection, conservation and utilization 1976–1986. PGRC/E, Addis Ababa.

Rouk, H. F. & Mengesha, M. H. (n.d.). An introduction to teff (*Eragrostis abyssinica* Schad.). A nutritious cereal grain of Ethiopia. *Debre Zeit Agricultural Research Centre Bulletin No. 26*. Alemaya University of Agriculture, Dire Dawa, Ethiopia.

Seyfu, K. (1983). Studies of lodging, floral biology and breeding techniques in teff (*Eragrostis tef* (Zucc.) Trotter). PhD thesis, University of London.

Stewart, R. B. & Degnachew, Y. (1967). Index of plant diseases in Ethiopia. *Debre Zeit Experimental Station Bulletin No. 30*. Alemaya University of Agriculture, Dire Dawa, Ethiopia.

Tadesse, E. (1975). Teff (*Eragrostis tef*) cultivars; morphology and classification. Part II. *Debre Zeit Agricultural Research Centre Bulletin No. 66*, pp. 1–73. Alemaya University of Agriculture, Dire Dawa, Ethiopia.

Tareke, B. (1975). A breakthrough in teff breeding technique. *FAO Information bulletin on cereal improvement and production 12*, 11–13.

Tareke, B. (1981). Inheritance of lemma color, seed color and panicle form among four cultivars of *Eragrostis tef* (Zucc.) Trotter. PhD thesis, University of Nebraska, Lincoln, Nebraska.

27

Pulse crops of Ethiopia: genetic resources and their utilization

HAILU MEKBIB, ABEBE DEMISSIE AND ABEBE TULLU

Introduction

Ethiopia is known as a centre of diversity and/or origin of numerous cultivated crop plant species. This was first recognized by N. I. Vavilov in the late 1920s and later confirmed by several other scientists. Vavilov (1951) indicated that some 38 crop plants have their centre of diversity in the Ethiopian region. Zohary (1970) mentioned 11 crop species which had their centre of diversity in Ethiopia. Primitive cultivars or landraces and wild relatives of some of the world's major crops are found in the country. Pulse crops form a significant portion of the available genetic resource base for plant breeding programmes.

In this chapter an attempt is made to describe the situation for the most important pulse crops cultivated in Ethiopia regarding their diversity and the germplasm kept in the national collection, and their conservation, evaluation and utilization.

Collection

Owing to the richness and potential of the biological resources of the country, numerous plant expeditions have been undertaken by scientists in the past. However, it was only after the establishment of the Plant Genetic Resources Centre/Ethiopia (PGRC/E) in 1976 that systematic collecting missions were launched.

The total holding of pulse accessions by PGRC/E is about 4300. The bulk of the germplasm was acquired from field collecting (*ca*. 2900) on the basis of a well defined strategy, and some was acquired through repatriation and acquisition from national and international sources.

The sampling procedure and techniques are based on a field collecting manual (Hawkes, 1980), which advises that seeds from up to 50 individual plants, and certainly not more than 100, should be collected non-selectively and bulked to obtain an optimum sample. Whenever rare types, i.e. plants which show characters not included in the random sampling, are noticed, a selective sampling technique is adopted. Such a sample is given a different collection number.

Conservation

Generally, the pulse crops under review are conserved in temperature-regulated storage units. Seeds intended for both base and active collections are dried to 3–7 per cent moisture content and maintained at −10 °C in laminated aluminium foil bags. Accessions which are too small to meet the sample size required for long-term storage are maintained in paper bags at 4 °C and 35 per cent relative humidity (Feyissa, 1988). The plans for *in situ* conservation of yeheb nut (*Cordeauxia edulis*) are based on the absence of factual data on its storage behaviour and need for its immediate conservation programme.

In Ethiopia farmers play a pivotal role in the conservation of landraces as they hold the bulk of genetic resources. Peasant farmers always retain some traditional seed stock for security even at difficult times unless circumstances dictate otherwise (Worede, 1987). This strategy of conservation is the second major option considered in Ethiopia.

Distribution and diversity

The distribution and the degree of genetic diversity of pulse crops in different agro-ecological zones of the country has not yet been adequately studied. A modest programme of work on crops such as *Lathyrus* and *Vigna* has been initiated with a view to identifying areas of maximum diversity. This will help in subsequent rational planning of collecting missions. An estimate of the diversity of some Ethiopian pulse crops is presented in Table 1 which was compiled by pooling available data (Mengesha, 1975) with data generated during collecting and preliminary evaluation activities.

Both intra- and infraspecific diversity in legumes are relatively large in Ethiopia. Crops such as faba bean, field pea, chickpea and lentil have their (secondary) centres of diversity in Ethiopia. There are *ca*. 600 species of legumes recorded in Ethiopia (Thulin, 1983) of which only a handful, namely, peas, lentil, chickpea, common bean,

Table 1. *Estimate of crop diversity in Ethiopia*

Administrative region	Estimate of crop diversity								
	Faba bean	Field pea	Chickpea	Lentil	Grass pea	Fenugreek	Lablab	Pigeon pea	Common bean
Arsi	H	M	M	L	—	L	—	—	—
Bale	M	M	L	L	—	L	—	—	—
Eritrea	M	M	M	L	M	M	—	—	—
Gamo Gofa	M	L	T	L	—	—	—	H	M
Gojam	M	H	H	M	H	M	M	—	—
Gondar	H	H	H	H	H	H	L	—	—
Harerge	M	L	M	L	—	M	—	—	H
Ilubabor	L	—	—	—	—	—	—	—	M
Kefa	L	T	—	—	—	—	—	—	M
Shewa	H	H	H	H	M	M	—	M	H
Sidamo	L	L	M	—	—	—	—	M	M
Tigray	M	M	H	M	M	M	—	—	—
Welega	M	M	—	—	—	M	M	—	L
Welo	H	H	M	H	M	H	—	L	M

H, M and L, high, medium and low diversity; T, trace; —, not enough data available.

Based on Mengesha (1975) and field observations of the authors.

faba bean and cowpea are largely grown as grain legumes for human consumption.

Faba bean (*Vicia faba* L.)

The origin of faba bean is so far not clearly established. In spite of claims by Abdella (1979) that faba beans originated in Egypt, most recent studies have indicated that the crop perhaps originated in west (Cubero, 1974) or central Asia (Ladizinski, 1975), but both the progenitor and place of origin remain uncertain. It is believed that faba beans were cultivated at an early date in the Nile Valley as far south as the Ethiopian highlands and as far east as Afghanistan. In both these regions primitive forms with small and sometimes black seeds are still cultivated (Hawtin & Hebblethwaite, 1983). Furthermore, Westphal (1974) indicated that the origin of *V. faba* is in the Mediterranean region or south-western Asia where it has been cultivated for centuries. *V. pliniana*, which grows wild in Algeria, is said to be its close relative.

Faba bean is one of the most common and major pulse crops in the highlands of Ethiopia, occupying 6 per cent of the total area under cultivation by the major crops. It was collected from areas ranging in altitude from 1600 to 3200 m above sea level; a few accessions were made between 3800 and 4000 m, although the majority come from 2200 to 2800 m. It is generally planted in June when the main rainy season commences and harvesting is carried out in December–January.

Faba bean is usually planted alone. It is sometimes cultivated together with field pea for support and it is usually incorporated in a rotation scheme, coming immediately after cereals.

Utilization

Varietal improvement efforts are at rather an early stage. The first national yield trial was initiated only in 1972 and had to be discontinued in 1974 due to inadequacy of the working collection (Telaye, 1988).

Recently, 349 faba bean germplasm accessions have been evaluated for grain yield and other agronomic characters at several locations. Of these, 28 high-yielding accessions were identified and selected for further multilocation yield testing in the national programmes (Ghizaw *et al.*, 1986). Yield trials have shown promising results and new potential cultivars are being selected and recommended for various agro-ecological zones within the country.

PGRC/E has collected 744 accessions of faba bean from various agro-ecological zones. About 450 of these accessions have been evaluated for a number of agro-morphological characteristics on the basis of International Board for Plant Genetic Resources (IBPGR) descriptors, synthesized by the PGRC/E in collaboration with the national crop breeders. A summary of the evaluation data for faba bean is presented in Table 2.

Field pea (*Pisum sativum* L.)

Field pea was perhaps domesticated in central or western Asia, spread to China and India, and reached Africa and the mountain regions of Ethiopia before the arrival of the Europeans. In recent work (van der Maesen *et al.*, 1988) four possible centres of diversity are mentioned, namely, the Near East, the Mediterranean, central Asia and Ethiopia.

Extensive areas of the central and northern Highlands of Ethiopia are cultivated with field pea. Although it is a typical plant of high elevation, some varieties thrive well at lower elevations. Field pea has been collected in areas as low as 1560 m and as high as 3560 m, but the bulk of the PGRC/E collection comes from areas between 2160 and 2760 m. The total holding of field pea is close to 1060 accessions.

There are two main cultivar groups, *abyssinicum* and *sativum*. The former has leaves with one pair of leaflets, the flowers are small and pink or purple, the seeds are globose, brownish or grey, often with blotches and a black hilum. The latter is usually more robust but less hardy, the flowers are white, pods and seeds are larger and the seeds are yellowish round and smooth or wrinkled. The two forms of cultivated pea are sometimes regarded as separate species, *P. arvense* and *P. sativum* respectively, but there is little taxonomic ground for treating them so. They are genetically similar and interbreed readily, producing fertile progeny. In Ethiopia, the *abyssinicum* type is predominantly grown.

Utilization

Since the early days of research efforts in Ethiopia, agronomy work has been largely based on local landraces procured from the local markets or farmers' fields. As a result cultivars such as CS 436, FP EX DZ, and G 2276–2c were released (Telaye, 1988). Recently the Ethiopian field pea germplasm collection has been evaluated for several morphological and agronomic characters. It has shown considerable diversity for the characters considered (Table 2).

Table 2. *Summary of evaluation data for chickpea, field pea, faba bean, lentil, grass pea, fenugreek and common bean*

Crop	*N*	Max.	Min.	Mean	STD	CV
Chick pea						
Days to flowering	639	63.0	36.0	46.5	4.7	10.1
Primary branches	640	3.0	1.0	2.0	0.2	11.2
Secondary branches	639	6.2	1.8	2.9	0.7	26.7
Days to maturity	639	123.0	84.0	97.1	5.8	6.0
Plant height	638	48.4	16.0	29.5	6.1	20.6
Pods per plant	639	141.0	10.0	33.8	14.7	43.5
Seeds per pod	332	2.4	1.0	1.5	0.2	15.9
Field pea						
Days to flowering	1063	180.0	32.0	74.0	12.9	17.4
Days to maturity	1052	183.0	92.0	131.8	17.9	13.5
Seeds per pod	876	11.0	1.0	5.5	0.9	16.3
Seed weight	203	227.4	10.8	123.3	18.5	15.0
Faba bean						
Days to flowering	452	88.0	33.0	62.9	7.9	12.6
Flowers per plant	410	87.0	20.0	42.5	14.9	35.1
Days to maturity	392	182.0	95.0	141.5	19.1	13.5
Plant height	452	195.0	50.0	100.7	24.2	24.0
Pods per plant	452	50.0	3.0	14.3	5.9	41.5
Seeds per pod	452	5.0	1.0	2.3	0.5	24.1
Seed weight	97	521.6	208.3	317.1	73.2	23.1
Lentil						
Days to flowering	683	79.0	33.0	53.5	8.0	15.0
Primary branches	516	6.0	2.0	2.6	0.5	21.9
Secondary branches	516	28.0	2.0	7.4	5.8	78.0
Days to maturity	670	132.0	78.0	98.8	13.2	13.4
Pods per plant	516	159.0	13.0	49.8	21.8	43.9
Seeds per pod	516	2.2	1.0	1.8	0.2	14.8
Grass pea						
Days to flowering	114	73.0	30.0	52.4	9.1	17.3
Days to maturity	114	160.0	92.0	116.5	27.3	23.4
Plant height	114	94.8	37.6	59.1	11.3	19.1
Pods per plant	75	121.0	14.2	43.8	25.7	58.8
Seeds per pod	75	4.0	1.0	2.4	0.6	24.6
Thousand seed weight	40	681.6	61.8	253.2	113.2	44.7
Fenugreek						
Days to flowering	170	55.0	31.0	42.5	5.0	11.9
Days to maturity	169	137.0	87.0	99.5	6.0	6.0
Plant height	170	40.6	9.0	19.7	8.3	42.2
Pods per plant	110	47.4	4.7	16.7	10.4	62.4
Seeds per pod	168	33.4	5.8	16.9	9.0	53.6
Thousand seed weight	71	22.6	11.7	16.1	2.6	16.6
Common bean						
Days to flowering	153	102.0	40.0	54.0	7.5	13.9
Days to maturity	153	151.0	77.0	88.1	13.9	15.8
Plant height	114	50.0	17.6	31.7	6.0	19.0
Pods per plant	134	83.0	3.8	17.6	8.6	48.1
Seeds per pod	140	15.0	0.4	5.5	1.6	29.2

Chickpea (*Cicer arietinum* L.)

Chickpea is believed to have originated in south-west Asia, although the wild form has never been found. An escape in the wild state in the Mediterranean region has been reported. The crop has been in cultivation in India, the Middle East and Ethiopia for centuries (Westphal, 1974). Ethiopia and India are centres of diversity for the cultivated chickpea (van der Maesen *et al.*, 1988). A wild related species (*C. cuneatum*) is reported to occur in northern Ethiopia.

Chickpea is an important pulse crop, ranking second among the pulses in Ethiopia. The annual average production reaches *ca.* 150 000 tonnes. There are basically two types of chickpea, Desi and Kabuli. The former is predominantly grown in Ethiopia. Ethiopia is a treasury of variability for chickpea and is considered to be a secondary centre of diversity.

Chickpea is sown after the main rainy season, in September–October on residual moisture, and harvested in January–February. In general it follows teff (*Eragrostis tef*) or wheat, or precedes wheat in the case of red clay soils (Westphal, 1974). It is to a large extent a monoculture crop, but it is sometimes found mixed with safflower, niger seed or noog (*Guizotia abyssinica*), sorghum or maize, depending on the region. Chickpea germplasm is collected from areas ranging in altitude from 1200 to 3000 m. However, it is largely grown between 1400 and 2300 m where the annual precipitation ranges from 700 to 2000 mm. The crop is grown mainly on black clay soils of pH between 6.4 and 7.9 (Murphy, 1963).

Utilization

Evaluation data of chickpea revealed the existence of wide genetic diversity (Table 2). The national crop improvement programme has taken advantage of the local germplasm and developed some cultivars such as Dubie, DZ-10-11 and DZ-10-4 by direct incorporation of the local landraces in the selection programme. Germplasm enhancement and utilization efforts have been initiated on a sizeable number of accessions and production of new cultivars is already under way. Recently Debre Zeit Agricultural Research Centre and the International Crops Research Institute for the Semi-Arid Tropics (ICRISAT) initiated a collaborative germplasm enhancement programme with 1000 accessions (900 accessions of Ethiopian origin) for subsequent utilization in the breeding programme in areas where the material proves to be useful.

Lentil (*Lens culinaris* L.)

Lentil is one of the oldest leguminous crops, believed to be indigenous to south-western Asia and the Mediterranean regions. From these areas the crop spread east to India, south to Ethiopia and north to Europe (Purseglove, 1968). It is now widely cultivated in temperate and subtropical regions as well as at higher elevations in the tropics. Recent collecting work revealed the existence of *L. ervoides* in central Ethiopia (A. Demissie, unpublished).

In acreage it is one of the major pulse crops of Ethiopia, surpassed only by faba bean, field pea and chickpea. Lentils are cultivated under rainfed conditions from 1500 to 3500 m. About 80 per cent of the PGRC/E collections were assembled from areas ranging in altitude from 2100 to 2900 m and receiving an annual rainfall of 950–1500 mm.

Generally, lentil is sown in a pure stand though sometimes it is grown in association with linseed, which can be separated at harvest. Lentil greatly resembles chickpea in habit and cultural requirements. It is sown in late June to early July in poor soil and in August–September in vertisols where double cropping is sometimes practised in certain regions.

Utilization

Lentil spread from its primary gene centre in south-west Asia to Europe, China, India and Ethiopia (Zeven & Zhukovsky, 1975). Important genetic variation has developed in the secondary centres of diversity. National crop breeders have recently taken full advantage of the genetic diversity occurring in lentil. As a result of direct exploitation of local landraces, one accession (EL-142) has been released as a variety. Recently, efforts have been initiated to evaluate 200 accessions of lentil landraces with the ultimate objective of identifying high-yielding lines and genetic stocks with acceptable disease/pest resistance or tolerance.

The work on lentil is mainly on cultivar development by combining desirable traits with useful genetic characters brought in from other regions through international organizations. A summary of the evaluation data is presented in Table 2.

Common bean, haricot bean or kidney bean (*Phaseolus vulgaris* L.)

Phaseolus vulgaris is believed to have originated in the New World and has been cultivated throughout North, Central and South America since ancient times. The common bean was introduced into

Europe by the Spaniards and Portuguese in the 16th century and was later brought to Africa. Now it is widely cultivated in the tropics, subtropics and temperate regions. The crop has achieved major importance in Ethiopia.

The total holding by PGRC/E is close to 300 accessions. It is collected from a wide range of soils (from light to heavy clay and loam soils) and climatic conditions. The altitude of collecting sites ranges from 600 to 2230 m with a high frequency of occurrence in areas of altitude 1700–1900 m. It can be seen from this how wide the ecological amplitude of the crop is. Several types, differing in seed size and colour, habit, flower colour, pod colour and size have been recorded. The common bean collections have been evaluated for some agro-morphological characters in appropriate agro-ecological sites (Table 2).

Utilization

The haricot bean is one of the important Ethiopian pulse crops, both as a protein source for local consumption and for export earning. The research work is based on introduced Mexican cultivars and those developed from exotic sources. A few local cultigens (Red Wolayita and Black Dessie) are important at subsistence farming level.

Though an introduced species, *P. vulgaris* has developed wide variation in a number of agro-morphological characters in Ethiopia. The PGRC/E collections assembled from various agro-ecological zones are currently incorporated in the breeding programme.

Minor pulse crops

Fenugreek (*Trigonella foenum-graecum* L.)

Fenugreek is indigenous to southern Europe and Asia and its cultivation now extends from the Mediterranean to western India and China, and south as far as Ethiopia; it is also found on the west coast of the USA (Westphal, 1974). It has been cultivated around Saharan oases since very early times.

To date a total of 248 accessions of fenugreek have been assembled from various regions, particularly from Gondar and Welo where the soils are predominantly black. The altitude of the collecting sites ranges from 1520 to 2750 m. Seeds are sown in August and harvested 3–4 months later. Plants are uprooted and dried for a few days before threshing and storing.

Utilization

Evaluation work has been initiated on the germplasm collected by PGRC/E. A total of 88 accessions have been characterized and the summary of the data is presented in Table 2. There is no major work on breeding and yield improvement aspects.

Grass pea (*Lathyrus sativus* L.)

The cultivated species belongs to the large genus *Lathyrus*, with about 130 species, which are widely distributed in the Northern Hemisphere and South America (Purseglove, 1968), with a few species in Africa. In Ethiopia five species, including *L. sativus* and *L. pratensis*, have been identified. The latter appears to be indigenous (Thulin, 1983).

A total of 245 samples has been assembled from fields, farm stores and village markets. Of these, 177 were collected from defined sites. The altitude of the collecting sites ranges from 1685 to 2700 m though the bulk of the material was collected from an area with the altitude ranging from 2200 to 2600 m.

Preliminary evaluation work has recently been initiated by PGRC/E. The study indicated the presence of significant diversity for the characters considered. The results of the evaluation work are summarized in Table 2.

Utilization

Grass pea is endowed with many properties that make it an attractive pulse crop in drought-stricken areas where soil quality is poor and extreme environmental conditions prevail. Several accessions collected by PGRC/E have been put under observational trials for screening and subsequent varietal development suitable to moisture stress areas. Prospects for selection of superior components within the landraces and identification of non-toxic strains are under consideration.

Cowpea (*Vigna unguiculata* (L.) Walp.)

To date the origin of *Vigna* is not equivocally established. In the opinion of earlier scientists, the cultivated cowpea originated in the Indian sub-continent (Vavilov, 1951); however, more recent studies (Faris, 1965) have gathered evidence to indicate an African origin for this pulse crop. Steele (1976) proposes a solely Ethiopian centre of origin followed by subsequent evolution predominantly in the ancient farming systems of the African savannah (Duke, 1981).

A total of 47 accessions of cowpea landrace material have been collected by the Ethiopian genebank for conservation and subsequent utilization.

Utilization

The national crop improvement effort of the cowpea programme utilizes exotic material that has come through the various international institutions, despite the existence of remarkable diversity in the indigenous material. The potential for varietal development for local germplasm is high. The passport data which include the altitude and soil characters are indicative of the existence of a wide ecological tolerance and this can possibly be equated with the existence of a wide range of diversity in cowpea. This apparently provides the basis for selecting suitable types with desirable traits for breeding programmes.

Groundnut, peanut (*Arachis hypogaea* L.)

Although the cultivated groundnut is widely found in tropical and warm temperate regions throughout the world, it is native to South America. The migration route of groundnut to Ethiopia is not known for certain. There is little, if any, worthwhile evidence for any pre-Columbian introduction of groundnut to Africa (Smartt, 1976).

Groundnut, though a legume crop, is categorized as a lowland oil crop in Ethiopia, usually grown at low to mid-altitudes in the warm regions. It is mostly intercropped with cereals and if planted alone is assigned to marginal lands. The cultural practices and varieties used by farmers are basically traditional types.

Although not much germplasm has been collected by PGRC/E, efforts have been made to assemble materials from certain localities. As a result, 14 accessions have been gathered during the course of general collecting operations from the eastern lowland areas with altitudes ranging from 500 to 1890 m.

Utilization

The research work on groundnut is based on introductions and local collections. A few local collections, e.g. Bisidimo, Olole and Sartu have shown good agronomic performances at a number of research sites, under both irrigated and rainfed conditions.

Lupin (*Lupinus albus* L.)

Lupinus as a genus has two centres of distribution, in both the Old World and the New World. The Old World centre is principally around the Mediterranean (Gustafsson & Gadd, 1964) and North

Africa as far as the mountains of Kenya and the Horn of Africa. *L. albus* belongs to the Old World centre and it is found in Sudan and Ethiopia.

In Ethiopia the cultivation of lupin is limited to the northern regions such as Gojam, Gondar and Tigray. Usually it is cultivated in marginal soils where other pulse crops do not perform very well. It is generally sown during the main rainy season (July–September) and harvested in December. Minimum tillage and cultural practices are followed for the cultivation of the crop.

The economic importance and agricultural potential of lupin is not widely recognized in the country except in the northern regions where the seeds are used to make 'katikala', a local drink. There are very few accessions conserved at PGRC/E.

Lesser known but potentially valuable pulse crops

The genetic resource base of pulse crops in Ethiopia is not limited to the species indicated above. There are several species which are lesser known but potentially valuable with significant regional importance; these are listed in Table 3.

General considerations

It is evident from this account that there is wide ecological amplitude for most of the pulse species in Ethiopia. Moreover, from this and from previous evaluation data (Mekbib, 1988) it can be seen that there is a high diversity for the characters recorded. The widely grown species have types well adapted to specific habitats. The recently introduced pulses such as *Phaseolus vulgaris* have become important in both acreage and production, while *P. lunatus* and *P. coccineus* are limited in distribution and acceptance.

Utilization of the available genetic diversity in the country is relatively low compared with cereals and oil crops. Work on the exploitation of the variability in pulse crops has only recently been initiated. The genetic exploitation of pulses differs from species to species and the effort initiated on utilization of cool season legumes appears to be encouraging. The joint utilization effort between national institutions and international organizations is a step in the right direction in maximizing the benefit of landraces and primitive cultivars available in the country.

Several of these pulse species possess specific attributes of utility in crop breeding programmes. These cultigens have desirable traits that can be crossed with high-yielding cultivars. Preliminary observations

Table 3. *List of pulse crops occasionally cultivated and rarely encountered in Ethiopia*

Scientific name	Common name	Genetic resources and utilization
Voandzeia subterranea (L.) DC	Bambara groundnut	Locally grown in south-western Ethiopia, few accessions are collected and conserved
Phaseolus coccineus L.	Scarlet runner bean	Rarely encountered as garden plant, few accessions are collected so far
P. lunatus L.	Lima bean	Fairly recent introduction, found in south-western regions
Canavalia ensiformis (L.) DC	Jack bean Sword bean	Recorded as found in south-west and part of Northern Ethiopia
C. *virosa* (Roxb.) Wight & Arnott	—	Probably sometimes cultivated in some regions, reported in Westphal (1974)
Mucuna pruriens (L.) DC	Velvet bean	Cultivated in Welega (Thulin, 1983)
Psophocarpus palustris Desv.	Goa bean	A rare occurrence, almost unknown (Westphal, 1974)
Vigna radiata (L.) Wilczek Syn.	Green gram Mung bean	Fairly recently introduced but becoming quite common, few accessions assembled
Cordeauxia edulis Hemsl.	Yeheb nut	Highly threatened species, endemic to Eastern Ethiopia and part of Somalia
Lupinus mutabilis Sweet	Tarwi	Recent introduction and sometimes cultivated for human consumption, mostly as a garden plant
Cajanus cajan (L.) Millsp.	Pigeon pea	Recent introduction and restricted cultivation
Glycine max (L.) Merr.	Soya bean	Recent introduction and grown in limited areas
Lablab purpureus (L.) Sweet	Hyacinth bean	Recent introduction and grown in limited areas

have shown that some accessions are tolerant to stress conditions and diseases and are sometimes preferred by local farmers. Being particularly adapted to diverse agro-ecological conditions and extreme climatic stress, they provide useful germplasm for introducing tolerance to drought, cold and diseases.

In general, with the exception of recently introduced species, the major improvement work, particularly of the cool season legumes such as faba bean, field pea, lentil and chickpea, is largely based on indigenous germplasm. There are still untapped resources of genetic material available for breeding work. This considerable genetic wealth has to be fully exploited in national efforts to ameliorate the farming situation.

It will be essential to establish more comprehensive collections of the different pulses, in order to represent the existing diversity in Ethiopia in these collections. So far pulse crops have only enjoyed lesser priority in terms of collecting. Field exploration/collection, survey and basic diversity studies are essential to identify high diversity areas for subsequent collecting operations and conservation.

Evaluation is exceedingly important and constitutes a prerequisite for the utilization of landraces. In-depth studies of the material to hand, with respect to characters of adaptation and resistance to stress conditions such as frost, cold, heat, drought and to adverse soil conditions, are required and should be strengthened. Landrace improvement endeavours on pulses, such as the exemplary undertaking on durum wheat, should be looked at as one of the facets of improvement programmes for pulse crops.

References

Abdella, M. M. F. (1979). The origin and evolution of *Vicia faba* L. *In*: *Proceedings of the 1st Mediterranean Conference of Genetics*, pp. 713–746.

Cubero, J. I. (1974). On the evolution of *Vicia faba*. *Theoretical and Applied Genetics*, **45**, 47–51.

Duke, J. A. (1981). *Handbook of Legumes of World Economic Importance*. Plenum, New York.

Faris, D. G. (1965). The origin and evolution of the cultivated forms of *Vigna sinensis*. *Canadian Journal of Genetics and Cytology*, **7**, 433–52.

Feyissa, R. (1988). Germplasm conservation at the Plant Genetic Resources Centre/Ethiopia. *In*: J. M. M. Engels (ed.), The conservation and utilization of Ethiopian germplasm. Proceedings of an international symposium, Addis Ababa, 13–16 October 1986, pp. 49–57 (mimeographed).

Ghizaw, A., Tilaye, A., Berhe, A. & Beniwal, S. P. S. (1986). Evaluation and maintenance of faba bean germplasm in Ethiopia. Paper presented at the 7th Annual Coordination meeting of ICARDA/IFAD Nile Valley Project on faba bean held in Addis Ababa, 23–27 September 1986 (unpublished).

Gustafsson, A. & Gadd, I. (1964). Mutations and crop improvement II. The genus *Lupinus* (Leguminosae). *Hereditas*, **53**, 15–36.

Hawkes, J. G. (1980). *Crop Genetic Resources Field Collection Manual*. Department of Plant Biology, University of Birmingham. IBPGR/Eucarpia.

Hawkes, J. G. (1985). Report on a consultancy mission to Ethiopia for GTZ to advise PGRC/E on germplasm exploration, conservation, multiplication and evaluation. Birmingham (mimeographed).

Hawtin, G. C. & Hebblethwaite, P. D. (1983). Background and history of faba bean production. *In*: P. D. Hebblethwaite (ed.), *The faba bean (*Vicia faba *L.). A base for improvement*. Butterworths, London, pp. 3–22.

Ladizinski, G. (1975). On the origin of broad bean *Vicia faba* L. *Israel Journal of Botany*, **24**, 80–8.

Mekbib, H. (1988). Crop germplasm multiplication, characterization, evaluation and utilization at the Plant Genetic Resources Centre/Ethiopia. *In*: J. M. M. Engels (ed.), The conservation and utilization of Ethiopian germplasm. Proceedings of an international symposium, Addis Ababa, Ethiopia, 13–16 October 1986, pp. 170–8 (mimeographed).

Mengesha, M. H. (1975). Crop germplasm diversity and resources in Ethiopia. *In*: O. H. Frankel and J. G. Hawkes (eds), *Crop Genetic Resources for Today and Tomorrow*. Cambridge University Press, Cambridge, pp. 449–53.

Murphy, H. F. (1963). Fertility and other data on some Ethiopian soils. *Imperial Ethiopian college of agriculture and mechanical arts, Experimental Station Bulletin no. 4*. Dire Dawa, Ethiopia.

Pundir, R. P. S. & Mengesha, M. H. (1982). Collection of chickpea germplasm in Ethiopia. *Genetic resources progress report 44*. ICRISAT, Patancheru.

Purseglove, J. W. (1968). *Tropical Crops: Dicotyledons, vols 1 and 2*. Longman, London.

Smartt, J. (1976). *Tropical Pulses*. Longman, London.

Steele, W. M. (1976). Cowpeas. *In*: N. W. Simmonds (ed.), *Evolution of Crop Plants*. Longman, London, pp. 183–5.

Telaye, A. (1988). Cool season food legumes in East Africa. *In*: R. H. Summerfield (ed.), *Proceedings of the International Food Legume Research Conference on pea, lentil, faba bean, chickpea*. Spokane, Washington, USA, 6–11 July 1986, pp. 1113–24. Kluwer, Dordrecht.

Thulin, M. (1983). Leguminosae of Ethiopia. *Opera Botanica*, **68**, 1–223.

van der Maesen, L. J. G., Kaiser, W. J., Marx, G. A. & Worede, M. (1988). Genetic basis for pulse crop improvement: collection, preservation and genetic variation in relation to needed traits. *In*: R. J. Summerfield (ed.), *Proceedings of the International Food Legume Research Conference on pea, lentil, faba bean, chickpea*, Spokane, Washington, USA, 6–11 July 1986, pp. 55–66. Kluwer, Dordrecht.

Vavilov, N. I. (1951). The origin, variation, immunity and breeding of cultivated plants. *Chronica Botanica*, **13**, 1–366.

Westphal, E. (1974). *Pulses in Ethiopia, their Taxonomy and Agricultural Significance*. PUDOC, Wageningen.

Worede, M. (1987). Overview: genetic erosion. PGRC/E, Addis Ababa (unpublished).

Zeven, A. C. & Zhukovsky, P. M. (1975). *Dictionary of Cultivated Plants and their Centres of Diversity*. PUDOC, Wageningen.

Zohary, D. (1970). Centres of diversity and centres of origin. *In*: O. H. Frankel and E. Bennett (eds), *Genetic Resources in Plants: Their Exploration and Conservation*. Blackwell, Oxford, pp. 33–42.

28

Oil crop germplasm: a vital resource for the plant breeder

HIRUY BELAYNEH

Introduction

According to Seegeler (1983), 328 oil plant species are known to exist in Ethiopia. Of these, 15 are cultivated and the rest may have uses other than for oil and may be cultivated or wild. Oil-bearing plants having oil contents in excess of 10 per cent, but which are not yet cultivated commercially, have been catalogued by Goshe & Hamito (1983) (Table 1).

Ethiopia is known to be either a centre of origin or a centre of diversity for many cultivated oil crops. Several of the cultivated oil-seed crops play an important role in the nutrition of the Ethiopian population and in foreign exchange earnings. The oil crops currently in production in the country are niger or noog, rapeseed, Ethiopian mustard or gomenzer, linseed, sunflower, sesame, groundnut, safflower and castor bean.

The overall objective of the research programme is to increase the production of oil seeds for food and to provide raw materials for agro-industrial development. This can be achieved by the development of high-yielding, stable cultivars with the necessary package of practices required for sustained high yields. The programme therefore falls into three sections:

- the improvement of noog, linseed, sesame and safflower which possess a wide range of variability and a wealth of unutilized indigenous germplasm;
- the improvement and popularization of oil seed *Brassica* and groundnut for which a wide range of indigenous germplasm is also available;
- the popularization of the introduced sunflower crop, for which probably no indigenous germplasm exists.

Table 1. *The oil or fat content on dry seed basis of some selected Ethiopian tree and shrub species*

Scientific name	Common or local name	Oil content of seeds/kernels (%)
Gossypium spp.	Cotton	10–24
Argemone mexicana	Prickly poppy	38.7
Ricinus communis	Castor bean	51.7
Jacaranda spp.	Jacaranda	36.6
Schinus molle	Pepper tree	10.9
Ximenia americana	Inkoy (Amharic)	49[a]
Schefflera abyssinica	Ketema (Amharic)	10.3
Maesa lanceolata	Kelewa (Amharic)	22.3
Trichilia roka	Ethiopian mahogany	44–56
Bersama abyssinica	Azamir (Amharic)	35.8
Allophylus abyssinicus	Imbus (Amharic)	26.4
Sterculia africana	Fua (Gambella)	30.8
Pittosporum mani	—	23.4
Sapium ellipticum	Gancho (Sidamo)	55[a]
Terminalia macroptera	Kokora (Oromo)	52–64[a]
Balanites aegyptica	Desert date	10–50[a]
Salvadora persica	Hadia (Tigray)	19–34
Melia azedarach	Neem tree	33–45[a]
Myrica salcifolia	Shinet (Amharic)	20
Erithyrina abyssinica	Red-hot-poker tree	15–20
Trema guineensis	Sendo (Amharic)	28
Acanthus spp.	Kosheshila (Amharic)	23
Cocos nucifera	Coconut	65–68

[a] Represents oil content of kernels.

Source: Goshe & Hamito (1983).

Germplasm collection

Ethiopia is known to be a gene centre for many cultivated crops. Much has been done in the past to exploit this germplasm wealth. The collection of oil seed crops has been carried out with the objective of enriching the existing gene pools of those crops utilized in breeding and selection programmes.

The collection and characterization of indigenous germplasm were initiated in the early 1940s, but lost momentum before being resumed in the early 1970s. A few Italian documents exist showing the work done on noog landraces between 1940 and 1946. The period 1974–85 saw an intensification of effort in the search for accessions with the required characteristics, with the intention of exploiting promising local cultivars to the fullest extent. The first step in this approach was

Table 2. *Oil crop germplasm collections maintained at PGRC/E*

Oil crops	Total number of accessions	Number of accessions characterized
Oil seed Brassica (*Brassica carinata, B. nigra, B. oleracea* and *B. napus*)	964	910
Cultivated niger or noog (*Guizotia abyssinica*)	939	826
Wild noog (*G. scabra*)	7	—
Sunflower (*Helianthus annuus*)	20	—
Linseed (*Linum usitatissimum*)	1832	470
Safflower (*Carthamus tinctorius*)	132	99
Castor bean (*Ricinus communis*)	350	49
Sesame (*Sesamum indicum*)	376	195
Groundnut (*Arachis hypogaea*)	19	16

to establish a collection of the local germplasm before it was irretrievably lost as a result of changes in land use, drought, etc.

Initially, the collection of oil crop germplasm was undertaken by extension agents and oil crop subcommittee members, most of the collections being samples from farmers' stores or local markets. In recent years, collections have been made primarily from standing crops in farmers' fields. However, many of the expeditions were confined to the roadsides. The oil crop team members and invited experts have participated in several expeditions which substantially increased the available germplasm, particularly of Ethiopian mustard, linseed, noog and sesame. The collecting trips were made in collaboration with the Plant Genetic Resources Centre/Ethiopia (PGRC/E). The germplasm holdings collected and preserved at PGRC/E so far are presented in Table 2.

Traditional utilization

In Ethiopia, oil seeds are the main source of oil in the diet of the majority of the population. Noog is the prime supplier of cooking oil and usually commands a premium over other available oils. Sesame and safflower oils are appreciated in regions where the crops can be produced. Linseed oil is used when other oils are in short supply. Mustard seed oil is also produced commercially. The cake remaining after oil extraction is rich in protein and can provide a valuable livestock feed.

In the local cuisine, dried and/or ground seeds from linseed, noog and mustard are used to prepare a stew or 'wot'. Mustard leaves are an important leaf vegetable in the highlands where they are boiled and served as 'gomen wot'. Roasted and crushed seeds of noog, linseed and safflower may be mixed with water to prepare beverages. Noog and sesame flours are sometimes mixed with flour for baking bread and the whole seeds are often sprinkled on top prior to baking.

Seed oil from castor, linseed and mustard plants can also be used for tanning leather and for varnishes. The use of mustard, safflower and castor oils for lighting is widespread and they are also commonly used to grease the 'metad' (pan) before 'injera' (teff bread) is baked.

Preparations from noog, mustard, sesame, linseed and castor plants are traditionally used for one or more of the following medical problems:

- to treat diarrhoea (sesame, linseed, niger);
- as a diuretic (linseed);
- to treat eye irritation due to dust (linseed, castor, sesame);
- as a purgative (linseed, safflower);
- for birth control (mustard, niger);
- to treat sores and rheumatism (safflower, sesame);
- to treat syphilis (niger);
- to treat coughs (niger);
- to treat an upset stomach (mustard).

Oil crops also have many other uses in Ethiopia, e.g. as condiments, as snack food (when roasted), as a lubricant, as an additive for soap and paints and as a fertilizer.

Variety development

The task of the oil crop breeders is concentrated on the development of high-yielding varieties, taking into consideration resistance to major diseases and improved quality. Depending on the adaptability of the species, the indigenous germplasm was initially characterized for agronomic and morphological characters at the Holetta, Melka Werer, Awasa and Debre Zeit Research Centres of the Institute of Agricultural Research (IAR).

Further evaluation tests were carried out at several sites. Large oil seed introductions from different countries have also been evaluated at a number of sites. The multilocation trials have resulted in the release of several improved varieties of oil crops. The names of the released and/or recommended varieties, their desirable characteristics and other information are presented below on a crop basis.

Table 3. *Number of indigenous entries in different variety trials, 1986*

Crop	Micro-trial	Pre-national variety trial	National variety trial	Extension yield trial
Niger	19	7	7	4
Ethiopian mustard	10	8	2	–
Linseed	21	9	3	–
Sesame	14	5	4	–
Groundnut	–	2	3	–
Safflower	2	2	–	–

Niger or noog (*Guizotia abyssinica*)

There is evidence that noog orginated in the highlands of Ethiopia, north of 10° N latitude (Baagoe, 1974). Harlan (1975) considers noog to be among the earliest of the crops domesticated in Ethiopia. It may have orginated from the wild species *Guizotia scabra* through selection by Ethiopian farmers over thousands of years. The crop has always been one of the most important oil crops in Ethiopia in terms of both area and production. All cultivars are local and raised under rainfed conditions.

The objective of the noog breeding programme is to develop improved cultivars which produce consistently and give a high yield of good quality. Special attention is being given to lodging resistance, uniform ripening, minimal shattering, frost tolerance and resistance to *Septoria* disease.

As the research effort in noog has not been as extensive as in other oil seeds, only one variety (Sendafa) has been released to growers. However, recent accessions have been evaluated for desirable characters and promising, indigenous lines have been advanced to the different levels of replicated variety trials (Table 3).

Two selection programmes on noog landraces are in progress now. The mass selection programme aims at improved plant type through usual selection while the half-sib recurrent selection programme uses an evaluation of progeny performance as a basis of selection. Each method is geared towards developing early and medium maturing composites with the desirable characteristics.

Oil seed Brassica (*Brassica carinata, B. nigra, B. napus* and *B. oleracea*)

Ethiopian mustard or gomenzer (*Brassica carinata*) is believed to have originated in Ethiopia from the natural crossing of *Brassica*

Table 4. *Oil crop varieties that have been released from introduced germplasm*

Crop	Variety	Year of release	Country of origin
Rape seed	Target	1976	Canada
	Tower	1984	Canada
	Pura	1984	W. Germany
	Tower Sel_3	1986	Canada/Ethiopia
Linseed	Victory	1978	North Dakota, USA
	Concurrent	1978	The Netherlands
	CI-1525	1984	France
	CI-1652	1984	Ireland
Sunflower	Russian Black	1974	Yugoslavia/Russia
	Hesa	1974	W. Germany
	Pop 158	1974	W. Germany
Groundnut	Shulamith	1976	Israel
	NC 4X	1986	USA
	NC 343	1986	USA
Sesame	T-85	1976	Uganda/India
	S		Uganda
	E		Uganda

nigra (senafitch) and *Brassica oleracea* (Gurage or Wollamo gomen). Ethiopian mustard is the fourth most important oil crop after niger, linseed and sesame and is widely grown only in Ethiopia. It is also used extensively as a vegetable.

The objective of the breeding programme is to develop improved cultivars of the oil seed Brassica, especially rape seed and mustard. Resistance to diseases, especially to *Alternaria* leaf spot, low erucic acid in the oil and low glucosinolate and fibre concentrations in the meal are receiving special attention.

Evaluation of Ethiopian landraces has led to the recommendation of five *Brassica carinata* varieties (S-67, S-71, S-115, Awasa population and Dodolla-1) for the highlands of Ethiopia. Four *napus* varieties, namely Target, Tower, Pura and Tower Sel_3, were identified from introductions and released for large-scale production (Table 4). The latter three varieties have low erucic acid and low glucosinolate and have replaced Target, which has high erucic acid and high glucosinolate similar to the recommended high-yielding Ethiopian mustard varieties.

The major effort in the crossing programme has been towards incorporating earliness, low erucic acid and low glucosinolate characters into local mustard selections. Interspecific crosses have been achieved and are being used to broaden the genetic basis of Ethiopian mustard.

The existing germplasm lines and selections of Ethiopian mustard are being screened for low erucic acid and glucosinolate and evaluated for oil yield (Getinet, Rakow & Downey, 1986). Low erucic acid oil and low glucosinolate meal are now essential if an international trade is to be developed and the meal is to be fed to animals.

Linseed (*Linum usitatissimum*)

Linseed is also an important oil crop occupying a wide production area. The crop has built up considerable diversity in Ethiopia after its early introduction from Asia. The objective of the breeding programme is to develop improved cultivars of linseed, and disease resistance to wilt, powdery mildew and *Septoria* are important.

A large number of local collections have been evaluated and characterized (Table 2). Single plant selections were made in the nurseries to capture 'within' plot variation and uniform lines were produced. A number of varieties and lines are being advanced through a four-stage hierarchy of yield trials (Table 3).

Varietal testing of introduced linseed germplasm has resulted in the release of four varieties, Victory, Concurrent, CI-1525 and CI-1652. The latter two bold-seeded varieties were released in 1984 after fulfilling the pre-release requirements. A hybridization programme based on the pedigree method and using early, wilt resistant and high-yielding lines is in progress. The existing local linseed collections, as well as introductions, will be screened for low linolenic acid content. Reduction in linolenic acid content in linseed should overcome problems like rapid flavour deterioration, low nutritive value, etc.

Sunflower (*Helianthus annuus*)

The purpose of the sunflower breeding programme is to identify well adapted cultivars of this crop, also giving consistent, enhanced yields of oil per hectare. Resistance to disease, especially to downy mildew, as well as to *Sclerotinia* and rust require more attention.

More than 300 varieties of sunflower have been introduced, mainly from North America, the USSR and Eastern Europe, and evaluated while in quarantine. A number of promising lines were advanced through a four-stage hierarchy of yield trials. Before 1976, three varieties of sunflower, Russian Black, Hesa and Pop 158, were recommended for general release. Of the varieties tested in the last five years, the long-maturing Argentario was identified as being widely adaptable. The two promising early maturing lines in the programme are Charnianka × Gene pool I and Gene pool II.

Sesame (*Sesamum indicum*)

Ethiopia is considered as either the centre of origin or a centre of diversity for sesame (Seegeler, 1983), hence the collection and characterization of local germplasm is very important in the overall improvement programme for the crop.

The purpose of the breeding programme is to identify the most adaptable sesame varieties for the various agro-ecological zones through selection and hybridization. Development of varieties with high oil content, resistance to bacterial blight disease, uniform ripening and minimum shattering characteristics, receives special attention.

Of the landraces assembled and evaluated, some have proved to be useful and were advanced through a four-stage hierarchy of yield trials (Table 3). Selection of progenies was also carried out and some have shown promising performances under multilocational tests. From the landraces collected earlier, 'Kelafo 74', a variety collected from the lower Wabe Shebelle Basin of Kelafo district, has proved to be among the best yielders, especially under irrigated conditions at Melka Werer, Tendaho and Gode. One progeny from a collection of the western region, SPS-111519, is under consideration for release in the rainfed zones.

Several exotic sesame germplasm accessions have been tested and three varieties, T-85, S and E (introduced from Uganda), were recommended for general release and are under production at several sites.

A hybridization programme to evolve lines with partial shattering characteristics, disease resistance and high seed and oil yield, is in progress at Melka Werer Research Centre. Preliminary results show that some lines with uniform stand, good capsule setting, partially opening capsules at harvest and bacterial blight resistant types are attainable.

Groundnut (*Arachis hypogaea*)

In Ethiopia, groundnuts are consumed roasted or are crushed for the production of edible oil. The overall target of the breeding work is the increase of the productivity of the crop for local consumption, agro-industrial projects and export. Furthermore, the identification of the most suitable groundnut varieties for the different agro-ecological zones is an important objective. Special attention is being given to disease resistance, in particular to leaf spot and rust.

From earlier local accessions, 'Asmara' and 'Dire Dawa', which were collected from the respective regions, were recommended for general release. Of the few recent local accessions, Bisidimo, Olole and Sartu have shown good performance at Melka Werer, Bisidimo and Babile.

Evaluation of earlier introductions has led to the release of Shulamith. This variety is under cultivation over a wide region. Of the recent introductions from the USA, two Virginia type varieties, namely NC 4X and NC 343, were released in 1986 for irrigated areas of the Awash Valley. The two varieties had shown a maximum yield of 7000 kg/ha at the irrigated site of the Middle Awash Valley and 2500 kg/ha at the marginal rainfall sites of Bisidimo and Babile. Moreover, moderately disease-resistant and early maturing varieties have been obtained from the recent introductions from the International Crops Research Institute for the Semi-Arid Tropics (ICRISAT).

Safflower *(Carthamus tinctorius)*

Ethiopia is considered to be the centre of origin and/or of diversity for safflower. It is grown as a companion crop with cereals, mainly teff. Research work on the crop was discontinued for some time and resumed only a few years ago. The objective of the breeding programme is to develop improved cultivars of safflower for low, mid- and high-altitude areas. Disease resistance to *Ramularia* and *Alternaria caratami* is important.

About 100 new landraces have been collected and characterized for morphological and agronomic performance and some outstanding lines have reached the micro-trial and pre-national variety trial stages (Table 3). A few introductions of safflower lines have been received from Indore, India.

Some 600 progeny selections were made from the earlier collections to get pure and uniform lines. Outstanding progenies have been advanced to the pre-national variety trial.

Castor bean *(Ricinus communis)*

The castor bean research programme started two years ago, after a long period of interruption. The purpose is to evaluate indigenous and exotic varieties mainly for adaptability and high seed and oil yield.

So far, about 200 local collections and 90 introductions have been evaluated and characterized. Some outstanding lines were observed in the preliminary observation nursery and have been advanced for further studies.

References

Baagoe, J. (1974). The genus *Guizotia* (Compositae). A taxonomic revision. *Botanisk Tidsskriff*, **69**, 1–39.

Getinet, A., Rakow, G. & Downey, R. K. (1986). Seed colour and quality characteristics in Ethiopian mustard (*Brassica carinata*). *PGRC/E–ILCA Germplasm Newsletter*, **12**, 12–15.

Goshe, B. A. & Hamito, D. S. (1983). Preliminary survey of oil bearing plants in some regions of Ethiopia. *Ethiopian Journal of Agricultural Science*, **5**, 89–96.

Harlan, J. R. (1975). *Crops and Man*. American Society of Agronomy, Madison, Wisconsin.

Seegeler, C. J. P. (1983). *Oil Plants in Ethiopia, their Taxonomy and Agricultural Significance*. PUDOC, Wageningen.

29

Significance of Ethiopian coffee genetic resources to coffee improvement

MESFIN AMEHA

Introduction

Among the economic species of coffee grown in more than 50 countries in different parts of the world, *Coffea arabica* L. is the only tetraploid species of the genus. It contributes over 80 per cent of the world's coffee production. In many scientific reports Ethiopia is considered to be the centre of origin and diversification of coffee (Sylvain, 1958; Fernie, 1966; Food and Agriculture Organization, 1968; Carvalho *et al.*, 1969; Ameha, 1980; Rodrigues, 1981; Worede, 1982; Watkins, 1985). The question of whether the south-west mountain moist evergreen forest, the farmer's field or the low altitude river banks is the primary habitat is not discussed here, although the issue is of primary interest to geneticists and breeders, for conservation purposes and in the search for primitive genes in the wild progenitors.

Arabica coffee is an evergreen shrub of variable size. The tree grows up to 14 m in height and about 2 m in width under forest strata and up to 6 m in height and about 12 m in width in canopy under farmers' holdings (a tree this size was observed in Wanago near Dilla in Sidamo administrative region). Naturally, it has a dominant central orthotropic stem and horizontally growing plagiotropic branches with pairs of secondary, tertiary, etc. branches originating from preceding branches in the hierarchy. The leaves are borne in opposite pairs along the side of the branches. The flowers emerge as inflorescences on all forms of lateral branches in each leaf axil of the nodes. Every flower normally develops into a two-seeded berry. The central stem gives rise to a number of orthotropic stems when stumped, wounded or bent.

Climate and soil

Coffee is a hardy plant which thrives well in almost all types of soils under a wide range of climatic conditions. In Ethiopia, it grows in almost all administrative regions in conditions ranging from the semi-savannah climate of the Gambela plain (550 m above sea level) to the continuously wet forest zone of the south-west (2200 m). It generally grows on sloping land of different gradients.

The soil varies from sandy loam to heavy clay while the general soil types, which are acidic (pH 4.2–6.8), are red to reddish-brown lateritic loams or clay loams of volcanic origin (Sylvain, 1958; Fernie, 1966). Annual rainfall varies from 750 to 2400 mm. The upper limit for day maximum temperature recorded is 36 °C and the lower limit for night minimum is 7 °C, with varying ranges of diurnal temperature.

Botanical cultivars

Ethiopian coffee production generally comes from botanically unidentified cultivars of Arabica populations. In the early years the local names of this coffee reflected either the area of production or the names of influential landlords or aristocrats and, even today, these local names are widely used.

Chevalier (1947) and Ciferri (1940) were the pioneers in grouping Ethiopian coffee by bean characteristics. Sylvain (1958) grouped them into Ennarea, Jimma, Agaro, Chochie, Yirgalem, Dilla, Arba Gugu, Harer, Loulo, Wolkitte and Welayita types using bean shape and colour, leaf size and colour, leaf tip colour and other obvious characters. Recently, a number of distinct morphotypes have been recognized. In the Harer type, for example, Abadiro, Buna Guracha, Buna Kella and Shimbre, among others, are quite different (Watkins, 1985). From the accessions conserved in the National Coffee Collection and in the collection of coffee berry disease (CBD) resistant types, different coffee strains can be distinguished both qualitatively and quantitatively. The Mettu types (74110, 7412, 74148, etc) possess very narrow shiny green leaves, somewhat compact in nature. Accessions like 2370, 3170, 7440, and 741 are robust and of a spreading type (Institute of Agricultural Research, 1971–84). This considerable variation requires a systematic approach by a taxonomist and geneticist in order to develop a practical classification.

Variation in coffee

In Ethiopia, research findings over 18 years have revealed the presence of enormous genetic variation for different agronomic traits (Brownbridge & Gebre-Egziabher, 1968; IAR, 1971–84; Ameha, 1980, 1983; van der Graaff, 1981; Worede, 1982; Ameha & Belachew, 1986a,b; Ameha, Belachew & Shimbr, 1986; Belachew & Ameha, 1986). Yield was almost normally distributed with mean yield per accession varying from 10 to 840, 55 to 895, 165 to 720, 150 to 775 and 365 to 1250 g per tree clean bean at Melko, Mettu, Agaro, Wonago and Gera, respectively. Similarly, when the yield variation of individual accessions over a number of years of production was analysed, the coefficients of variation ranged from 34 per cent to 211 per cent.

Frequency distributions for resistance to CBD, leaf rust and liquoring quality were skewed but all in favourable directions (IAR, 1971–84; Ameha, 1980, 1983; Worede, 1982; Ameha & Belachew, 1984; Belachew & Ameha, 1986). The negative bimodal skewedness observed for bean grade is a typical character of Arabica coffee. Generally, there are more discard and remainder beans in the processed coffee as a result of Man's preference for well shaped, large beans of AA, A and B grades. When bean sizes were measured, most frequent dimensions varied from 4.8 to 14.5, 4.0 to 10.5 and 2.5 to 8.5 mm for length, width and depth, respectively. The bean colour is classified into green olive-green, green yellow, green bluish, yellow amber and pale yellow and the bean shape as small oval, almost round, large rectangular and large oval (Ciferri, 1940; Sylvain, 1958; FAO/IBP, 1973; Belachew & Ameha, 1986). The degree of variation for other characters, such as tree shape, branching habit, resistance to diseases and pests, persistent sepals, etc. is also striking. As expected, it appeared that most of the traits are quantitatively heritable and, in general, considerably influenced by environment (Tadesse & Engels, 1986).

Several researchers in the Western Hemisphere have attributed the limited variation there to mutation and restricted genetic recombination. The coffee varieties in Latin America evolved from a few seeds which were derived from only a few trees grown in the Botanical Garden of Amsterdam in the middle of the 18th century. However, here in Ethiopia, nature has played a vital role in the selection, distribution and adaptation of Arabica coffee and it was only 20 years ago that Man entered the picture with his ideas of systematic selection. What is being grown now is the result of tens of thousands of years of natural selection, selection that has occurred *in situ*. It is from this background that the hybridization and heritability study for CBD revealed new findings in the predominantly self-fertile coffee species (Ameha & Belachew, 1984; 1986a). Three to five recessive genes controlling resistance to CBD and a hybrid vigour for yield and yield components were observed in the indigenous coffee. These recent findings also indicate that Arabica genotypes are generally location-specific. No one cultivar performed consistently well across locations for yield and vigour, suggesting how specific Arabica 'ecotypes' are in their adaptability.

Loss of genetic diversity

In the last 40 years, a significant reduction of genetic diversity has occurred in the Ethiopian coffee. According to FAO reports almost 90 per cent of the Ethiopian forest cover had vanished by 1965 (FAO, 1968, 1973). Added to this deforestation are the effects, in subsequent years, of the development of new roads bringing expanded agriculture and forest utilization, particularly in the rainy forests of the south-west where coffee occurs naturally in association with forest. Consequently, coffee genetic erosion has gone far beyond the point of no return. Many coffee forests are no longer intact. Persistent drought in the last 15 years has further aggravated the situation. Hundreds of hectares of coffee forest have been replaced by food crops since farmers want more food for their families. Furthermore, food crops pay farmers more than coffee does when production per unit area is compared in times of drought. Even under normal conditions, when food is abundant, the income from coffee is discouraging. This, together with the drought, has led to the destruction of the coffee forests.

With the advent of CBD in 1971 and the subsequent identification of CBD-resistant selections, the distribution of resistant cultivars resulted in the retention of relatively invariable individuals in some typical coffee forests where they were replanted after forest clearing. This has definitely caused significant losses in genetic diversity. It has been estimated that between 25 000 and 35 000 hectares of semi-forest coffee have so far been replaced by CBD-resistant cultivars, leading to at least 10 per cent loss.

Observations suggest that if the present pace of agricultural development, forest utilization and population growth continue, by the year 2000 about 120 000 hectares of the estimated 350 000 hectares of semi-forest coffee will be replanted with the advanced cultivars and about 80 per cent of the remaining 230 000 hectares will be lost through other factors. The loss of the heterogeneous coffee populations, which represent the gene pool for hundreds of agronomic traits, will be catastrophic. The well known quality coffee of Limu, Nekemte, Gimbi, Harer and Yirga Chefe, which fetches a high premium, will no longer exist unless immediate ways are found to preserve it. The conservation work of PGRC/E which has started at Chochie, Kefa, in cooperation with the Jima Research Centre of the Institute of Agricultural Research, is highly appreciated. It needs to be strengthened with better facilities and more manpower thus allowing more accessions to be conserved. In the same manner, and as a matter of considerable urgency, collection and *in situ* conservation of Harer coffee must resume. Watkins (1986) and the IAR team concluded that at present 80 per cent of the coffee from Habro Awraja, which produces 40 per cent of the Harer coffee, is rapidly declining in production. Prompt action is required to save the germplasm of this coffee which has a tremendous, worldwide reputation for quality and fetches twice the price of coffee from other regions.

A project with comprehensive conservation strategies including (a) ecosystem conservation for the wild and semi-wild coffee and (b) collection and *in situ* conservation, was prepared and submitted for implementation some years back. PGRC/E, together with other governmental institutions and organizations affiliated with coffee, must be encouraged to facilitate and implement those strategies as a matter of top priority, by law if necessary.

References

Ameha, M. (1980). Yield assessment of indigenous coffee collection grown at Jima Research Centre. *Ethiopian Journal of Agricultural Science*, **11**, 69–77.

Ameha, M. (1983). Variabilities of indigenous coffee collection to rust resistance. Simposio Sobre Ferrugens do Caffeeiro, Oeiras, Portugal.

Ameha, M. & Belachew, B. (1984). Resistance of the F_1 to coffee berry disease in six parent diallel crosses in coffee. *Proceedings, First Regional Workshop on Coffee Berry Disease*. Association for the Advancement of Agricultural Sciences in Africa, Addis Ababa, pp. 167–77.

Ameha, M. & Belachew, B. (1986a). Field evaluation of resistance to stress conditions in crosses and their parents of coffee, *Coffea arabica* L. *First Ethiopian Coffee Symposium, Institute of Agricultural Research, Addis Ababa, August 1986*. IAR, Addis Ababa.

Ameha, M. & Belachew, B. (1986b). Genotype–environmental interactions in coffee, *Coffea arabica* L. *First Ethiopian Coffee Symposium, Institute of Agricultural Research, Addis Ababa, August 1986*. IAR, Addis Ababa.

Ameha, M., Belachew, B. & Shimbr, T. (1986). Yield assessment of CBD (coffee berry disease) resistant progenies of coffee under different environments. *First Ethiopian Coffee Symposium, Institute of Agricultural Research, Addis Ababa, August 1986*. IAR, Addis Ababa.

Belachew, B. & Ameha, M. (1986). Variation among national coffee collections for some agronomic characteristics. *First Ethiopian Coffee Symposium, Institute of Agricultural Research, Addis Ababa, August 1986*. IAR, Addis Ababa.

Brownbridge, J. M. & Gebre-Egziabher, E. (1968). The quality of some of the main Ethiopian mild coffees. *Turrialba*, **18**, 361–72.

Carvalho, A., Ferwerda, F. P., Frahm-Leliveld, J. A., Medina, D. M., Mendes, A. J. I. & Monaco, L. C. (1969). *In*: F. P. Ferwerda and F. Wit (eds), *Outlines of perennial crop breeding in the tropics*. Miscellaneous Papers 4, Landbouwhogeschool, Wageningen, pp. 189–241.

Chevalier, A. (1947). III. Systématique des caféiers et faux-caféiers. Maladies et insectes nuisibles. *Cyclopedia diologique, No. 28*, Paris.

Ciferri, R. (1940). Primo reporto sul caffe nell – Africa Orientale Italiana. Firenze, Regio Instituto Agronomico per l'Africa Italiana. *Relazioni e monografie agrario coloniali, No. 60*.

Food and Agricultural Organization (1968). *Coffee Mission to Ethiopia 1964–65*. FAO, Rome.

FAO/IBP (1973). *Survey of crop genetic resources in their centres of diversity. First report*. FAO, Rome.

Fernie, L. M. (1966). *Impressions of Coffee in Ethiopia*. Coffee Research Station, Lyamungu, Tanzania.

Institute of Agricultural Research (1971–84). *Coffee Progress Report* (annual). Jima Research Centre, IAR, Addis Ababa.

Rodrigues, C. J. Jr (1981). *Coffee Leaf Rust in Ethiopia* (consultant report, Eth. 78/004). FAO, Rome.

Sylvain, P. G. (1958). Ethiopian coffee – its significance to world coffee problems. *Economic Botany*, **12**, 111–39.

Tadesse, D. & Engels, J. M. M. (1986). Phenotypic variation in some fruit characters in coffee collected from Chora wereda. *PGRC/E–ILCA Germplasm Newsletter*, **12**, 2–8.

van der Graaff, N. A. (1981). Selection of Arabica coffee types resistant to coffee berry disease in Ethiopia. *Mededelingen Landbouwhogeschool Wageningen, 11-1*. University of Wageningen.

Watkins, R. (1985). Coffee *Coffea arabica* L. genetic resources and breeding. Consultant Report, Wye College, University of London.

Watkins, R. (1986). *Proposed Actions for Maintenance and Production of Harer Coffee*. Coffee Improvement Project, Addis Ababa.

Worede, M. (1982). Coffee genetic resources in Ethiopia: conservation and utilization with particular reference to CBD resistance. *Proceedings, First Regional Workshop on Coffee Berry Disease*. Association for the Advancement of Agricultural Sciences in Africa, Addis Ababa, pp. 203–11.

30

Use of Ethiopian germplasm in national and international programmes

J. G. HAWKES AND MELAKU WOREDE

Introduction

It will have become evident from the previous chapters in this book that the crop genetic resources of Ethiopia are very diverse and constitute an invaluable base for plant breeding both within and outside the country. Ethiopia is one of the world centres of diversity, identified by N. I. Vavilov some 60 years ago. Not only does it possess important diversity in crops domesticated elsewhere, such as wheat, barley, grain legumes and several oil plants; it also has developed its own indigenous cultigens, such as teff, sorghum, niger seed (noog), ensete, Ethiopian mustard and coffee, many of which are now of great international importance. Ethiopian breeders have taken full advantage of the crop genetic diversity in their own country, combining it with useful genetic characters brought in from other regions.

Clearly, the importance of Ethiopian crop diversity has not gone unnoticed amongst world breeders. Vavilov, who visited Ethiopia in 1927, pointed out the value, particularly to wheat and barley breeders, of the Ethiopian landraces and their extraordinary morpho-agronomic variation (Vavilov, 1931).

In this final chapter we shall attempt to summarize the value of Ethiopian crop genetic diversity both nationally and internationally.

Wheat

Ethiopia is unique in containing a very wide diversity of tetraploid wheat, but very little hexaploid wheat diversity; this latter was probably introduced in recent times.

Improvement of indigenous landraces is described by Tesfaye Tesemma (Chapter 22) in terms of higher productivity, and stem and leaf rust resistance; promising lines are often crossed with exotic varieties in order to improve yield and quality. Prospects for the selection of superior components within landraces are also under consideration.

Breeding for resistance to *Septoria tritici* is being dealt with by Hailu Gebre-Mariam (Chapter 23), using groups of bread wheat varieties from Ethiopia and other sources. Useful resistance was found in these durum varieties checked for leaf rust under field conditions. In general, wheat breeders in Ethiopia recognize the need to involve Ethiopian varieties and landraces in their programmes. Most of this material is, of course, provided by the Plant Genetic Resources Centre (PGRC/E).

Concerning the international value of Ethiopian wheats, we have already mentioned Vavilov's 1927 visit and his collections of wheat, which were used widely by breeders in the Soviet Union. Much of this material was also made available, we understand, to breeders in Germany and other European countries.

The genebank base collection at the International Centre for Maize and Wheat Improvement (CIMMYT) earlier possessed 16 bread wheat and 69 durum wheat accessions (Sencer, 1988), and latterly 1800 Ethiopian wheat entries were introduced from genebanks in the USA, Germany and Italy. Six Ethiopian durum wheats are currently being used in the CIMMYT breeding programme.

Perrino (1988) also reported on Ethiopian wheats collected on three separate expeditions from Italy in the early 1970s and currently stored in the Germplasm Institute at Bari. These wheats were mainly tetraploid, and amounted to over 400 accessions, in which high levels of diversity were found.

Barley

Nearly 5000 Ethiopian barley accessions are stored at PGRC/E. Their high diversity was confirmed by Engels (Chapter 9), with the diversity fairly evenly spread throughout the areas where the crop is grown. Breeding for this crop in Ethiopia has turned away from the exclusive use of exotic materials and towards a greater emphasis on the autochthonous germplasm (Hailu Gebre and Fekadu Alemayehu, Chapter 24). Current breeding concentrates on population improvement of landraces by mass selection and other appropriate methods. The evaluation of Ethiopian barleys reveals useful characters such as resistance to barley yellow dwarf virus, powdery mildew, net blotch and loose smut as well as high protein quality, high tillering quality, tolerance to marginal soil conditions and vigorous seedling establishment. Breeding trials have shown promising results, and new potential cultivars are being recommended. Both malting and food barley have received attention among Ethiopian breeders and the prospects are highly satisfactory.

On an international scale, the Ethiopian barleys became famous ever since the expeditions of H. V. Harlan in 1923 (see H. V. Harlan, 1957), N. I. Vavilov in 1917 and E. L. Smith and C. Thomas in 1963–4. Qualset (1975) has described the high degree of resistance to barley yellow dwarf disease in Ethiopia, particularly in the area just north of Addis Ababa where the concentration of resistance alleles is very high, according to screening carried out in the USA.

According to Witcombe (1983) more than one-third of the total barley collections available worldwide originated in Ethiopia – a total of nearly 4500 accessions. Barley materials are noted also for CIMMYT (Sencer, 1986) and Bari, Italy (Perrino, 1988). Furthermore, Somaroo & Holly (1988) report that the International Center for Agriculture Research in the Dry Areas (ICARDA) possesses nearly 2500 barley collections from Ethiopia, mostly 6-rowed and *deficiens* types; these have shown considerable promise in the search for earlier heading and maturing characters, as well as high protein content.

Recent barley collections were described by Toll (1980, 1981), totalling 675 samples, of which duplicates were to be sent to Braunschweig, West Germany, and the rest stored at PGRC/E.

Thus the quality of Ethiopian barleys both in terms of total diversity and, most importantly, the types of resistance and adaptation genes found, shows their extremely valuable importance to breeders in Ethiopia and in the international community.

Sorghum and millets

The primary centre of origin and diversity of sorghum is assumed to be the Sudan and Ethiopia (see also Doggett, Chapter 10), and it is thus not surprising that useful genetic traits have been found among Ethiopian landraces.

Sorghum breeders in Ethiopia have found sources of cold resistance, high protein (lysine), good grain quality, resistance to stalk borer, downy mildew, smuts, bacterial streak, anthracnose and *Striga*, as well as useful agronomic and kernel characters (Yilma Kebede, Chapter 25). Landrace selections from the sorghum collections have been valuable and five entries are on the recommended list, two for high elevations, two for intermediate and one for low elevation areas. It is confidently expected that more screening and more germplasm exploration in Ethiopia will reveal a wider range of valuable traits than is known at present.

Ethiopian sorghums are also well known internationally. Thus Mengesha & Remanandan (1988) report that nearly 4500 accessions are stored at the International Crops Research Institute for the Semi-Arid Tropics (ICRISAT). These authors draw particular attention to the sorghum line E 35-1 which has been selected from a Zera-zera landrace from Ethiopia, introduced into West Africa for direct cultivation and also used in several African breeding programmes. Other important characteristics, which were identified in a recent expedition to south-west Ethiopia (Ahluwalia *et al.*, 1987) in Ethiopian sorghum (cultivated, wild and weedy types of forage sorghum), include lodging resistance, stem sweetness and high tillering capacity.

Nearly 300 accessions of Ethiopian minor millets also figure in the ICRISAT collections. Characters of interest include morphology, height, number of spikes per inflorescence, size and shape of the fingers, resistance to lodging and tillering capacity.

Since Ethiopia is the only country that grows teff as a cereal crop this section will be concerned only with Ethiopia. Its importance in terms of area cultivated is greater than any other cereal crop in Ethiopia (Seyfu Ketema, Chapter 26) and improvement has been initiated through selection from landraces and mutation breeding. The main aims are to develop lodging resistant, high-yielding stable varieties; so far, higher yielding materials have been obtained through pure line selection and hybridization (for further details see Seyfu Ketema, Chapter 26).

Grain legumes (pulses)

Of 13 million hectares currently under cultivation in Ethiopia, pulses occupy about 13–14 per cent of the cultivated area, cereals *ca.* 83 per cent and oil crops *ca.* 4 per cent. Faba bean is the major pulse crop, occupying about 6 per cent of the total area under major crops, followed by field pea, chickpea and lentil.

Ethiopia has high inter- and infraspecific diversity of pulse crops. Several species of pulse crops have been identified (Westphal, 1974), and their genetic resources and utilization are described by Mekbib, Demissie & Tullu (Chapter 27). Recent work indicates that Ethiopia is an important centre of diversity for some cool season legumes, namely, field pea and chickpea (van der Maesen *et al.*, 1988), and Ethiopia is considered the secondary centre of diversity for *Lens culinaris* (Zeven & Zhukovsky, 1975).

Varietal improvement efforts at national level are at rather an early stage with regard to pulse crops. However, a considerable number of accessions of faba bean, field pea, chickpea, lentil, etc. are being incorporated in crop improvement programmes. Breeding for pulse crops in Ethiopia is largely dependent on the use of indigenous germplasm. Evaluation of various pulse crops reveals useful characters such as earliness, high number of pods and tolerance to some adverse soil conditions. Further details are provided in Chapter 29 (Hailu Mekbib *et al.*).

On an international dimension the Ethiopian grain legume germplasm has received considerable attention. ICRISAT stores over 900 accessions of Ethiopian chickpea and 14 of pigeon pea. These are currently being screened for resistance characters (Mengesha & Remanadan, 1988).

The ICARDA genetic resources unit contains a total of 375 Ethiopian lentil accessions which, when screened in Syria, showed early flowering and maturing characters. These lines were small-seeded, with black or dark brown testa (Somaroo & Holly, 1988).

Perrino (1988), describing the Italian expeditions, reported 96 entries of *Pisum*, 38 of faba bean and 68 of other legumes from Ethiopia. Screening data are not mentioned.

Oil crops

Ethiopia is considered to be the centre of origin or diversity of a number of important oil crops, such as niger seed (*Guizotia abyssinica*), Ethiopian mustard (*Brassica carinata*), safflower (*Carthamus tinctorius*), linseed (*Linum usitatissimum*) and castor bean (*Ricinus communis*). There are national breeding schemes for all of these and several others of slightly lesser importance (Hiruy Belayneh, Chapter 28).

Oil-crop breeders are concentrating on developing varieties with higher yields, improved quality and disease resistance. For *Guizotia*, in addition, special attention is being given to lodging resistance, uniform ripening, minimal shattering, frost tolerance and resistance to *Septoria* disease. Many indigenous lines are doing well in variety trials, and landrace selection is also in progress.

Brassica carinata, *B. nigra*, *B. rapa* and *B. oleracea* are all under consideration for the production of better cultivars with low erucic acid, low glucosinolate and fibre concentrations and resistance to diseases, especially *Alternaria* leaf spot. Useful selections of *B. carinata* have already been derived from Ethiopian landraces, and a crossing programme is under way.

Linseed breeding is focussing on oil yield as well as the general characters mentioned above, with resistance to wilt, powdery mildew and *Septoria*. Promising selections have already been made from the national germplasm collection.

Safflower (*Carthamus*) work includes the selection of promising lines from the national collection and progeny selections.

The castor bean (*Ricinus*) programme is still only two years old but already some outstanding lines have been observed from the national germplasm collection and have been advanced for further study.

Finally, work on the non-indigenous oil crops, sunflower, sesame and groundnut, is in progress.

The Ethiopian oil plants have engendered considerable interest internationally. There are collections stored in a number of seed banks in Europe (Holland, Sweden, UK, etc.) and also in other continents (Kebebew, 1988). Some of these collections have recently been 'repatriated' to Ethiopia. A collection of 76 *Brassica* accessions was made recently by Astley, Haile Giorghis and Toll (1982), of which a duplicate set of samples was sent for cytotaxonomic screening to the Vegetable Gene Bank of the Institute of Horticultural Research, Warwick, UK. These include nearly 60 samples of *B. carinata*.

Coffee

Coffee (*Coffea arabica*) originated in Ethiopia and for that reason its genetic diversity is considered to be higher in that country than in other regions where it was introduced in more recent times.

The significance of Ethiopian coffee genetic resources in coffee improvement is dealt with in detail by Mesfin Ameha (Chapter 29) and thus need not be described at length in this chapter. It is worth reiterating, however, that coffee in Ethiopia has become adapted to a very wide eco-climatic range and with excellent traits for yield, quality and resistance to diseases and pests. However, grave alarm is being expressed at the rapidly diminishing wild resources caused by forest destruction and other changes in land use.

International interest in Ethiopian coffee has been high since the first decades of this century; the works of Chevalier (1929) and Carvalho (1956, 1959) especially, have drawn attention to the need for further international investigations on this crop, particularly as some resistance to *Hemileia* was reported in Ethiopian collections (see also Meyer, 1965, and Vayssière, 1961, for further details).

Forage grasses and legumes

Work on Ethiopian forage genetic resources at the International Livestock Centre for Africa (ILCA) is described by Hanson and Solomon Mengistu (Chapter 15) and by Lazier and Alemayehu Mengistu (Chapter 21). Although African grasses are of considerable importance, in general the emphasis in Ethiopian germplasm work lies with the legumes. Evaluation in plot and field trials of *Trifolium* and *Vicia* species for higher altitudes is in progress, with certain *Argyrologium, Macrotyloma, Eriosema, Neonotonia, Indigofera, Crotalaria* and *Stylosanthes* accessions also showing promise. Among the grasses only *Melina minutiflora* and a *Zornia* species performed well.

In feeding trials several *Acacia* species and *Sesbania sesban* have seemed to perform reasonably well.

Many accessions of highland and lowland forage species have also been collected, but it seems that no landraces exist, since cultivation has been in progress for only 50 years. Thus collecting has been restricted to wild populations, and in the last six years some 1700 accessions were collected and entered into the seed bank as well as being set out in observation trials and field plots. No breeding and selection work has yet been attempted, as far as we are aware.

Final remarks

In this chapter we have attempted to set out some of the highlights of Ethiopian germplasm utilization in national programmes within Ethiopia. Clearly, much more detailed information is given in the appropriate chapters.

The information on the use of Ethiopian germplasm internationally is not so precise, since it has been impossible to scan through the plant breeding literature of the whole world in the hope that one or two relevant pieces of information might be revealed. We have thus relied chiefly on materials provided from international and national centres at the 1986 conference (Engels, 1988) which were not featured in the present book. There are also a limited number of easily accessible references on collecting in Ethiopia by foreign missions, germplasm exchange data and general papers and books by Vavilov, H. V. Harlan, Chevalier and others known to us in the literature.

Despite this limited amount of information from international sources it is quite clear that Ethiopian genetic resources have been valuable in the past, are valuable in the present, and, it is hoped, will become even more valuable in the future. Future development will only be possible, however, if the free exchange and flow of germplasm internationally between Ethiopia and other countries, carried out to mutual advantages, is maintained and strengthened (see also Worede, Chapter 1). Such a movement of germplasm would help to develop screening and utilization programmes, to the ultimate benefit of both Ethiopia and the world at large.

References

Ahluwalia, M., Dabas, B. S., Seme, E. N., Demissie, A. & Nafie, N. A. (1987). Exploration and collecting landraces of cultivated wild and weedy types of forage sorghum in Kenya, Ethiopia, Sudan. *IBPGR Mission Report No. 87/43*.

Astley, D., Mehateme, Haile Giorghis & Toll, J. (1982). Collecting Brassicas in Ethiopia. *Plant Genetic Resources Newsletter*, **51**, 15–20.

Carvalho, A. (1956). O cafe selvagem da Abissinia. *Boletin Superintend. Serv. Cafe, São Paulo*, **31**, 13–15.

Carvalho, A. (1959). Preliminary information on the genetics of Ethiopian coffee. *Nature (London)*, **183**, 906.

Chevalier, A. (1929–47). *Les Caféiers du Globe, 3 vols*. Paris.

Engels, J. M. M. (ed.) (1988). The conservation and utilization of Ethiopian germplasm. Proceedings of an international symposium, Addis Ababa, 13–16 October 1986 (mimeographed).

Harlan, H. V. (1957). *One Man's Life with Barley*. Exposition Press, New York.

Kebebew, F. (1988). Germplasm exchange and distribution by PGRC/E. *In*: J. M. M. Engels (ed.), The conservation and utilization of Ethiopian germplasm. Proceedings of an international symposium, Addis Ababa, 13–16 October 1986, pp. 276–84 (mimeographed).

Mengesha, M. H. & Remanandan, P. (1988). The gene bank at ICRISAT and its significance for crop improvement in Africa with special reference to Ethiopian germplasm. *In*: J. M. M. Engels (ed.), The conservation and utilization of Ethiopian germplasm. Proceedings of an international symposium, Addis Ababa, 13–16 October 1986, pp. 333–49 (mimeographed).

Meyer, F. G. (1965). Notes on wild *Coffea arabica* from southwestern Ethiopia with some historical associations. *Economic Botany*, **19**, 136–51.

Perrino, P. (1988). Country report on plant genetic resources in Italy. *In*: J. M. M. Engels (ed.), The conservation and utilization of Ethiopian germplasm. Proceedings of an international symposium, Addis Ababa, 13–16 October 1986, pp. 377–89 (mimeographed).

Qualset, C. O. (1975). Sampling germplasm in a centre of diversity: an example of disease resistance in Ethiopian barley. *In*: O. H. Frankel and J. G. Hawkes (eds), *Crop Genetic Resources for Today and Tomorrow*. Cambridge University Press, Cambridge, pp. 81–96.

Sencer, H. A. (1988). CIMMYT's wheat germplasm bank and its significance for crop improvement in Africa with special reference to Ethiopia. *In*:

J. M. M. Engels (ed.), The conservation and utilization of Ethiopian germplasm. Proceedings of an international symposium, Addis Ababa, 13–16 October 1986, pp. 353–62 (mimeographed).

Somaroo, H. B. & Holly, L. (1988). The significance of plant genetic resources for crop improvement at ICARDA with special reference to Ethiopian barley and lentil germplasm. *In*: J. M. M. Engels (ed). The conservation and utilization of Ethiopian germplasm. Proceedings of an international symposium, Addis Ababa, 13–16 October 1986, pp. 340–52 (mimeographed).

Toll, J. (1980). Collection in Ethiopia. *Plant Genetic Resources Newsletter*, **43**, 36–9.

Toll, J. (1981). Collection in Ethiopia. *Plant Genetic Resources Newsletter*, **48**, 18–22.

Van der Maesen, J. J. G., Kaiser, W. J., Marx, G. A. & Worede, M. (1988). Genetic basis for pulse crop improvement: collection, preservation and genetic variation in relation to need traits. *In*: R. J. Summerfield (ed.), *Proceedings of the International Food Legume Research Conference on pea, lentil, faba bean, chickpea*, Spokane, Washington, USA, 6–11 July 1986, pp. 55–66. Kluwer, Dordrecht.

Vavilov, N. I. (1931). The wheats of Abyssinia and their place in the general system of wheats. *Bulletin of Applied Botany, Genetics and Plant Breeding, supplement 51*, p. 233.

Vayssière, P. (1961). L'Ethiopie, pays d'origine du caféier d'Arabie. *Café, Cacao, Thé*, **5**, 77–81.

Westphal, E. (1974). *Pulses in Ethiopia, their Taxonomy and Agricultural Significance*. PUDOC, Wageningen, p. 261.

Witcombe, J. R. (1983). A provisional world list of barley expeditions. *Plant Genetic Resources Newsletter*, **53**, 25–40.

Zeven, A. C. & Zhukovsky, O. M. (1975). *Dictionary of Cultivated Plants and their Centres of Diversity*. PUDOC, Wageningen, p. 219.

Index